America and the Sea

America and the Sea

A LITERARY HISTORY

EDITED BY

Haskell Springer

The University of Georgia Press

Athens and London

© 1995 by the University of Georgia Press
Athens, Georgia 30602
All rights reserved
Designed by Betty Palmer McDaniel
Set in ten on thirteen Ehrhardt
by Tseng Information Systems, Inc.
Printed and bound by Maple-Vail Book Manufacturing Group
The paper in this book meets the guidelines for
permanence and durability of the Committee on
Production Guidelines for Book Longevity of the
Council on Library Resources.

Printed in the United States of America

99 98 97 96 95   C   5 4 3 2 1

Library of Congress Cataloging in Publication Data
America and the sea : a literary history /
edited by Haskell Springer.
p. cm.
Includes bibliographical references and index.
ISBN 0–8203–1651–2 (alk. paper)
1. American literature—History and criticism.
2. America—Discovery and exploration—Historiography.
3. Sea stories, American—History and criticism. 4. Sea poetry,
American—History and criticism. 5. Seafaring life in literature.
6. Ocean travel in literature. 7. America—In literature.
I. Springer, Haskell S.
PS169.S42A49 1995
810.9'32162—dc20   93–44385

British Library Cataloging in Publication Data available

Contents

Contents

Contents

Preface

Except for certain allowances on the subject of Herman Melville, readers have not been particularly attentive to the role of the sea in American literary history. While, for example, the vanished western frontier has been repeatedly so studied and invoked as to make it a hoary cliché, that other, and permanent, American frontier, the sea, hardly registers today in our cultural consciousness as setting, theme, metaphor, symbol, or powerful shaper of literary history. Our collective attempt here is to survey this vast and relatively little-known expanse, from the days of New World exploration and settlement to 1990, charting the seas found there.

In effect, we argue that from the renowned canonical writings of the nineteenth century, through the fiction and poetry of many of our best contemporaries, the sea has been source, scene, and symbol for writings by the most highly regarded American authors. Adding other powerfully expressive writers whom most readers seldom if ever think of as engaged with the sea, though they were; further adding colonial-era sermonizers, writers of personal narratives during three centuries, and a host of popular eighteenth-, nineteenth-, and twentieth-century poets, novelists, and romancers, we try to make clear that to understand the history of American literature one must attend to its persistent concern with the sea.

The year 1985 saw the posthumous publication of Willard Bonner's *Harp on the Shore: Thoreau and the Sea,* and in 1988 appeared Bert Bender's *Sea Brothers: The Tradition of American Sea Fiction from Moby-Dick to the Present.* These recent efforts are the only meaningful book-length studies complementing the important work done previously by Thomas Philbrick in *James Fenimore Cooper and the Development of American Sea Fiction* and by Jeanne-Marie Santraud in *La Mer et le Roman Américain dans la Première Moitié du Dix-Neuvième Siècle.* No one has yet looked at the sea's pertinence to American literary history in general.

Doing so now, all the contributors to this book have had to keep in mind that telling the tale is not as straightforward a matter today as it was even a few years ago. Competing theories of language, of history, and of literary history politicize our efforts and guarantee dissatisfaction in some quarters no matter how successful we may be by other standards. Some even claim (justly, perhaps) that literary history cannot be written. Aware of such issues, but determined to tell a meaningful story, we have tried to be sensitive to them in several ways. Our chapters recognize the inevitability of canons, traditional or otherwise, while they avoid uncritical valorization of certain texts or authors. The importance of genre, gender and race, community, class and competition, audience, purpose, text and context, tradition, and innovation is subsumed throughout. But as editor, I have insisted that each chapter be written so as to be accessible to a diversely educated audience rather than merely to academic literary specialists. That decision (like any other on this matter) has obvious benefits and shortcomings. But it is predicated, ironically, on what has happened, in contemporary consciousness, to the language of the ship and the sea.

For fairly obvious historical and geographical reasons, most Americans no longer have an active sense of the sea's presence in their language. Though newspapers still print headlines such as "Town alters course, sails out of sea of red ink," and presidential spokesmen still tell audiences that the chief's hand is firmly on the helm—that the president is steering a steady course, the clichéd use of such metaphors cannot conceal that sea language is either a foreign tongue or invisible to most readers of all but the saltiest books. Misuse, therefore, is common. "Leeway" is now sometimes corrupted to the nonsensical "leadway," "tack" appears as "tact," and the plural of "craft" (vessel) is thought to be "crafts." Is the "steerage" a place or a mechanism? What, exactly, is a "helm"? What is the difference between a bulkhead and a bulwark—implicitly as well as literally?

Because of such cultural change, many readers of *Moby-Dick, The Sea-Wolf, Sabbatical,* and *Middle Passage* have only the faintest notion of what is literally happening on certain pages. But lacking comprehension of the factual and literal, a reader may find figurative meaning out of reach. Without trying to reeducate our audience in nautical terminology, our chapters make that point in various ways. If sea literature has a common focus or intent, that focus is on meaning itself, expressed in the facts and language of sea and ship. The sea experience, literally and in mind, as fact and as still-potent metaphor, is the generative center of these chapters.

This book was conceived a number of years ago in eastern Kansas, on the shore of a prehistoric sea. Despite long delays produced by severe and unexpected personal trials for several of its participants, its gestation has produced and profited from the professional development of its contributors, who in the process have become colleagues and in some cases friends. That analogue to the archetypical all-in-the-same-boat experience central to much sea literature deserves recognition. My own debt to fellow contributors is great, and gratefully acknowledged here.

Writing the story of American literature and the sea has been a joint effort; each contributor wrote his or her own chapter, and that piece of the whole was then read by the editor and often by at least one other contributor. Occasionally an outside reader was called on for advice. Certain paragraphs, sections, and critical positions in some chapters are traceable to the editor, though no chapter expresses opinions unacceptable to its author. The title is the editor's. It reflects the fact that "America" has most often meant the land that is now the United States—and still does, of course, in common parlance. Though I recognize that such usage may, to some readers, imply a cultural imperialism, and that the name properly applies to both American continents, I use it here in its usual sense, to denote this book's treatment of literature applying to the United States, including the territories and colonies that preceded it.

The citation system in these chapters is fuller than that usually found in literary histories, but (partly in recognition of our intended audience) does not follow the common academic pattern in other literary studies. Between the note on the text page and the listing in the bibliography, any reader ought to be able to find a passage or item of interest. Since more than one edition or other form of a work is frequently listed in the bibliography, and since most readers will not have access to original publications or to the particular edition cited, we most often do not cite page numbers of primary works, but instead refer to chapters, stanzas, sections, and so on. An exception is those works not divided into parts at all, or only into daunting ones; there we cite by page or line number.

I am glad to give thanks to the Kansas University English Department, particularly Michael Johnson, its chairperson. For advice and friendly assistance, Gail Coffler, Jim Millinger, Phillip Paludan, and Daniel Vickers have my gratitude. Mary Lacey contributed readings to the chapter on modern poetry. Pam LeRow saw that the word processing was done efficiently and effectively.

America and the Sea

Introduction

THE SEA, THE LAND, THE LITERATURE

The very winds that move over the lands have been cradled on [the sea's] broad expanse and seek ever to return to it. The continents themselves dissolve and pass to the sea, in grain after grain of eroded land. So the rains that rose from it return again in rivers. In its mysterious past it encompasses all the dim origins of life and receives in the end, after, it may be, many transmutations, the dead husks of that same life. For all at last return to the sea—to Oceanus, the ocean river, like the ever-flowing stream of time, the beginning and the end.

RACHEL CARSON, *The Sea Around Us*, ch. 14

I. History

One begins, of course, at the beginning, but where, in literature, is the beginning? Though the sources of meaning are likely to be obscure, they are accessible by one means or another—science, myth, psychology, religion, history. . . . Within the current cliché that ours is a water planet (however much we resist acting on the implications of that truism) is the fact that water—both fresh and salt—is life. Science and myth agree that on this planet there was first water and then dry land. Not only was water the precondition of an earthy Earth, it has also been the necessary condition first of planetary life, then human life, then civilization—essential to food, drink, cleansing, transport, trade, and defense. The oasis, the well, the pond, the stream, the river, the lake, and the sea: water is the joiner of human beings and the center of their communities.

But it is also the separator, the border, the dangerous boundary. As it has been the facilitator so it has been the forbidder; if savior, then also destroyer. This instability, changeableness, even paradoxicality of water accounts for

[1]

much of the human relationship to it. On this hand the pond and the well, on the other, the flood; it is life itself, but a little bit too much is death. Despite the fact of earthquakes and volcanic eruptions, no comparable paradoxicality attaches to the land.

We are told that we come from dust and to dust we shall return; but what first animates that dust and then reproduces it is water. Coming from the protective womb waters, we proceed throughout our lives to dry up and die. Finally, we physically give up our fluids and become earth, as the water outside invades our remains and converts us fully into the not-us. Water, that is, makes possible the integrity of the self, and water changes self to other. It is the element of individuation and the element of dissolution. In between, as universal ritual testifies, it is the symbol of becoming and of conversion: from the waters of baptism to the washing of the corpse, water is present in rites signaling change of state and in stories of such change. So it is when, in biblical narrative, the children of Israel symbolically cross a sea early in the process of their becoming a separate and self-defined people—an analogy that has not been lost on American culture, from the writings of the Puritans to those of African Americans.

It is little wonder, then, that water should have such a prominent place in the epics, myths, and living dreams of the human race and of Americans in particular—and that we should take it largely for granted under conditions of calm or control, but be in awe of its powerful fluidity when it rises and flows. Little wonder too that, in winter as well as summer, people seek the shore and look out on the ocean. Such water-gazers (Melville's term) are responding to energy, size, movement, and mystery. E. A. Robinson said it memorably: "An ocean is forever asking questions and writing them aloud along the shore" ("Roman Barthalow").

In their westward movement, starting before there was even an "America" in human consciousness, Europeans encountered the seas that led to the New World and then later helped to define it. The Atlantic, the world's stormiest ocean, was, very early in modern European history, an economic and cultural focus—as it had long been for the fishing and whaling "Indians" on the other side. Next met, though we tend to forget them, were the freshwater seas of the Great Lakes, each of which, Rudyard Kipling remarked, is a "fully accredited ocean." This phenomenon, like the salt sea, inspired much legendary and mythological literary expression from Native Americans, as it later astonished and attracted European adventurers, settlers, and their own imaginations. Then (with its own indigenous North American sea cultures) came the immense Pacific, 25 percent larger than all the world's land masses combined—a challenge to trade and exploration, but also the liquid wall against which the westwardly realizing United States found the literal and symbolic end of its land frontier—until Alaska and Hawaii once again added new shores to the American map.

These bodies of water have been powerfully influential on the history of this continent and this country. The literary historian must keep in mind the discovery, exploration, exploitation, and settlement of North America by the English, Spanish, French, and Dutch, among others, and the subsequent experiences of three hundred years of immigration to them (both voluntary and forced) by ship. Comprised in this history are the American industries of the sea such as fishing, whaling, shipbuilding, and transportation; indispensible commerce with other countries and among the states; extensive exploration by sea, both governmental and private; naval actions; and the more lurid smuggling, slaving, and piracy. The following historical overview leans toward seeing that history in terms of its intersections with American literary expression.

The maritime heritage of the United States had its origins in New England. Most of the early settlers apparently had had more than enough of the ocean during their passage to the New World, though some had turned that passage into narratives, sermons, and poems emphasizing the glory, the horror, and the well-nigh eschatological power of the sea as seen from the perspective of those in its "bosom" or conversely its grip. But the difficulties of transplanting English agricultural life in stony New England and the long experience of their culture with maritime endeavor meant that some turned, in one way or another, back to the sea. For many decades thereafter, while settlement hugged the Atlantic shore, the economies of the colonies and the states that succeeded them were heavily, though to varying degrees, dependent on the sea: on the shipbuilders who got from the thick northern forests the materials for an industry that sold its products both domestically and abroad; on the import-export merchants, among whose accomplishments was the clever Triangle Trade (the first of several similarly named patterns), in which the American colonists began to declare their independence by evading British laws; on the whalers who kept the oil lamps fed; on ships and seamen, mostly from New England, who took materials to market and imported goods from Europe; on the fishermen who supplied larger and larger markets for fresh, dried, and salted fish. Fishing, the oldest of New England's maritime enterprises, was particularly important, as testified to by the name of Cape Cod, of course, and by the wooden codfish that has hung in the Massachusetts legislature since 1784. Later, in the nineteenth century, aided by icing and the development of railroad transport, fresh mackerel, haddock, halibut, herring (sardines), and shellfish, brought in by New England boats, joined cod as fundamental products in the region's economy. The many active New England ports of the period, with their web of sea-related enterprises, stood in contrast to the few more southerly ones, testifying to the region's primacy in maritime endeavor. Here developed, along the shores as well as among seafarers themselves, a maritime culture—always predominantly male, but incorporating over time the work and the words of women too.

The end of the colonial era and the beginning of the national period, 1775–1815, has traditionally been called by maritime historians the "heroic age." For a third of it the United States was at war with Great Britain; for most of the rest of it other wars made the Atlantic unsafe far beyond the risks of its formidable natural hazards and the absence of aids to navigation—a combination that in retrospect seems well summed up in Philip Freneau's formulation: "dread Neptune's wild, unsocial sea." Thomas Jefferson's Embargo of 1807, which prohibited foreign trade by American ships, though modeled on two previous American trade embargoes and seen by the Jeffersonians as the only feasible alternative to war, created disastrous conditions for the American marine and caused political upheaval in Federalist New England. In the War of 1812 the British navy effectively interfered with most American nautical endeavor, including coasting and fishing. Even John Jacob Astor's attempt to dominate the trade in furs with China by establishing a depot in the Pacific Northwest (the history of which was later written by Washington Irving in *Astoria*, 1836) failed because of the war.

The difficulties of the period led to Americans' using many ruses, and to their seeking new cargoes, new trading partners in exotic countries. Their efforts to overcome the formidable military and political obstacles, in addition to the hazards of unknown coasts, new languages, and strange cultures, meant that for many the maritime enterprise of the period required heroic effort. The sea novels of James Fenimore Cooper are now our most accessible literary windows into some nautical aspects of this era, but many other narratives, fictional and otherwise, were popular at the time. For example, there was Archibald Duncan's *Mariner's Chronicle, Being a Collection of the Most Interesting Narratives of Shipwrecks, Fires, Famines, and Other Calamities Incident to the Life of Maritime Enterprise,* a four-volume work originally published in 1804, which one Richard Manning, around 1832, gave his nephew Nathaniel Hawthorne (whose own sea-captain father had died of disease when on a voyage), and Hawthorne then gave his friend Herman Melville.

Within this forty-year heroic age occurred the period of "neutral shipping" (1793–1807), during which the long-lived U.S. role as shipper to the world really began. War between England and France had resulted in a blockaded

Continent. American shipowners found ways around the restrictive rules of the time, and as neutrals in the war, made large profits supplying such goods as sugar and coffee. Increasing trade called for more and larger ships, as well as larger quantities of supplies and provisions, so ports and secondary maritime activities all along the eastern seaboard grew and throve.

The "golden age" of American maritime activity is the romanticized term generally given to the period from 1815 to 1860. In the years following the War of 1812, all aspects of American maritime activity developed, expanded, and in some cases boomed. For example, packet ships of the 1820s, 1830s, and 1840s sailed for England and Europe on schedule regardless of weather. Charles Dickens, who crossed on the *George Washington* in 1842, wrote that "the noble American vessels have made their packet service the finest in the world." A larger number of packets, steamers, and tramps plied the extensive coastal routes from Bangor to New Orleans, trading such commodities as lumber and cotton. The fishing fleets continued to expand, harvesting the rich resources not only off their coasts but also, in the case of the sealing and whaling industries, at great distances from their home ports. American ships also searched for business virtually all over the world, competing for cargoes of pepper from Sumatra, figs from Smyrna, and trying new enterprises—ginseng to China, furs and sealskins for many markets, ice to South America, India, Asia. In a midcentury story by Harriet Beecher Stowe, "Deacon Pitkin's Farm," the possibilities of seafaring for young America are imagined this way: A young man without prospects runs off to sea on the *Eastern Star,* East Indiaman out of Salem. A friend gives him fifty dollars to invest, confessing that he wanted to go himself, but his wife would not hear of it. The sweetheart left behind says, "Oh that horrid, horrid sea! It's like death—wide, dark, stormy, unknown. We cannot speak to or hear from them that are on it." But after seven years, and the reported loss of the ship, the bronzed and matured adventurer comes back rich to save the family farm and marry the sweetheart. So went the ideology of youthful American enterprise in its maritime form.

Not surprisingly, the reality was rather different, and more complex. Seeking hides to supply the Massachusetts leather goods industry, ships went all the way around the Horn to Mexican California—as did the Boston brig *Pilgrim* in 1832, with Richard Henry Dana, Jr. among its crew. The profits of the voyage included Dana's improved health and *Two Years Before the Mast,* which created a virtual literary genre—the "voice from the forecastle" narrative—in which the American history of seafaring intersects with the literature of the United States under the heading of labor and management. The tales told by sailors such as Dana, followed by Melville and many others, have proved to be far more engaging reading than the usually prosaic and self-justifying, little-known accounts by officers in command of ships, military or civilian—and accord with the long story of labor exploitation usually told by the historian. Well into the twentieth century it was a legal maxim, generally agreed with even by civilian seamen, that unquestioning obedience to authority was essential to life and safety at sea. But under several pressures, particularly those for profits during the "golden age" and afterward, that authority was repeatedly abused. The numerous stories told by these authors, of unfair and brutal conditions of employment such as deprivation, bad food, denial of contractual rights, arbitrary authority, and physical punishment (including flogging) for even minor failings, turn out to be essentially true—highlighted by notable exceptions. The merchant seaman, for most of American history, lived in a state somewhere between indentured servitude and outright slavery. A green hand might ship out filled with the romance of seafaring, but the euphoria probably disappeared in almost no time—as it does (fictionalized) in Melville's *Redburn:*

> Yes! yes! give me this glorious ocean life, this salt-sea life, this briny, foamy life, when the sea neighs and snorts, and you breathe the very breath that the great whales respire! Let me roll around the globe, let me rock upon the sea; let me race and pant out my life with an eternal breeze astern and an endless sea before!

> Miserable dog's life is this of the sea! commanded like a slave, and set to work like an ass! vulgar and brutal men lording it over me, as if I were an African in Alabama. Yes, yes, blow on, ye breezes, and make a speedy end to this abominable voyage! (ch. 13)

Legislation to remedy this virtual slavery was late, and weak in effect. It lagged far behind the efforts of seaman-organizers such as James Williams, whose African-American heritage no doubt exacerbated his indignation at the wage-slavery under which working seamen like himself were held. Not until the La Follette Act of 1915 were common sailors given legal freedom. Collective bargaining ultimately benefited many seamen, but even those efforts contributed in their own way to the general decline of the national marine. More recently, the failures of government, industry, and labor, as well as certain obvious developments such as air travel, have perpetuated the problems. Nowadays, as we can see by reading John McPhee's *Looking for a Ship,* a few seamen can make good money in an apparently moribund industry.

During that so-called golden age, American shippers continued to transport new immigrants, particularly Irish and Germans. But not all of those who arrived in American ports were there by choice. Slavers went on plying the "Middle Passage" as they had been doing since the early seventeenth century, illegally supplying the demand for forced labor, while Frederick Douglass, who later wrote of the galling contrast between the freedom of ships before the wind and his own enslavement, escaped from bondage disguised as a free sailor.

If we can say that the American seaman was for many years a virtual slave at sea, that comparison is made possible by our knowledge of the real thing, and of what made *it* possible. Slaving—the details of which make a horrifying chapter in American maritime history—was a business in which up to 50 percent of Africans being transported to America died at sea. The first African slaves came to North America in 1619, brought by a Dutch ship to the Jamestown colony. The *Desire* of Salem, the earliest-known American slave ship, took a cargo of Pequot Indians to the West Indies in 1638 and returned with some Africans. During the remainder of the colonial period, importation steadily increased, primarily to fill the need for labor on tobacco, rice, and indigo plantations, and then for harvesting cotton—until the slave population of the colonies totaled about half a million. For most of its history, slaving was not the specialty or sole business of certain shady American merchants or captains, as one may perhaps imagine today. Rather, large numbers of ships, widely owned, often carried partial cargoes of slaves. Though slavery was concentrated in the South, the deep involvement of New England shippers in buying and transporting cargoes of Africans cuts away any putative Northern moral high ground.

After independence, opposition to the trade in people increased—along with the numbers imported. Responding to the growing revulsion, even the merchants of Rhode Island, who had led the colonies in ownership of vessels engaged in the slave trade, shifted more and more to other enterprises. By 1808 importation was illegal; by 1820 transporting slaves was defined as piracy. But American enforcement of these salutary laws was lax, and some slaving continued up to the Civil War. Only one man was ever executed for violating the antitrafficking laws, while many other ship captains and owners managed to avoid the prohibitions against interstate transport of slaves. Though many African slaves were directly imported, perhaps a larger number came to the American colonies from the West Indies, where they were "seasoned"—beaten or otherwise forced into submission. The voyage from Africa to the Caribbean (frequently the second leg of a triangular trade that began in New England with a cargo of rum for Africa and concluded with a passage from the islands with sugar and molasses destined for New England distilleries) became known as the "Middle Passage," a term that has more recently been taken to mean the forced trip from Africa to America by whatever route. It is memorialized in some slave narratives, and in Alex Haley's *Roots*, but brilliantly evoked in Robert Hayden's poem "Middle Passage" and imagined most recently in Charles Johnson's National Book Award–winning novel of the same name.

One reason slavers were able to go about their awful business as long as they did is that they employed fast vessels that, after the belated outlawing of their trade, frequently outran the navy ships pursuing them. Fortunately for national pride, though, the American reputation for speedy ships does not rest on slaving. Beginning with its packet ships, reaching an apogee in the midcentury clipper ship, and concluding with the large Down-Easters of the later nineteenth century, the United States developed and retained a reputation for building and sailing many of the fastest windships in the world—though it lagged behind others in seeing the inevitable triumph of iron and steel over wood, and of steam over sail. Before the Civil War, the development of the admired and imitated American clipper ship, largely for the China and California trades, brought great wealth to canny merchants, reputation to the national merchant marine, fame and fortune to some very skillful builders and ship captains—and a wealth of experience (as many later proudly wrote) in the school of very hard knocks for the deep-water sailor.

The epitome of the search for speed under sail was the full-rigged clipper ship (so named, supposedly, because it traveled at a fast "clip," or because it "clipped" the waves rather than pushed through them, or simply because the word denoted streamlining), whose racy lines, daringly lofty masts, and astonishing spread of canvas declared its intention. Responding to the economic demands of the China tea trade, and the California and Australia gold rushes, it was at the height of its glory in the 1840s and 1850s—though building primarily for speed did continue up to the 1870s. The clipper, most notably from the Boston yard of Donald McKay, comes down to us in the words and pictures of contemporaries as a model of beauty and elegance (Samuel Eliot Morison called it our Rheims and our Parthenon), and has often been seen—probably romanticized—as the embodiment of "splendid overreaching." Though steamships had preceded the clippers, they could not carry enough coal to make the longest voyages; nor, until many years after the clipper era, reach the best speeds set by windships such as *Champion of the Seas*, which maintained a never-equaled twenty-knot average during one twenty-four-hour period, or the other ships that set records on a variety of passages and whose vaunting names included *Westward Ho, Flying Cloud, Stag Hound, Comet*, and *Young America*.

The clippers, sailed around the ferocious Capes Horn and Good Hope to the limits of their physical endurance (and that of their crews) by hard-driving skippers, did not last long under such strain. Furthermore, when demand for speed slackened, they could no longer command the high freight rates that justified their costs of construction and operation, and the reduced carrying capacity of their sharp hulls. Finally, the Suez Canal, opened in 1869, definitively ended the clipper domination of the China tea trade. Because of the reputation of those great ships, though, the name "clipper" retained enough potency to make it a seductive appellation of American commercial airliners—meant to connote the beauty and speed of airships.

Not all sailing ships strove for speed, of course. Among those with bluff, rounded bows, and limited spread of canvas, one class was easily distinguishable to a casual eye by the boats hung in davits along their bulwarks. These were whalers. Though fishing probably mattered more to more people than whaling ever did, and though scholars suggest that Banks fishing was more hazardous than whaling, it is the whale hunt that lives in the American imagination as the distinctive national maritime endeavor of the nineteenth century. The fact that whaling is now probably the best-known maritime activity of the past is probably because of *Moby-Dick* alone, and that book may also be the source of today's inaccurate, romantic picture of whaling, a picture in which men seek out the whaling ports, abandoning the dull and dulling life in the stony fields, behind the shop counters, or in the factories of New England, to strike out for the eternal sea, where they pit themselves, puny human beings, against the leviathan, in an exciting epitome of The Hunt. In fact, most whalers went to sea as untried youths, and if they too had such romantic notions, they probably soon lost them. The life they led was at best monotonous, dirty, brutal, and very low-paid—so much so, that except for the officer class, young white New Englanders tended to make only one voyage before giving up on whaling, leaving the field more and more to a mixture of free blacks, Indians, and the foreign-born. At worst, it was a hellish existence from which an unlucky or improvident seaman could return after three or four years at sea actually owing money to his employers.

What made whaling a large and often hugely profitable industry (for the owners, captains, and merchants, that is) was the importance of whale products, mainly for illumination, but for other widespread uses as well. The best of the refined oil produced a clear, bright lamplight; spermacetti, a component of sperm whale oil, made the very best candles; machinery of all sorts in an industrializing America depended on whale oil for lubrication; "whalebone" from the baleen plates of the right and humpback supplied corset stays, buggy whips, and umbrella ribs; and ambergris from the intestines of certain diseased whales, a large lump of which could be more valuable than a full hold of oil, was the base of fine perfumes.

When Herman Melville sailed aboard the *Acushnet* of Fairhaven in 1841, the United States was the world's foremost whaling nation. It remained so up to the Civil War. One measure of that preeminence is the fact that American whale ships found and put down on the charts more places than did the Great U.S. Exploring Expedition of 1838–42. In midcentury, New Bedford, with more than three hundred ships, became the world's most active whaling port, eclipsing Nantucket (the *Pequod*'s home), which had, more than a century earlier, learned how to make a profitable business of sperm whaling, and which, as early as 1775, had a fleet of 150 whale ships—though during the Revolution and then the War of 1812 the independent Quaker whalers of the island saw most of their ships destroyed. Despite such depredations, the fishery boomed from 1830 to 1860, but by the time of the Civil War, overfishing had caused a strong decline. The industry was really doomed by the combined effects of the war, disasters (such as the destruction of thirty-two ships of the Arctic fleet by ice in 1871), and, mainly, the commercially successful extraction and refining of petroleum beginning in 1859—though the last whaling voyage in a wooden ship, the schooner *John R. Manta*, began, from New Bedford, in 1925. The last modern American factory-ship whaler did not go out of operation until after World War II. In the decades since then, the United States has played a rather inconsistent role in the world as both a defender of the great cetaceans against extinction and a politically cautious defender of the harvesting rights of whaling countries.

Throughout the nineteenth century, and even after the end of the age of sail, many published narratives of whaling and commercial voyages and other records of personal maritime experience (large numbers still in manuscript) enriched the American historical and literary record. For one thing, inspired by the extraordinary accomplishments of Captain Cook's three voyages (1768–79), by national pride, and by economic hopes, the United States sent out several notable exploring ventures, public and private, published accounts of which were very popular. (Edgar Allan Poe attempted to capitalize on that popularity with his *Narrative of Arthur Gordon Pym.*) Not only did unprecedented numbers of educated men go to sea and then seek to tell their tales to a large audience apparently much interested in them, and not only did even the minimally educated attempt to speak from the quarterdeck as well as from before the mast, but women too left us their visions of sea and self. Usually the wives of sea captains, they accompanied their husbands on uncounted whaling voyages and many commercial ones as well, writing accounts in which we can read variant versions of nautical experience and in which we can see the social and literary construction of gender under circumstances seldom found ashore.

While sail was supreme, the ports of Boston and New York, aided by river and canal networks, produced those great cities (not coincidentally the nation's literary centers). The well-known maritime foundations of their growth are evident in the metaphorical language of writers such as Thoreau, Emerson, James, Wharton. In the Deep South were Savannah, Charleston, and New Orleans, cultural centers in their own right, largely shipping cotton. In tonnage owned, though, right after Boston and New York came Philadelphia and Baltimore. But mere rank does not give an accurate picture of relative commercial importance: cargoes between New York and Liverpool alone exceeded all others. Washington Irving sailed early in the century as a cabin passenger on that route, embroidering the experience in his essay "The Voyage." Melville sailed it before the mast, and *Redburn* was the result.

Ports, and the fortunes of their cities, waxed and waned during the period. Samuel Eliot Morison notes that Salem, long active in the China and Africa trade (though not slaving), left the roster of important seaports only in 1845, the year Nathaniel Hawthorne was appointed to his custom house post there. Under the conditions of declining commerce, Hawthorne, having little official work to do, made good use of his time in conceiving *The Scarlet Letter*. West Coast shipping, fishing, sealing, and whaling did develop during the period, though some of those activities matured only later in the century. The importance of the maritime in national thinking was demonstrated by the acquisition of the Oregon territory, sought in part because of access to the Pacific and trade with the Orient. San Francisco, though, with its huge natural advantages as a harbor and its soon-developed port facilities, dominated on the West Coast. Its growth from such activities explains its ability to support the work of journalists such as Bret Harte and Mark Twain, as well as pointing to the subject matter repeatedly chosen by its native Jack London.

As the long, long age of sail waned, the era of the windjammers succeeded that of the packets and clippers. Much larger square-rigged ships, usually of iron and steel and foreign-built (except for the wooden, Maine-built Down-Easters), generally limited in routes to the very longest and cargoes whose value would not permit the expense of coal for steam shipping, they were the last of the great windships, and shared the seas with steamers well into the twentieth century (a fact reflected, for example, in the plays of Eugene O'Neill and in Mark Helprin's "Letters from the *Samantha*"), though steam became the mode of virtually all shipping late in the nineteenth. Apparently originating as a term of denigration suggesting that these ships could not effectively trim to the wind but had to "jam" themselves into it, "windjam-mer" became a designation of pride for ships whose size and strong con-struction produced many impressive passages.

The inevitable end of windships for military and commercial purposes had been signaled long before the windjammer era: regular Atlantic crossings under continuous steam power had been available in the 1840s. Economics, aided by traditional thinking, however, kept sailships in the van until the weight of technological advances guaranteed first the ubiquity of steamships, and later their replacement by diesel-powered vessels. The Panama Rail-way (1855), the effects of the Civil War, the Suez Canal (1869), the Trans-continental Railway (1869), and the opening of the Panama Canal (1914), accompanied great improvements in marine power plants and, ultimately, worldwide availability of fuel for them. These factors definitively changed American maritime experience. For one thing, the huge waves of immigration from southern and eastern Europe came by steamship. Those immigrants had a shorter, safer voyage than their Irish, German, and Scandinavian pre-decessors. They were probably not aware of the difference or thankful for it, but the merchant seaman *was* aware, and, if we can believe those who have written most convincingly on it, not thankful at all. One view of that change appears in O'Neill's "*The Hairy Ape*," where Paddy says, " 'Twas them days men belonged to ships, not now. 'Twas them days a ship was part of the sea, and a man was part of the ship, and the sea joined all together and made it one."

While O'Neill could still write of the "beauty and singing rhythm of the sea," to him and to others the ship propelled by mechanical force controlled the sailor, rather than vice versa, and they saw in that a representation of in-dustrial America. In the minds of many who knew the sea, the possibilities for self-realization, for the enactment of voyages of self-discovery and matu-ration were terribly diminished. But these historical changes also motivated a resurgence of sea fiction by working seamen.

Meanwhile, as the twentieth century approached and fewer and fewer Americans made their livings by the sea, another sort of sea-focused lit-erature was being written—along the Atlantic shores for the most part— as resorts and vacation communities drew many urban business people and intellectuals, most notably women of letters, to the economically depressed seaside. There is where we find, in fact and in many of their writings, Kate Chopin, Elizabeth Stoddard, Sarah Orne Jewett, Harriet Beecher Stowe, and others. Of course, as the twentieth century progressed (though inter-rupted by two wars and a severe economic depression) the shift of the sea context from business to pleasure only increased, until now the phrase "the sea" evokes for many Americans the beach, the marina, the yacht club, the cruise ship, and sport fishing. These too have had their own literary expression, which continues to reach a popular audience.

Until the Civil War, then, the American maritime enterprise had grown and prospered for several reasons, including favorable conditions for shipbuilding, certain natural advantages, entrepreneurial boldness, immigration, national expansion to the West Coast, boom times such as the gold rush, and general growth of worldwide shipping in the mid-nineteenth century. In the years just before the war, American-owned ships carried close to 75 percent of the country's foreign trade. But by the end of that war, the figure had dropped to 25 percent. The merchant marine story since then has been seen by historians as rather dismal: some call the period from the Civil War to World War I the "dark age," and except under the impetus of war, the peaceful American marine has diminished ever since. The largest reason for the decline was probably economic self-sufficiency: plenty of native raw materials, steadily increasing domestic manufacture of goods formerly imported, domestic consumption (by a rapidly growing population) of most farm products. Then, too, favorable opportunities for landward investment, especially in industry, drew capital away from shipping. So the country "focused inward," as one maritime historian says, relying more and more on foreign vessels for its foreign trade. The United States, though still heavily dependent on the ship (oil imports come readily to mind), is now, in ownership of vessels, one of the lesser maritime nations in the world—except for its redoubtable navy, of course.

In all the periods of American maritime history, the strength or weakness of the country as a naval power has not only been significant to its general history, but also has found its way into the national literature. The exploits of John Paul Jones, the defeat of pirates (Barbary and others), the long history of Old Ironsides, the "opening" of Japan, the exploring expeditions commanded by naval officers, the destruction of the *Maine*, the Great White Fleet, Pearl Harbor, the Normandy invasion. . . . From Cooper's naval novels and Melville's *White-Jacket* and *Israel Potter*, through Richard McKenna's *Sand Pebbles*, Edward L. Beach's *Run Silent, Run Deep*, Herman Wouk's *Caine Mutiny*, Marcus Goodrich's *Delilah*, to William Brinkley's *Last Ship* and Thomas Clancy's *Hunt for Red October*, a naval story, often based on the author's experience, has frequently had well-received (or belatedly praised) literary treatment. In presenting their military subject matter, these works frequently engage troubling questions of rank, privilege, and competence as seen in the context of a democratic nation. Looking at history nonfictionally, an entirely different sort of book, Alfred Thayer Mahan's *In-*

fluence of Sea Power upon History (1890), had a powerful influence of its own on American policy, justifying the imperialistic extension of our hegemony to Cuba, the Philippines, and further.

Boom or bust, war or peace, to remember the role of the sea in the economic and cultural life of the United States is fundamental to examining literary seafaring. Looking back at this general history, for example, we see that after the Civil War, while American ownership steadily declined, fewer and fewer American men went into the merchant service. Look at the nationalities of the crew in O'Neill's sea plays for a literary reflection of this fact. Following O'Neill as a seaman-author there was Langston Hughes; in recent years, though some fairly obscure voices have gone on speaking, who among well-known writers (other than Alex Haley, perhaps) can be pointed to as an American sailor whose subject relates to that life and experience? The decline in commercial engagement with the sea has no doubt contributed to the similar decline in direct, extended authorial involvement. Experience in the navy (Richard McKenna, Herman Wouk, William Brinkley), in the merchant marine (Claude McKay, Langston Hughes), on commercial fishing boats (Peter Matthiessen, John Casey, Paul Watkins), or on pleasure craft (John Barth, Susan Kenney) has been the background since the early twentieth century for most of those authors (the numbers are not large) whose writings reflect their own seafaring. The paucity of such experience among American authors, though, seems not to have depressed the market in other ways, since the literary exploitation of the sea and the American seagoing past has not faded in proportion to the now minor status of the United States as a maritime nation.

That literary survival may seem rather paradoxical from another perspective, for despite the economic reliance of the United States on shipping, who today in the potential reading audience is conscious of ships, their functions, their comings and goings? Oceans now hardly seem to exist for most Americans except as locales for pleasure and entertainment—though the mind is sharpened somewhat when storms periodically pick up the sea and fling it against the shore's puny defenses. Then the audience of the evening news takes a serious, if brief, look. Yet perhaps these sentences are too pessimistic. Though in the eyes of most Americans the sea may lack some of the magnetic potency investing it up to at least the twentieth century, that loss may be making possible its protection from human stupidity and cupidity. When nothing could harm the all-powerful sea, anything could be done to it. But now, organizations such as Greenpeace and the Cousteau Society, ironically aided by recent oil spills, have managed to focus the attention of many individuals, and a number of governments as well, on the oceans' susceptibility to incessant human attack. Ecological concerns seem to be far better understood or at least respected today than in the recent past, and the sea is generally admitted to be vital to the life and health of economies as well as of organisms. Other developments are notable: the maritime past is a fascination of many here and abroad; volunteers are cleaning sea birds and protecting turtle eggs; governments are becoming less and less tolerant of pollution, overdevelopment, and exploitation of marine resources—which are now seen as undeniably limited. It is just barely possible, then, that we ourselves may see in the interactions of humans and oceans "hope and history rhyme" (Seamus Heaney).

II. Holistically Speaking

When did American sea literature itself begin? The oral literature of Native Americans preceded any European expression, of course. Widespread among the tribes, from east to west, were myths of creation out of the great waters—often the myth of the earth-diver who brought up from the bottom of the deep and ubiquitous sea the tiny bit of earth or sand from which the dry land was formed. In addition, among the coastal peoples (as might be expected) tales tended to reflect their environment, their location by "the river with one bank"; so allusions to sea creatures, as to whaling and fishing, were apparently common.

The exploits of Koluscap (or Gluscap), culture-hero of the Micmacs, Passamaquoddies, and other eastern tribes, include supplying the pipe the whale smokes, and tricking a particular whale into bringing him ashore, though she strands herself. On the Oregon coast the trickster Coyote is the subject of many stories, including those about his establishing the salmon-fishing rituals. There is the Tillamook tale of South Wind marrying the daughter of Ocean, "the chief of chiefs"; the Chinook story of the monster that causes the roaring out beyond the surf; the Coos tale of the woman who married a merman . . . (*Coyote Was Going There*).

Not all Native American oral literature (no matter how local its subject) is completely indigenous, while some stories, east and west, explicitly tell of contact with others from across the sea. Showing possible Scandinavian influence, for example, is a Passamaquoddy tale of two sisters who became mermaids (Leland, *Algonquin Legends*). A particularly interesting contact account is the Tillamook tale of "The Journey Across the Ocean," suggesting visitors from Asia in a sailing ship, possible return visits by Native Americans, and an intermarriage. And another story from the Oregon coast records, from the perspective of the inhabitants, the visit of the first ship to the land of the Clatsop people. They destroyed it (*Coyote Was Going There*).

In looking for the beginnings of American sea literature from the dominant

European perspective, we can point to the journals, sermons, narratives, and poems of the earliest settlers and would-be settlers of the new western world, often written at sea, and powerfully marked by their experience of water without end and tempests of apparently eschatological strength. But in some ways more appropriate is an earlier story. Homeward bound after 160 days at sea, including those spent in the unwitting discovery of a new world, Christopher Columbus ran into a full gale south of Flores, westernmost of the Azores, and very nearly foundered. Fearing less for his personal safety than for his reputation, the admiral wrote an account of his voyage on parchment, begging the finder to deliver his narrative to the king and queen of Spain. Rolling the parchment in a waxed cloth, he ordered up a large wooden cask, sealed the manuscript in it, and hove it overboard. Columbus's parchment narrative may be the first item in American sea literature—lost, appropriately, at sea.

More important than any claim to precedence is the figurative power of his action. As it happens, Columbus's compulsion to tell his tale is one of the paradigms of sea voyage literature. Though the example that springs most readily to mind is *The Rime of the Ancient Mariner*, many men and women have told their own stories of the sea because the experience is often terrible and almost always altering. In nearly all such sea literature the narrators have been transformed by their trials and are driven to repeat the story. From Odysseus's voyage, to Ishmael's "'I only am escaped alone to tell thee,'" through innumerable personal and fictional accounts to Paule Marshall's *Praisesong for the Widow* and Charles Johnson's *Middle Passage*, many works conform to the myth of the protagonist returned from trial at sea enlightened with transforming knowledge.

Our receptiveness to sea literature rests in part on our sympathetic response to this archetypal journey, speaking as it does to truths that often transcend differences of culture and of gender. In this voyage-centered literature the sea assumes a double role: it is the field of action on which the separation and transformation are played out, and it is the thing itself, the heart of mysterious knowledge to which the protagonists aspire and with which they return. The vastness, loneliness, and power of the ocean change them forever and drive them to write about it, but always with the metaphorical understandings particular to the times in which they experience it: the expanding cosmos of the Renaissance, the eighteenth-century Sublime, the symbolic disorder of the Romantic era, the Darwinian naturalism of the fin de siècle. . . . The experience may be elemental; the telling is shaped by language and culture, which are always changing. T. S. Eliot appropriately reminds us that "The sea has many voices, / Many gods and many voices" ("The Dry Salvages," I).

To most minds, as to Eliot himself, those voices are heard from the sea's verge and even from far inland rather than from the deck of a ship. Therefore, this book, recognizing the metaphorical implications of water-gazing from the shore, takes as its province more than what is popularly thought of as sea literature (*Two Years Before the Mast, Moby-Dick, The Sea-Wolf, The Old Man and the Sea*); it includes the sea's appearance in American literary discourse as symbol, concept, and figure of speech as well. So we note the justice of James Russell Lowell's rhetoric in 1871, when, looking back thirty years to "The American Scholar," he says, "We were socially and intellectually moored to English thought, till Emerson cut the cable and gave us a chance at the dangers and glories of blue water." We also ponder the implications of a common store of surviving analogy in a perception of the Nebraska prairie in book 1, chapter 2 of Willa Cather's *My Ántonia* (1918): "the grass was the country, as the water is the sea," as we do in Conrad Richter's *Sea of Grass* (1937), set in Texas: "I saw a wave of antelope flowing inquisitively toward the buggy far ahead, a wave rusty as with kelp, rising and falling over the grassy swells and eventually turning in alarm, so that a thousand white rumps, whirled suddenly into view, were the breaking of the wide prairie wave on some unseen reef of this tossing upland sea" (ch. 5, echoed in ch. 15). This "sea of grass" trope is well-nigh ubiquitous in writings about the Great Plains, say scholars of that subject—pointing in addition to prairie "schooners" and towns with names such as "Westport." Meant at first to convey, to an audience familiar with the sea, the strangeness of a land that seemed, at times, hardly land at all, it became and remained an expected metaphor implying horizontal immensity.

In addition, the domestication of sea tropes, for example by Louisa May Alcott in *Work* (1873), as well as the focus on the sea's symbolic, social, and psychological import in a domestic novel as powerful as Elizabeth Stoddard's *Morgesons* (1862), or in Ntozake Shange's *Sassafrass, Cypress & Indigo* (1982), points to more ways, gendered ways, in which "the water comes ashore" in American literature, well beyond the context of what those words suggest in Frost's "Neither Out Far Nor in Deep." Literary engagements with the sea by the shorebound, men and women, can be variously understood in terms of psychological, cultural, and historical factors that may differ greatly from those pertinent to the literature of seafaring, or may prove to be congruent with them. In any case, as Chapter 8 notes, taking Emily Dickinson's poetry for its example, "a powerful literature of the sea does not depend upon direct experience on the high seas, but upon consciousness."

The consciousness reflected in the writings of shore-based watergazers and those of seafarers may be of different sorts, but thematic and conceptual similarities abound. From Thoreau, Melville, and Whitman, for example, through London, Hemingway, Matthiessen, and others, biology has often supplied the context for American literary probes into the meanings of life on Earth—probes, at once literal and metaphorical, which turn to the great waters for fact and symbol. That context, biological fact and interpretation, with their inevitable philosophical consequences, gives immense substance to my epigraph from Rachel Carson.

Much of sea literature also says (with T. S. Eliot) that the sea has a voice—and speaks to those prepared to hear it. The restless, dangerous, awe-full, beautiful, destructive, creative, deceptive, alluring, forbidding, infinite, profound sea is common to both groups of writers: its variety and paradoxicality is itself a frequent subject, as is the human response to those qualities separately and together. Until recently, however, only those who had lived the experience wrote powerfully of the special character and complexities of life at sea—physical, meteorological, linguistic, sartorial, social, occupational, psychological; class-defined, tradition-ruled, in rhythm and method distinct from life ashore. The special structures of the shipboard community, the diurnal consequences of watery isolation, the leaving and the returning—these were largely the literary province of the seafarer, almost always male. What is *not* found, though, even in such reports (before this century) is the sexual component of life aboard ship. Melville hinted at it in *White-Jacket* and *Moby-Dick*, and came closest to embodying it in plot in *Billy Budd, Sailor;* others steered clear. Now, though, both men and women, seafarers and lubbers alike, plunge into even that culturally complex subject.

Beyond divergences and similarities such as these in the literature, a reader finds overarching them a greater commonality—an engagement, often in fascination, with the ocean's perceived differences from the land. Those contrasts are contained in or implied by several primary, overlapping features of the sea itself. Chief among them are eternal motion, boundlessness, and obscuring depth—the sources of its power to image or embody, in imaginative writing based on whatever sort of encounter, physical or intangible, a range of human, earthly, and cosmic truths.

First of all, the ocean water, never completely without motion (though it may sometimes appear to be), is defined by that motion. Its calmer movement often suggests anthropomorphic connections, as in the rather frequent mention of a "heaving bosom." That association, in relation to others such as its tides, its teeming life, and its apparent "moods," usually gender it female and feminine to seamen and, sometimes, to male writers. Women are apparently less agreed about its gender, some exalting or identifying with the sea's generativeness, others, such as Kate Chopin or the diarist Dorothea Balano, evoking its apparently masculine seductiveness, and still others seeing its threatening dangers as masculine aggression. Recognizing and denying the traditional gendering of the sea by men, for example, Mary Mackey's 1976 poem cries, "Don't Tell Me the Sea Is a Woman."

In any case, the sea's femininity is different from that so often seen in the "virgin" (or the "raped") land by American writers. The sea can be lover, mistress, mother, spouse, but not also virgin unless the scene somehow strongly evokes the land: "These are the times, when in his whale boat the rover softly feels a . . . land like feeling toward the sea; that he regards it as so much flowery earth. . . . The long drawn virgin vales; the mild blue hill sides . . ." (*Moby-Dick*, ch. 114).

Walt Whitman, alone among the canonical "major" American writers, makes much of the sea as sexual female, as he does in "Song of Myself," "Out of the Cradle Endlessly Rocking," and other poems. This general absence of a feminine sea in our well-known literature, at least to the early twentieth century, is consistent with the paucity of writing on mature sexuality, especially by those in the traditional canon. The sexual avoidances in Irving, Cooper, Thoreau, Twain, and lesser authors, the evident neuroses in Poe, the male fears apparent in a story such as Hawthorne's "Rappaccini's Daughter" or Melville's *Pierre* illustrate the point. The danger or dangerous mystery of women to so many male writers and protagonists is concomitant with the threat of the sexual sea. We may reasonably speculate that the complex sexual symbolism of the sea, unlike the simpler virgin-and-mother nexus of the land, was psychologically forbidding to these men. And a further sexual threat, considering the mixed homophilia and homophobia in the literature written by many American men, is the ever-present possibility, at sea, of a metamorphosis from a perceived "feminine" calm to an aggressive "masculine" turbulence.

Actually, in our literature the sea is seldom beneficently calm. There are, so to speak, no Indian summers on the waves, no sea pastorals. And the absence is not merely a reflection of natural fact: Just as the seascape, the starkest of all scenes, leads the mind to ponder essentials, so the sea's threatening motion images the condition of human life, stripped of illusion. Faced with the puissance of the sea, men—particularly men—realize that their own force, which on land can make such changes, is negligible. Their minds, as many writings testify, focus on *its* power, rather than their own. In "The Open Boat," Stephen Crane, with terse, ironic humor, describes the general ocean situation:

A singular disadvantage of the sea lies in the fact that after successfully surmounting one wave, you discover that there is another behind it, just as important and just as nervously anxious to do something effective in the way of swamping boats. (pt. 1)

Even from the apparent security of the land one feels the threat of the sea. When it "comes ashore" (in deceptive Frostian truism), it has astonishing power. In another Frost poem, "Sand Dunes," the dunes

> . . . are the sea made land
> To come at the fisher town
> And bury in solid sand
> The men she could not drown.

Similarly, Melville's poem "Misgivings," whose subject is the coming Civil War, begins, "When ocean clouds over inland hills / Sweep storming . . . ," suggesting the irresistible, destructive power of anticipated events by comparing them to a sea storm come ashore. And Thoreau's *Cape Cod* shares the watergazer's feeling of awesome power in the sea, memorably describing and evoking its irresistibility while pointing out that the dead emigrants of the wrecked *St. John* had been seeking the New World, but had been carried by the sea, with supra-geographical force, to a newer yet. Emily Dickinson, who wrote scores of sea-conscious poems, also expressed this power in one of her riddles: "An everywhere of Silver / With Ropes of Sand / To keep it from effacing / The Track called Land" (#884).

Melville contended that "however baby man may brag of his science and skill, and however much, in a flattering future, that science and skill may augment; yet for ever and for ever, to the crack of doom, the sea will insult and murder him, and pulverize the stateliest, stiffest frigate he can make. . . . No power but its own controls it" (ch. 58). This perpetual, threatening power, much less diminished by technological advances than most think, has permitted many authors visions, usually denied by the acquiescent land, of human possibility larger than life. The personalities and accomplishments of Captain Ahab, Wolf Larsen (*The Sea-Wolf*), Santiago (*The Old Man and the Sea*), Raib Avers (*Far Tortuga*), are all what they are in large part because of their conjunction with the always-moving, always-dangerous sea.

Its unrestrained might gives the sea its ability to do away with limits, and is therefore deeply involved in the implications of its second major characteristic, boundlessness. Through much of American history, a perception of limitless space invested both the sea and the land. However, though the sea's "limber margin" (Yvor Winters) forms an undeniable end to the land, the reverse is not true: the sea stretches, uninterruptedly, around our world. And so while, under the force of western settlement, that image of the land, with its inspiration of expansion and of national and personal grandeur gradually lessened in the course of the nineteenth century, the sea's various appeals of boundlessness remained fully potent—as we can see in the actions of Kate Chopin's Edna Pontellier or earlier in Dickinson's lyric of pleading and demand: "My River runs to thee— / Blue Sea! Wilt welcome me? / . . . / *Say— Sea—Take Me!*" (#162).

Pip, the cabin boy of the *Pequod*, discovers that one danger of the "shore-less ocean" is an intolerable, maddening lonesomeness. Bulkington, of the same ship, returns to the "howling infinite" of the sea of thought, ultimately to perish there. In Poe's "MS. Found in a Bottle," the protagonist sails, perhaps, beyond time. The physical and psychological dangers of infiniteness are plentiful and real; human beings, to survive whole, need certain boundaries to time, to human relations—and to the salt sea of their own blood. So humanity shrinks from pure boundlessness; but at the same time it is intrigued and stimulated by the opportunities of that limitlessness because, as William Blake said, "the bounded is loathed by its possessor." Therefore, when America, which should have itself been boundless in reflection of its ideology, stretched from sea to shining sea, Walt Whitman still invoked his soul in "Passage to India" to "farther, farther, farther sail" on an endless and finally metaphysical sea. Whitman, like some others, was doing the work of a culture that, ideologically, refused to recognize limits.

Thoreau too, for all his exalting the properties of the land rightly regarded, invokes the sea for his comments on human potential. In *Walden* it is the source and context of his metaphorical treatment of such possibility; in the "Conclusion" he exhorts his reader to leave the stranded ship of obsolescence, skirt the reef of reality, and "explore the private sea, the Atlantic and Pacific Ocean of one's being alone." Everyone lives on the shores of this private sea, he says, but no one "has ventured out of sight of land, though it is without doubt the direct way to India." The metaphorical state of exploring at sea is for Thoreau what Whitman calls "more than India." In the boundless sea Thoreau found the right analogy; its infinity floods his belief and floats it.

Its liberating infinitude, then, as well as its motion, helps explain the sea's attraction to American writers—in further part because the limited and bounded is, conceptually speaking, potentially tame and trivial. As such, it can also be repressive because of its conventionality. Huck Finn decided to escape that repression by lighting out for the Indian territory, but the later absence of a West of personal liberation left only the sea—to which, in fact, Ken Kesey takes his crew of West Coast mental hospital inmates in part 3 of *One*

Flew Over the Cuckoo's Nest (1962) on their Pacific fishing-trip escape from momism and "the Combine." Not surprisingly, Kesey's free and dangerous sea does its job, temporarily restoring to the men their long-lost feelings of confidence, joy, and independence. Making explicit the potent sea/lost land context of the scene, Chief Bromden says, "I was feeling better than I'd remembered feeling since I was a kid, when everything was good and the land was still singing kid's poetry to me." More recently Rachel Ingalls, setting her *Mrs. Caliban* (1983) in a coastal California wasteland, reintroduces meaning into the life of her depressed housewife protagonist in the form of a manlike creature from the depths of the mysterious sea.

Finally, the sea's horizontal limitlessness is augmented by a vertical one: the unknown and untold possibilities of its obscuring depth. The very word "deep" has been a common literary synonym for sea at least from the King James translation of Genesis, and is still evocative enough in our era to function well as a Peter Benchley title suggesting hidden threat, mystery, awe.

One of the most common uses of "deep"—to mean "mentally profound" —probably owes the terms of its expression to a recognized analogy to the sea. And because much of what the deep thinker is contemplating is commonly thought incomprehensible, such a person is often feared or at least distrusted—as is the profound and alien sea. But to those for whom depth of mind is exciting, the sea metaphor is enriching. To some minds, unsolved mysteries are more valuable than those that can be resolved. For many writers that is precisely the case with the inscrutable ocean. Like some others, however, Emily Dickinson, contemplating the analogy between human and oceanic profundity, at least once denied that the mysteries of the world exceed our ability to penetrate them by asserting that the brain "is deeper than the sea—" (#632). In spite of their similitude, the brain has the God-like power to take the external into itself: "For—hold them—Blue to Blue— / The one the other will absorb— / As sponges—Buckets—do—." That's a brave assertion, from a brave poet, but it may also strike a less inland soul as naive.

The unseen and mysterious underwater world, in its obvious analogy to the human psyche, is also a powerful lure of seemingly infinite potential. On that fishing trip in *Cuckoo's Nest*, the psychologist (appropriately enough) hooks a huge fish on the ocean floor, and though unable to tell what it is, fights for hours to bring his "monster from the deep" to the surface. Similarly, hunting for the "ungraspable phantom of life" in the deep waters of the sea more than one hundred years earlier, Ishmael went whaling.

One need not be an Ishmael, though, or his driven shipmate, Bulkington, either, to respond to the eternal lure of the mysterious sea's dark depths:

in "The Slow Pacific Swell" Yvor Winters speaks as an enthralled "landsman" watergazer; Robert Frost sees (though perhaps with jaundiced eye) that people "cannot look out far. / They cannot look in deep. / But when was that ever a bar / To any watch they keep?" For Denise Levertov, in *The Jacob's Ladder*, the sea is always "turning its dark pages."

In all cases, that is, the ocean suggests the portentously unknown. Melville, probably the deepest-diving of all writers about the sea, realized that its obscuring depth, like its continuous motion and boundlessness, lure the mind in part by imaging that dark mind—and the even darker mind of God. In doing so it embodies a compelling mystery, human and cosmic.

III. These Chapters

To those from England who in the early seventeenth century sailed to Virginia, New England, or Maryland, the sea was of interest only as a barrier to be passed and survived. But that barrier was so formidable and fearsome that surviving it was a frequent subject of written contemplation. The sea-deliverance narrative became a particularly important form of personal, spiritual, literary expression at the time. Renaissance notions of nature as subject to human artifice were shaken or destroyed by the sea experience. At the same time, the suffering at sea was often seen as an undoer of complacency or error, preparing the Anglican or Catholic or Puritan voyager for regeneration in a Land of the Golden Age, Terrestrial Paradise, or Promised Land on the other side. In the sea-deliverance narratives they wrote, those who lived through desperate trials at sea often read their survival as a sign of God's special favor to the communal enterprise. As Chapter 1 concludes, from such seventeenth-century attitudes derive long-lived American tendencies to believe in an "orgiastic future" (F. Scott Fitzgerald), to have a sense of historical and divine mission. In the writings of William Bradford and the Mathers (Richard, Increase, and Cotton), among many others, the dangerous, mysterious, and transforming sea plays a vital part in the expressions of what were to become American literary and cultural traditions.

The Revolutionary and Federal periods of the eighteenth century (Chapter 2) saw changes in American attitudes toward the sea—based in part on technological improvements and the growth of maritime endeavors, but also on changes in philosophical and aesthetic ideas. The growth of commerce and the strengthening of nationhood were boasted of in sea literature, while reliance on the design of divine Providence became more and more a matter of literary convention than belief, as God's immediacy gradually gave way to the concept of a natural universe ruled by a distant Power. In factual and fictional written expression, the survivors of sea disasters no longer draw doctrinal conclusions from their deliverance, but simply return from the alien sea to the harmony of their social contexts. By the end of the century the sea comes to be seen as both a shield, protecting the new society from European intrusion, and as a reminder of the awesome power of uncontrolled nature. The disorder of the sea, its wild, unsocial character, is frequently imaged in sea literature of the time in terms of the natural Sublime, in which the vast and grand in nature both enthralls and appalls the puny human being.

In real-life narratives, sensationalism and sentimentality show that the aims of sea literature now included entertainment and creativity along with didacticism. The sea figures meaningfully in broadsides, plays, poems, and novels of the eighteenth century, in the productions of Susanna Rowson and Royall Tyler, among others. But Philip Freneau is the most successful and nuanced user of sea imagery. His poems, frequently patriotic, treat the sea as a field of America's projected influence, but also suggest the sea's superiority and invulnerability to all human endeavors, transient as they are.

Freneau's last published poems appeared in 1824, as did James Fenimore Cooper's *Pilot*. In his sea novels (twice as many volumes as in his famous Leatherstocking series, and seen by some readers at the time as his true forte), Cooper not only improved upon his American predecessors but also seriously reshaped the nautical fiction tradition handed down by Tobias Smollett and Sir Walter Scott. Having established the sea novel as a viable genre in American literature, and having written extensively on nautical concerns in other genres, he is the preeminent American writer of the sea before Melville, and therefore the main focus of Chapter 3. Cooper's early sea fictions (of the 1820s) locate themselves on the ocean's verge, in stories of smuggling, piracy, and captivity, expressing themselves in Byronic, patriotic tones. His eight works of the 1840s reexamine the romantic bases of that earlier work and develop illuminating subtleties, complicated oppositions—moral and otherwise—involving gendered conflicts between the shore and a farther sea, domesticity and independence, self and others.

America's most celebrated author of the time, Washington Irving, also wrote works with significant nautical content: in his histories of Columbus and his companions, in *Astoria*, and in several brief fictional pieces he too explored the relationship between this new-world nation and its seas. So did Nathaniel Hawthorne as the editor of nautical works. Furthermore, in the observing narrators of his stories, the sea raises issues of self-consciousness. And Edgar Allan Poe, not only in *The Narrative of Arthur Gordon Pym* but also in short fiction as well, exploited nautical, geographical themes and scenes for their speculative possibilities and metaphorical aptness in suggesting extreme states of mind.

A host of popularizers, including the famous "Ned Buntline" (who, like others, chose a nautical pseudonym), learned from Cooper and flourished during his own long career. They cheapened the currency, no doubt, but they were satisfying a public appetite for piracy, shipwrecks, sea serpents, and nautical character types. In turn, Cooper learned realism from Richard Henry Dana, Jr., whose immensely popular *Two Years Before the Mast* spawned imitations, mostly nonfictional. Against a background of popular nautical narrative in the earlier nineteenth century, "Cooper [navigated] the distance between Scott's shoreline sea of domestic sentiment and Melville's open sea of metaphysical speculation."

The background against which Cooper, Poe, and the other canonical authors stand out includes innumerable personal stories, in print and unpublished, of sea-based American experience. Chapter 4 discusses the literary significance of the many nonfictional narratives, journals, and diaries written from the time of the Revolution to the end of the nineteenth century, mostly by writers unknown to fame. They are expressions, directly and indirectly, of typically American concerns and values. Narratives of the Revolution and the War of 1812 overtly laud the cause of the emerging nation, while naval expeditions throughout the century sought further to expand and coalesce America's place in the world of scientific knowledge. Also evident in the narratives by both men and women is a democratic and humanistic concern for equitable treatment of sailors, tolerance for other peoples, and the development of moral and religious ideals—though some gender-based differences are apparent.

This nautical nonfiction also expresses a strong concern for the development of the individual. Some writers seek merely to portray accurately the difficulties and rewards of life at sea, while others stress how such a life can foster one's physical and psychological improvement. The personal sea narratives of the first century and a quarter of the nation's history chart the struggles and achievements of the maturing nation and its people.

The maritime concerns reflected in the fiction and nonfiction treated in previous chapters are also the subject of the hymns, chanteys, and sea songs discussed in Chapter 5. Early hymns addressed the relationship of colonists and settlers to the natural world, and the sea experience (so susceptible to allegorical reading) long remained current in hymnody. The nautical ballad of the new and growing republic reflected patriotic enthusiasm about the new country and also about its new navy; and then the rise of chanteying (originating in the rythmns of work at sea) made every maritime endeavor of the nineteenth century a subject for exploitation or celebration. The passing of the age of sail meant a shift in subject to the passenger liner, until, as the United States more and more lost touch with its maritime heritage in the earlier twentieth century, the subject rarely appeared in song. But a resurgence of interest in the nautical around midcentury produced songs focusing on the sea and its heritage as resources to be protected.

Chapter 6, "Poetry in the Mainstream," focuses primarily on the antebellum America that was essentially a coastal, sea-conscious culture with a developing literary tradition anchored in romantic impulses. Most of the sea poetry of the period mirrors these elements of American life by responding to a "call from the sea," a call with three distinctive voices. The first voice calls the listener to a realm of fantasy and enchantment, the second, to the exploration of death, and the third is a counterpoising voice of life. All three voices reflect the romantic preoccupation with the unusual and the mysterious—the first often taking the reader beneath the sea, the second emphasizing the sea's immense destructive power, ghost ships, actual historical disasters, and the fate of those lost at sea and those left behind. The third voice produced a richly suggestive poetry revolving around the motifs of creativity and ebb and flow, and the metaphors of the journey of life and the ship of state. All the poets from that time remembered today, from Bryant, Longfellow, and Whittier to Emerson, Whitman, and Tuckerman, contributed to this American poetry of the sea.

Herman Melville, deservedly America's best-known nautical writer, has a chapter to himself. From beginning to end of his forty-five years of authorship, Melville's sea-consciousness shaped his work, giving it a profundity and suggestiveness beyond that accomplished by any other American user of nautical themes, plots, or language. Working from his years of experience at sea and from his voluminous reading, he began by writing romances and narratives of exotic adventure and life at sea, quickly moving into symbolic and allegorical investigations of culture, psychology, and metaphysics in such astonishingly original books as *Mardi, Moby-Dick,* and *Pierre.* But attempting to be commercially successful, while also striving, more or less surreptitiously, to say what he knew would often be unacceptable to his audience, he could never fully realize his own high authorial goals.

Melville evokes the sea's beauty and its horrifying power, its subtlety and deceptiveness. He finds intolerable truths in landlessness, in the vast, terrifying indifference of the sea (supposedly God's) to human striving. He perceives its essence as biology, and also feels its metaphysical and sexual suggestiveness. In his own metaphor, he became, in book after book, a deep diver—ashore or afloat—and went on, using the sea largely as metaphor, to examine American culture, slavery, landscape, the Civil War era, other subjects, and his own self, in unconventional prose and poetry.

In his late poetry, particularly *John Marr and Other Sailors,* Melville's imagination again focused directly on the sea, drawing it as sharkish, indifferent, immovable, but yet somehow his own. Despite its undeniable destructiveness, he found there, in fact, memory, and symbol, a life-enhancing power: "Healed of my hurt, I laud the inhuman sea." He died in 1891, leaving in his writing desk the uncompleted *Billy Budd, Sailor*, a testament to the power and complexity of his thought and his language to the very end. That last work, set in the days "before steamships," is also, no less, an elegy for the unique species of human being—the sailor—about whom he had begun writing so long before.

Appearing at this point in the text, but pertinent to the book as a whole, the portfolio entitled "American Seascape Art" explores some of the interrelations between literary materials and the rich and varied visual tradition of American seascape. It is not intended as a minisurvey of masterpieces but is a selection of works that explore in various media what Melville called "the shock of recognition" between artists who inspire one another in various ways to re-imagine the meaning and the forms of the sea, the coast, and the voyage as an aesthetic language.

"Realism and Beyond," Chapter 8, examines the complex pattern of American sea literature in the later nineteenth century, after the great age that culminated in the writings of Melville. Shifts both in American maritime life and American values in these years undermined many of the assumptions of the prose romance and lyric poetry as literary forms, and the writings discussed in this chapter, by, among others, Dickinson, Howells, Twain, and James, adopt new stances: ironic, consciously antiromantic, and focused to some degree more on questions of consciousness than on the meticulous recounting of adventuring on some mythic quest. This period also witnessed a substantial development of literature by American women writers such as Stowe, Stoddard, Larcom, and Jewett on the meaning of the sea, the coast, and the voyage that focused especially on the life of the maritime community, a community (particularly in New England) in decay from its earlier economic prosperity.

Toward the end of the nineteenth century, as Chapter 9 explains, a number of Americans began to revive the time-honored tradition—established by Cooper, Dana, and Melville—of writing about the sea on the basis of their own experience as working seamen. By far the most prominent of these was Jack London; but several others—for example, Thornton Jenkins Hains, Morgan Robertson, Arthur Mason, Felix Riesenberg, Richard Matthews Hallet, and Archie Binns—followed this traditional pathway to literary careers in sea fiction. Produced between the 1890s and the 1930s, this body of fiction tended to honor the ordinary seaman, thus drawing on the example of Dana and Melville (well before the Melville revival). And it developed two essential themes that Melville had begun in his own last sea stories: an elegiac celebration of the last days of sail, and, relatedly, a dramatization of navigational crises that were, in fact, symbolic crises of faith.

By far the most powerful influence on American sea fiction after *Moby-Dick* was the Darwinian thesis that all species of life originated in the sea and are governed by the laws of natural and sexual selection. For this reason, even such saltless writers as William Dean Howells and Henry James were convinced that the civilized reality might be most clearly revealed "in the strong sea-light" (as in James's story, "The Patagonia"); and later writers with more specific interests in natural history and biology, such as Ernest Hemingway and Peter Matthiessen, would turn to the sea as the crucial setting in which they might more effectively study what, to Melville's pre-Darwinian sense, had been the "ungraspable phantom of life."

Providing both contrasts and similarities with the great salt seas, the Great Lakes, oceanic but drinkable waters in the heart of a continent, have produced literature little known to most readers. Chapter 10 explains that the fiction of the Great Lakes has its place in at least two major traditions: the ancient one of the unchanging sea and the more modern one of the conquest of a new world. The Great Lakes offer a challenge to comprehension because, since they are fresh water, it is hard to accept that they are like the ocean in their superiority to human control. Shipwreck and destruction are accepted traditions when it comes to the sea, but we have never stopped believing in our ability to control and subdue the Great Lakes. Our failure to do so—a failure expressed and understood in Indian stories long ago— questions our myths of control over nature in general, in ways missing from saltwater fiction. Human frailty, fear, and the perception of a nameless, ineffable, malevolent presence on the Great Lakes is common to its literature by both natives and settlers, and persists today.

The relationship between African Americans and the sea began historically with the Middle Passage, the experience of being transported in misery across the Atlantic from Africa to slavery in America. Yet Chapter 11 draws on African-American folklore and folk songs as well as autobiography, fiction, and poetry, by many women and men, to show the sea as presenting African Americans with actual possibilities for liberation during the eighteenth and nineteenth centuries, and with metaphysical and mythical possibilities for liberation in the twentieth. In the late twentieth century, African-American writers such as Haley, Hayden, and Johnson recreate the experience of the Middle Passage as a means not only of denouncing slavery but also of demonstrating the capacities of black people to transform and to transcend this attempted destruction of body, soul, and culture. In view of the arduous and persistent endeavors of black Americans to establish roots in the American land, most of their writing references rivers and swamps, farms and cities; nevertheless, the sea, appalling in its associations with the annihilating passage to this country and appealing in its associations with liberation from American racism, remains a constant in African-American literature.

With the decline of narrative and purely descriptive verse in the twentieth century, the sea has, for the poet, become primarily a fertile source for metaphors and symbols. This is detectable first in the poetry of the Imagist movement. Having identified and exemplified it there, Chapter 12 proceeds to survey its significance in the main currents of subsequent American poetry, following a broadly chronological sequence.

First come the major poets who established their reputations before World War II. The rich variety of uses to which the sea is put by those poets, together with what it represents for each of them imagistically, prevents the formulation of any neatly generalized summary. The sea's timelessness, energy, vastness, violence, and changeability, however, symbolize a force that may be hostile, sympathetic, or indifferent to human beings. Here the poet for whom the sea is especially important as a theme is Robinson Jeffers.

Next, the chapter examines the work of a wide range of poets emerging during and after World War II. It demonstrates that, for many of them, there is a significant shift in emphasis from the sea itself to the creatures living in or on it. The ways in which the existence of those creatures relates to or illuminates the human condition form the basis for a considerable body of speculative verse. Charles Olson is the great sea-poet of this period.

Summing up and connecting, as far as is possible, the variety of ways in which the sea has challenged the poetic imagination in this century, this discussion then shows how, for Hart Crane and Wallace Stevens, in specific poems, the sea is linked directly with the process of poetic creativity itself.

In the first half of the nineteenth century, the sea romances of Cooper, Melville, Dana, and Poe had pitted the individual against the natural world, in contests where human strength challenged physical (and metaphysical) nature in its severest manifestations. Regardless of the extremities to which their protagonists were driven, a coming-to-terms between humankind and nature seemed possible. In the years subsequent to that Romantic age, such exploits evolved literarily into staid Victorian contests of manners like those satirized in, for example, "The Love Song of J. Alfred Prufrock." That is where Chapter 13, "Modernist Prose and Its Antecedents," begins.

The fictions of the realist-naturalist period (from about 1870 to 1910) often gloomily portray the condition of the individual and of human culture under the influence of Darwin's evolutionary theories and of contemporary philosophy. Mark Twain's satiric works point to the kind of terms a realist could make with a changing, disheartening world. In particular, Twain used a youthful hero in his semblance of a sea romance, *Adventures of Huckleberry Finn*, to explore the possibility of realizing one's moral being, even in a corrupt society. Later in the century, Stephen Crane and Frank Norris, in both their land and sea works, portrayed the plight of the individual caught unprepared in a universe unresponsive to human effort.

Then, from the early twentieth century to about the death of Hemingway in 1961, the modernists dominated the American literary scene with their views of what biology, technological advancements, and changing social patterns meant to "modern" life. Eugene O'Neill, in advance of other American writers of the early century, chose the expressiveness of the sea and sea life for a group of one-act plays and several major dramas. Of those who followed suit by employing sea imagery and maritime metaphors to turn their views of modern life into art, Hemingway and Fitzgerald were foremost. Their works reflect their own geographical and moral meandering, and find symbols for the sense of alienation and isolation that characterized modern life for them. Hemingway's art most powerfully responded to the new age with nautical settings and symbols that capture Modernist predicaments but also express human possibility.

Our final chapter demonstrates that the traditional sea subjects, contexts, and symbolic meanings continue to be important in American fiction since the 1960s. Among them is the treacherous sea (*Jaws*, for example), the sea of life (*Sailing*), escape from civilization (*The Birth of the People's Republic of Antarctica, The True Sea*), the sea—in war or peace—as testing ground (*The Last Ship, Adrift*), the transformative sea (*Middle Passage, A Sea-Change*), and the loss of nautical traditions and ways of life (*Far Tortuga, Calm at Sunset, Calm at Dawn*). But in addition, a new attitude toward the sea emerges in some recent writing—a perspective that may be variously traced to newer scientific thinking, to the technological developments that have diminished certain threatening mysteries, to an idealistic romanticism, to disgust with civilization's failures. The old fear of the sea, the sense that it is at least alien, and often hostile to boot, often gives way to a sense of it as a place of possibility, as a potentially sustaining force for human life. Some of the books cited in this connection are Ursula Le Guin's *The Farthest Shore*, Lois Gould's *A Sea-Change*, Ted Mooney's *Easy Travel to Other Planets*, Rachel Ingalls's *Mrs. Caliban,* Ntozake Shange's *Sassafrass, Cypress & Indigo,* and even Joseph Heller's *Catch-22.* This change suggests a different kind of hopefulness from that adduced in several places above: for those whose other earthly frontiers (in the American sense of the word) seem long gone, perhaps the last, best chance has not really vanished. If so, the question now is what to make of it.

HASKELL SPRINGER

Chapter One

THE COLONIAL ERA

The sea experience, and particularly the storm, tested Renaissance notions of the primacy of art over nature and threw into question the possibility of establishing order out of chaos. It brought into sharp relief the inability of human power to influence or control nature. Sea literature of the colonial period offers a view of life in which Renaissance enthusiasm and exuberance collide with the stern reality of life's limitations and its potential for suffering and disaster. But it does more. In *The Tempest*, for example, Shakespeare creates out of such human suffering as the sea inflicts the vision of a harmonious cosmos to which disordered nature is reconciled. Sea literature of the Renaissance is thus often corrective, expiating ignorance or wrong thinking. Through suffering, misperception or complacency is overthrown and replaced by enlightened knowledge of the cosmos and our place in it.

In December 1619, John Donne preached a sermon (no. 72) at the Hague on Matthew 4:18–19 ("Follow me, and I will make you fishers of men"). In it he employed one of his favorite figures: "The world is a sea in many respects and assimilations. It is a sea as it is subject to storms and tempests; every man—and every man is a world—feels that." He then drew a number of comparisons, ending in this: "All these ways the world is a sea, but especially is it a sea in this respect, that the sea is no place of habitation, but a passage to our habitations."

Donne's phrase captures an essential feature of American sea literature. In American writing of the seventeenth and eighteenth centuries, the sea is by itself rarely the reason for the writer's being there. For the earliest writers, the place of habitation was the New World—Virginia, New England, Maryland, the West Indies. The alien world of the sea offered no haven, and Donne's extended simile became a commonplace description of spiritual pilgrimage as well as of the physical passage familiar to those who made the Atlantic crossing. The sea literature of the colonial period is almost entirely a sea-deliverance literature. Very few works consciously set out to be meditations on the sea. Inevitably, the poems and narratives of the period are written from the perspective of men pinned down by wind and wave, clinging to a spar, eating the soles of their shoes, and drinking their own urine after days and nights on a floating wreck without food or water. Few intellectualize the process, and there is little deliberate aesthetic contemplation of the natural beauty or power of the sea.

For the most part, the writers were persons who had somewhere to go by ship. On the way, disaster or misfortune threatened or overtook them, and afterward they wrote about it. The intensity and trauma of the experience drove them to write, and a number of times the result was genuine literary art, but almost none of the writers set sail with aesthetic purposes or preconceptions. Nobody set out to make something up.

Belief in Divine Providence was common to all Christian faiths at this time, but it had special meaning for American writers. For men and women of the Renaissance, Providence was a matter of natural fact as well as of theology. It was not a matter of debate that God held the cosmos together by divine will and operated in the world using bounty or famine, sunshine or tempest to communicate His pleasure or displeasure. Belief in Providence was not exclusively a Puritan conviction. In the narratives of the Virginians Strachey and Norwood, for example, both of whom express considerable dislike of Puritanism, the concept of Providence is accepted without question.

For the most part, too, the writers' beliefs concerning the role of Providence in human affairs guided their aesthetic values. American writers of the seventeenth century looked at sea experience not as an object of literature but as a metaphor for the Christian life and as a sign of special favor pointing to the historic mission of their enterprise. Art is frequently the result, in early American literature, of the writers' use of sea experience—and many writers were conscious craftsmen; but art for art's sake is unknown.

Thus, in the literature of the New World, the sea voyage was a familiar metaphor for national and spiritual pilgrimage, manifested in individual lives and in the development of the community. Sea deliverance was not merely an adventure or the occasion for uncommon suffering. It was the dramatic record of an individual's trial and call to faithfulness. It was an experience charged with meaning for the audience who listened as well as for those who endured it. Those who read the narrative were called on to bear witness to God's Providence in the experience of the writer and in the life and history of the community.

Sea deliverance provided both narrator and audience a shared sense of theology and perception, and it gave them a common sense of community and purpose. That the narrative had a function beyond the individual exemplum was clear to a seventeenth-century audience. The special Providence of a sea deliverance is evidence of God's particular concern. For the Virginia writers, sea deliverance demonstrated God's special favor for the enterprise of English colonization of the New World. In a sermon preached to the Virginia Company in 1622, John Donne described a vision of the enterprise that combined a desire for English hegemony in the New World with the Kingdom of God: "You shall have made this island, which is but as the suburbs of the Old World, a bridge, a gallery, to the new; to join all to that world that shall never grow old, the kingdom of heaven, you shall add persons to this kingdom, and to the kingdom of heaven, and add names to the Books of our Chronicles, and to the Book of Life." Likewise, American Puritans looked to sea experience for evidence of God's revealed will, evidence that would complement the revelation in Scripture and confirm New England's special "errand into the wilderness." Puritan historians from John Winthrop to William Hubbard, William Bradford to Cotton Mather, wrote of the world as a vast ocean fraught with dangers for those whose voyage to America was part of the unfolding drama of sacred history. Similarly, the Catholics of Maryland saw in the glory of their exertions at sea the intercession of God and His Saints on behalf of their community. Sea deliverance provided the audience a sense of personal and communal participation in the unfolding drama of God's will working in history.

With or without an accompanying sermon, the sea-deliverance narrative offered the readers an exemplum and reinforced their sense of historical uniqueness and mission. The trial of the individual was also the trial (and triumph) of a chosen people. The sea deliverance was an artifice by which a tribal people like the Puritans of Boston or the Catholics of Maryland might respond to external dangers that threatened their well-being. The external threat was controlled by an act of imagination. Through literature the fears of a community were transcended, and the resultant "Music of the Waters" was made to express communal aspirations and ambitions.

For immigrants in the colonial period (and later times as well), the trans-Atlantic crossing was both the trial by which one began a new life and a metaphor for the transition into a life of grace. The Atlantic offered a rich source of figurative language through which the ocean traveler might organize and express the meaning of the crossing. Interpretation of a successful voyage as a sign of providential protection, of spiritual (later patriotic) fitness, did not end, though, with the colonial period. Rather it remained, in altered form, in the national consciousness as a metaphor for the manifest destiny of a nation. The ocean crossing, which seventeenth-century Americans saw as a metaphor for the soul's pilgrimage to the City of God or a trial of faith for a people especially chosen to do the Lord's work in the wilderness, became in the eighteenth century the rite of passage for Crèvecoeur's American, the New Man forsaking Europe and the past. This American, like succeeding generations of immigrants, is separated from the Old World literally by a trial by water, through which passage he is redeemed for the New. Finding no place of habitation in the ocean, Americans began to transform their sea experience imaginatively. The magic and redemptive qualities of Renaissance art are married to belief in the special mission of the American land. The challenge to order posed by the storm at sea is controlled by correcting the perspective of participant and audience to reveal the providential drama of American history.

II

Shakespeare, with other Englishmen, was attracted by the excitement surrounding the Virginia voyage of Sir Thomas Gates. Gates sailed from Plymouth on June 2, 1609, disappeared in a storm, was shipwrecked in the Bermudas, managed to escape to Virginia, and returned to England in the autumn of 1610. In the wake of all this, several pamphlets and at least one poem describing his shipwreck and deliverance were published.

The poem, "Newes from Virginia," by R. (presumably Richard) Rich, self-styled soldier and "one of thc Voyage," is twenty-two stanzas of lumbering fourteeners, the first few of which briefly recount the storm and shipwreck of Gates's *Sea Venture*, the building of two pinnaces, and the passage to Virginia. More of the poem is devoted to a defense of colonization in Virginia, and a description of the new land and its opportunities for trade and profit. But "Newes from Virginia" does summarize in a single stanza the basic cycle of sea-deliverance literature:

> The Seas did rage, the windes did blowe, distressed were they then:
> Their Ship did leake, her tacklings breake, in daunger were her men.
> But heauen was Pylotte in this storme, and to the Iland nere:
> Bermoothawes call'd, conducted then, which did abate their fear.
>
> (st. 4)

A True Declaration of the Estate of the Colonie in Virginia, a prose report officially commissioned by the Council of Virginia, was entered in the Stationers Register in November 1610, after Gates's return to England. Included are portions of Gates's report to the council and information from letters of colonists. Two more prose accounts of the travails of *Sea Venture* were written in

1610 by members of the expedition. Silvester Jourdain's *Plaine Description of the Barmudas* was published in London in October 1610. William Strachey's "True Reportory of the Wracke, and Redemption of Sir Thomas Gates" circulated in manuscript among Virginia council members and their friends, including Shakespeare, but was not published until 1625 when it appeared in Samuel Purchas's *Purchas His Pilgrimes.*

Jourdain's *Description* (about half of which is devoted to a description of the flora and fauna of the islands) includes the observation that the storm so terrified the sailors that "some of them having some good and comfortable waters in the ship, sought them, and drunke one to the other, taking their last leave one of the other" (10).

As Shakespeare discovered, Strachey's "True Reportory" is a compelling narrative, a minor masterpiece of early American literature. Elegant and bold, Strachey's prose is carefully organized and structured, its cadences stately and its rhythms skillfully orchestrated. The storm scene, whose echoes are heard in the opening scene of *The Tempest,* is colorful and dramatic, full of graphic details and powerful images:

> When on S. James his day, July 24. being Monday (preparing for no lesse all the blacke night before) the cloudes gathering thicke upon us, and the windes singing, and whistling most unusually, which made us to cast off our Pinnace towing the same untill then asterne, a dreadfull storme and hideous began to blow from out the North-east, which swelling, and roaring as it were by fits, some houres with more violence than others, at length did beat all light from heaven, which like an hell of darkenesse turned black upon us, so much the more fuller of horror, as in such cases horror and feare use to overrunne the troubled, and overmastered sences of all, which (taken up with amazement) the eares lay so sensible to the terrible cries, and murmurs of the windes, and distraction of our Company, as who was most armed, and best prepared, was not a little shaken. For surely (Noble Lady) as death comes not so sodaine nor apparent, so he comes not so elvish and painful (to men especially even then in health and perfect habitudes of body) as at Sea. (5–6)

The work of Strachey's better-known contemporary, Captain John Smith, includes two items of interest to students of American sea literature. *A Sea Grammar,* a seaman's dictionary and compendium, is full of robust, active prose and is today indispensible for understanding Renaissance nautical terminology. The wholly conventional "Sea Marke" uses the wrecked vessel in the poem—a buoy, or "Mark"—as a familiar metaphor for the vagaries of fortune. Smith's own career was in decline when the poem was written, and it appeared the year of his death, 1631, in his *Advertisements for the Unexperienced Planters of New England.*

Like Smith, Henry Norwood, seventeenth-century treasurer of Virginia, had a career marked by the vicissitude of politics and favor. A loyal supporter of Charles I, Norwood determined to leave following the king's execution. "A Voyage to Virginia" is genuine adventure, constructed and executed with great skill. Norwood's descriptions are often powerful and graphic, as for example his depiction of the famine aboard the lost and wandering *Virginia Merchant:*

> The famine grew sharp upon us. Women and children made dismal cries and grievous complaints. The infinite number of rats that all the voyage had been our plague, we now were glad to make our prey to feed on; and as they were insnared and taken, a well grown rat was sold for sixteen shillings as a market rate. Nay, before the voyage end (as I was credibly inform'd) a woman great with child offered twenty shillings for a rat, which the proprietor refusing, the woman died. (168–69)

Norwood's engaging cavalier spirit permeates his narrative. The sufferings of the passengers are ennobled by instances of courage, determination, and presence of mind, qualities all possessed in large measures by Norwood himself. He is courtly, brave, vigorous, disarmingly frank about his personal strengths and weaknesses. And he is quick to credit similar qualities in others.

The most fully realized sea-experience poem of the early colonial period was the work of John Josselyn, of England and briefly of Maine. While associated more with New England than with Virginia—he also wrote two books on the natural history of the region—Josselyn was no Puritan, and was in fact openly critical of those who were. In November 1639, while crossing from New England to London, Josselyn noted that "about three of the clock in the afternoon, the Mariners observed the rising of a little black cloud in the northwest which increased apace, made them prepare against a coming storm; the wind in short time grew boisterous, bringing after us a huge grown Sea; at five of the clock it was pitchie dark." The storm lasted three days. For a week afterward, Josselyn remembered, "all the while we saw many dead bodies of men and women floating by us." Later he preserved his memory of the storm in verse:

> And the bitter storm augments; the wild wind wage
> Wars from all parts; and joyn with the Seas rage.
> The sad clouds sink in showers; you would have thought

The high-swoln-seas even unto Heaven had wrought;
And Heaven to Seas descended: no star shown;
Blind night in darkness, tempests, and her own
Dread terrours lost; yet this dire lightning turns
To more fear'd light; the Sea with lightning Burns.
The Pilot knew not what to chuse or fly,
Art stood amaz'd in Ambiguity.

(30–32)

Like the Renaissance English who sailed to Virginia, the Puritans of New England found sea imagery both available and apt, and their diaries, sermons, and poems are replete with the language of ships and the sea. In their sea literature, however, Providence and the Novus Mundi are united to explain and extol their particular experience of a divinely chosen people.

William Bradford, for example, speaks of the ocean voyage across "The vast and furious sea" as uniting the Pilgrims in praise and thanksgiving for their deliverance, and Edward Johnson in his *Wonder-Working Providence of Sions Savior in New England* identifies the common purpose of the Puritans as their having been chosen by the Lord for special work.

Bradford's *Of Plimouth Plantation* contains a famous section identifying Providence in events on the *Mayflower*. A member of the crew, "a proud and very profane yonge man" who continually berates the Pilgrims, is struck down by sickness and his body thrown overboard. But when a virtuous young man, John Howland, is washed over the side in a storm, he manages to catch hold of a halyard and is saved. Bradford notes the hand of Providence at work in each instance. Howland's individual experience assumes communal meaning for Bradford; to him it is a symbol of the Pilgrims' enterprise as a whole. The trial of the ocean voyage is converted to the glory of the enterprise, the fortitude of the passengers made an evidence of divine favor. Having crossed "the vast ocean, and a sea of troubles" only to be greeted by "a hideous and desolate wilderness," the Pilgrims found the "Mighty ocean which they had passed . . . was not as a main bar and gulf to separate them from all the civil parts of the world." Yet the very grimness of the Pilgrims' circumstances serves to accentuate the providential nature of their enterprise, and the ocean that separates them from the rest of the world also emphasizes their favor.

One of the most heartrending of Puritan sea narratives is that of Anthony Thacher, who with his family arrived in New England in 1635 and was on his way from Newbury to Marblehead in a pinnace when a hurricane struck the coast on the night of August 14. The vessel was driven onto the rocks and broken up. The crew and passengers, including the Thacher children, drowned. Only Thacher and his wife survived. Thacher's subsequent letter to his brother in England is one of the most affecting expressions of a characteristic Puritan tension. The letter eloquently renders Thacher's inward struggle to accept emotionally what intellectually he knew to be a work of Providence:

> Now Called to my remembrance the time and manner how and when I Last saw and Left my Children and freends. One was severed from me Sitting on the Rocke at my feete, the other three in Pinnace. My Little babe (ah poore Peter) Sitting in hiss Sister Ediths arms Who to the utmost of her power Sheltered him out of the waters, My poore William standing Close unto her. All three of them Looking rufully on mee on the Rocke, there very Countinance Calling unto me to helpe them, Whom I Could not goe unto, neither Could they Come unto mee, neither Could the mercilesse waves aforde mee Space or time to use any meanes attall either to helpe them or my Selfe. Oh I yet See their Cheekes poor Silent Lambs pleading pity and helpe at my hands. (62)

Likewise in 1635, the Reverend Mr. Richard Mather of Dorchester, England, progenitor of famous sons and grandsons, sailed for a new life in Massachusetts Bay. After a voyage of several weeks, Mather reported his ship in the vicinity of Cape Anne. Sight of land brought promise of the journey's end. Two days later, however, as Mather recorded in his *Journal*, "ye Lord had not done with us, nor yet had let us see all his power and goodnesse which he would have us to take knowledge of; and therefore on Saturday morning about break of day the Lord sent forth a most terrible storme of raine and easterly wind, whereby wee were in as much danger as I think ever people were." The ship lost all three of its anchors trying to keep off a lee shore. The sails "were rent in sunder and split in pieces as if they had been but rotten ragges," and the ship was driven toward the rocks:

> In the extremity and appearance of death, as distresse and distraction would suffer us wee cryed unto the Lord, and he was pleased to have compassion and pity upon us; for by his overruling providence and his owne immediate good hand, he guided the ship past the rock, asswaged the violence of the sea, and the wind and raine, and gave us a little respite to fit the ship with other sayles, and sent us a fresh gale of wind . . . It was a day very much to bee remembered, because of the day the Lord graunted us as wonderful a deliverance as I think ever people had, out of as apparent danger as I thinke ever people felt. (28–29)

Mather's sea deliverance comes at a dramatic moment in his *Journal*. The ship is within sight of land, the voyage nearly over. The storm and deliverance occur as a final corrective that restores the passengers' perspective, making them worthy of their divinely appointed mission in the New World. For Richard Mather and his descendants, the ocean crossing was a regenerative experience that prepared the heart for personal salvation and communal mission.

A similar experience—both personally and mythically—befell Father Andrew White, an English Jesuit, who two years before Richard Mather's voyage to Boston in 1635 left the Isle of Wight in the company of a fleet bound for the new colony of Maryland. During the voyage Father White's company ran into a gale, and "About ten o'clock at night a black cloud rained down a direful tempest." The mainsail was ripped from the mast, and even the bravest sailors were terrified of the storm. "The tempest animated the prayers of the Catholics," Father White recalled, and like the voyage of his Puritan contemporary, Richard Mather, Father White's sea deliverance is dramatic and transforming, a sign of special Providence for a chosen people (10–12).

The importance of sea literature to the Puritans is indicated in Increase Mather's 1684 *Essay for the Recording of Illustrious Providences,* whose first chapter is devoted to sea-deliverance narratives. Mather's intention was twofold: to impress upon his readers the essential mystery of nature, in which God acts providentially, and to demonstrate to his audience the special Providence of New England. Anthony Thacher's letter heads the list of narratives; several others are repeated from the Englishman James Janeway's collection of sea-deliverance narratives, *Legacy to His Friends* (1674). Typical is the story of Captain Edward Gibbons of Boston, his ship delayed by contrary winds and then becalmed. Gibbons and his crew faced starvation and cannibalism. The desperate men drew lots, "but before they fell upon this involuntary Execution, they once more went into their Prayers, and while they were calling upon God, he answered them for there leapt a Mighty Fish into the Boat." Twice more they are down to lots when saved by Providence. First, a large seabird alights on a mast and is taken and eaten. Later, they are rescued by a passing ship. Similarly, in each of the *Essay*'s nine remaining narratives the principals are providentially saved from shipwreck, storm, or starvation at sea.

A generation later, the sixth book of Cotton Mather's *Magnalia Christi Americana; or, The Ecclesiastical History of New England* (1702) is devoted to "Illustrious Discoveries and Demonstrations of the Divine Providence." As in his father's *Essay,* chapter 1 of Cotton Mather's "Christus Super Aquas," is devoted to sea deliverances, and seven of the eleven are taken from Increase Mather's work. The sea deliverances illustrate the mission of America. Suffering and providential intervention glorify and affirm political destiny. Cotton Mather uses the history of New England, including its sea deliverances, as part of an unfolding divine drama. He also uses the empiricism of science to demonstrate Providence and confute its critics: "We will now write a book of rare occurences, wherein a blind *fortune* shall not be once acknowledged . . . all the *rare occurences* will be evident operations of the Almighty God." He also adopts the new literary fashion of the eighteenth century in diction and sentiment. God is now "the Great Governor of the World" (Newton's world, to be sure), and in recreating the sea experience, Cotton Mather consciously provides a narrative distance more appropriate to the age: "I will carry my reader upon the huge Atlantic, and, without so much as the danger of being made sea-sick, he shall see 'wonders in the deep'" (341–43).

Richard Steere's *Monumental Memorial of Marine Mercy*, published in the same year as Increase Mather's *Essay*, describes the storm-tossed, Boston-to-London crossing of the ship *Adventure*. Tossed "like a ball in Sport, / From wave to wave in Neptunes Tennis Court," *Adventure* nearly foundered, the passengers at one point using bearskins to stanch the leaks which multiplied throughout the ship. Steere's poem not only creates a vivid picture of the terrors of a ship caught in a violent storm at sea, but it also provides a perceptive psychological portrait of the attitudes of the passengers and crew. Steere makes use, too, of the familiar voyage metaphor to enlarge the reader's understanding of man's dependence on Providence; once again, the narrative structure parallels the spiritual passage. The ocean crossing serves as a metaphor for the reader's progress from ignorance to truth, from indifference to understanding. The several physical crises that confront the passengers and crew of the *Adventure* bring a better understanding of Providence. According to Steere, voyage without peril might well induce complacency, a condition of acute spiritual danger:

> Had we continu'd *thus* upon the Deep
> We had bin Charm'd into a drowsie sleep
> Of calme Security, nor had we known
> The Excellence of PRESERVATION:
> We had been Dumb and silent to Express
> Affectedly the Voy'ges good success.
>
> (40–45)

The phrase "Express Affectedly" is an important one. Steere's purpose is not only to recount the events of his experience but also to move and improve the audience. He must recreate the situation so that his audience experiences the spiritual progress of "the Voy'ges good success."

Benjamin Bartholomew's verse account of the perilous voyage of the ship *Exchange* en route from Boston to Barbados in 1660 is less felicitous than Steere's poem, but interesting for its portrayal of the captain who in the midst of panic among passengers and crew manages to keep the ship afloat during a tremendous gale. Despite its faults, Bartholomew's is an engaging work. Implored by the passengers and crew to lead them in prayer as the vessel appeared certain of foundering, the captain replied, "to pray I would agree / Yet for our lives Let all meanes used be: / Therefore I think tis best to hoyse foreyarde" (ll. 123–25).

The conscious humor of a restoration wit animates the anonymous verse narrative *A Description of a Great Sea-Storm, That Happened to Some Ships in the Gulph of Florida, in September Last; Drawn up by One of the Company, and Sent to His Friend at London* (1671). The author was emigrating to the New World when the fleet in which he was sailing was struck by a hurricane off Florida's Atlantic coast. Internal evidence suggests the writer was something of a rake and well-acquainted with the seamier side of London tavern life. His comic spirit is evident throughout the poem, and he observes all goings on (and his own poetic efforts) with a certain wry detachment and a well-developed sense of fun. A charming bit of verse, the *Description* celebrates human foibles and confusion (including the author's) in an amusing and engaging way.

The American sea literature of the first half of the eighteenth century is, like that of the seventeenth, predominantly a narrative literature in which numerous sea captains and sailors recorded their deliverance from storm, shipwreck, battle, and pirates.

The most famous eighteenth-century shipwreck in New England was the wreck of the *Nottingham Galley* of London on Boon Island on December 11, 1710. Three weeks after their ship was broken up on the Island where their only provisions were a few bits of cheese, some beef bones, seaweed, and a single sea gull, the crew began to eat their dead. A few days later the castaways were sighted from the shore and taken off the island. News of the tragedy shocked the coastal communities of New England and even created a stir in England. Cotton Mather promptly appended a version of the story to a sermon written in response to the events on Boon Island. Well aware that a market for his story also awaited them in England, Captain Dean hurried back to London, where two editions of his narrative were run off in 1711. After Dean's first London edition appeared, the first mate of the *Nottingham Galley*, Christopher Langman, and two crew members published their own version of the story. Langman and the others contradict Dean's account in several respects and blame the captain for the wreck. In several later editions Dean stoutly maintained his innocence.

In 1722 nineteen-year-old Philip Ashton of Massachusetts was kidnapped from Port Rossaway, Nova Scotia, by the buccaneers of the infamous pirate, Captain Ned Low. For nearly a year after his capture in Nova Scotia, Ashton was the captive of Low's swashbuckling crew. After a number of adventures, he finally escaped the pirates, only to be marooned on an island in the Bay of Honduras. There he endured sixteen months of privation, hunger, and loneliness before he was rescued and returned to New England. Once home, he was interviewed by the Reverend John Barnard, and three months later, Barnard's reconstruction of the Ashton story, together with a sermon, was published in Boston as *Ashton's Memorial: An History of the Strange Adventures and Signal Deliverances of Philip Ashton, Jr.* The narrative invites comparison with *Robinson Crusoe*, published six years earlier, though the young sailor from Marblehead had none of the fortuitously placed provisions and implements that enabled Crusoe to transform his island into a thriving economy. Probably Ashton was not so pious or prudish as Barnard sometimes makes him out to be. Barnard's acquaintance with the literary fashion of the age in England influenced the narrative's style, which is occasionally overwritten and sentimental. By and large, however, he renders Ashton's adventures with immediacy and skill.

In 1726 William Walling of Middletown, New Jersey, took passage on a small boat in order to sail home from Manhattan. Walling and his pilot were caught in a sudden storm and blown out to sea. After eight days, Walling's feet were frozen, and his pilot had gone mad, appearing on deck naked in the frigid weather, insisting that the devil had ordered him to dress in petticoats. Walling lost several toes before a passing ship rescued him. By that time his shipmate was dead. Walling renders all of this in a style that is understated but compelling, and makes the delirious antics of the deranged pilot positively eerie.

Briton Hammon, a black man of Marshfield, Massachusetts, sailed from Plymouth in December 1747. In turn, he was shipwrecked, captured by Indians, traded to the Spanish, imprisoned, and left to die. Later he escaped, was recaptured, escaped again—this time to the British navy. He served on several ships and was wounded during a battle with a French man-of-war. Discharged in London in 1760, he made his way back to Marshfield to the home he had left thirteen years before. Hammon's relation is attractive on its own merits, but it is also remarkable for its being the work of a slave, a black man owned by General John Winslow of Marshfield. But Briton Hammon was no ordinary man. Sensitive and pious by nature, he was literate and well-read, especially in the Scriptures. His narrative is a tale of patient suffering and perseverance, of hardship matched by gritty determination—a story told in a style that is terse yet emotionally charged. Hammon's narrative is also one of the more important works of black American writers in the colonial period, and students accustomed to the poetry of Jupiter Hammon (apparently no relation) and Phillis Wheatley will find considerable merit in Briton Hammon's sturdy prose.

Captain Joseph Bailey, of Antigua, left New York, on December 20, 1749, with a cargo of merchandise and weighed anchor for home. Bailey risked bad weather in the belief that he could outsail winter storms as he hurried southward for the West Indies. It was not long before he regretted his decision, but as he says, "the Wind being northerly, there was no going back." On Christmas Eve a howling gale overtook Bailey's brigantine, beating the vessel about for three days and ultimately oversetting her. For seven days Bailey and his crew clung to the battered vessel in the choppy waters of the Atlantic before they were picked up by a passing ship. Bailey's is a dramatic narrative, taut and well-executed. A wealth of detail and concrete images crisply related render indelible a succession of action-filled scenes such as this one:

A terrible hard Gale ensu'd that wash'd our Deck of every thing that was moveable, Lumber, Coops, and Water-Casks; and for some Time the People were up to their Necks in Water at the Pump. Night came on and the Weather increas'd; and about 8 Clock in the Evening a Sea poopt us, and drove in our two Larboard Lights in the Cabbin, which almost fill'd it and the Steerage, putting out all our Lights below, and spoil'd our Tinder; but by the Help of a Pistol with some Powder I got Light again. I found the Sea had carry'd away one dead Light, so that to supply the deficiency, was obliged to put my Bed into one of the Windows, and secur'd it in the best Manner I was capable. I then set the Boys to bailing the Water out of the Steerage with Buckets as fast as they could, and call'd for the Broad-Ax and plac'd it at the Cabbin-Door, so that I might readily find it in Case she Broached too, to have cut away her Main-Mast. (6–7)

On November 22, 1752, Nathaniel Peirce of New Hampshire sailed for Louisbourg with a cargo of lumber. Winter storms crippled his vessel and then reduced it to a floating wreck. One by one the nine-member crew was washed overboard until only Peirce remained. After twenty days on the wreck (seven by himself), the captain was sighted by a passing ship and taken to Portugal. Even in the company of narratives in which instances of great courage and endurance are common, Peirce's story is unusually compelling. A plainspoken ship's captain, Peirce writes in a style that is direct and yet rich in active detail. "But what will not a Man attempt to save Life?" he asked—and answered with his own example: "No Man knows what Hardship he can bear, 'till he is bro't to the Test." The captain discovered his own inner resources as he watched his crew succumb one by one to delirium and death:

> The 16th, Nathaniel Barns the Mate, by his Behaviour, shew'd that he was delirious; talking and behaving much as Brown had done; and for Fear that he should steal from them in the Night, as Brown had done, they with a Reef Plat took a Turn round his Body, under his Arms, and made the Ends fast to the Crutch, which was between the Master's Birth and Timothy Cotton; and before Night they tho't he was struck with Death by some Signs in his Countenance, and the Motion of his Body; but when Night came on, he lay still and soon died, without any Struggle, that they could perceive; and the next Day in the Afternoon they threw him into the Sea. (10)

For such writers as Captain Peirce, the sea experience was transformational. The earliest writers of sea literature in America described their sea experience using images inherited from centuries-old dreams of Novus Mundi and the Terrestrial Paradise. During the Renaissance, the sea was a magical barrier, and one who successfully crossed it overcame time and history in the Earthly Paradise. Not surprisingly, Puritan and Catholic writers of sea literature took these images and converted them to their own purposes. Indeed they stood them on their head. The Puritans transformed the Earthly Paradise into the Late Howling Wilderness and saw the ocean crossing as a regenerating, converting experience that made sacred what was otherwise secular space. The New World's ocean offered vast imaginative possibilities to Renaissance and Puritan writers alike.

In a sense both succeeded. The American tendency to deny history, to invest always in what Fitzgerald would later call the "orgiastic future" is traceable to the Renaissance belief in Novus Mundi, the Earthly Paradise, Land of the Golden Age, the Fountain of Youth. The American sense of mission and destiny in history comes largely from the Puritans' efforts to fructify their existence in a wilderness city on a hill largely ignored by a Europe "dead in sin." In both instances, the sea—vast, brooding, mysterious, and fraught with dangers—is the transforming crucible from which the American hero returns with the dream made whole. In the written responses to that experience lie the beginnings of American sea literature.

DONALD P. WHARTON

Chapter Two

THE REVOLUTIONARY AND FEDERAL PERIODS

During the first two of the four wars between England and France for control of the North American continent (King William's War and Queen Anne's War, 1689–1713), the American colonies became a maritime country. Seaborne commerce increased in spite of the deprivations of war. Pirates and privateers flourished, but so did trade. Conflict on the seas forced merchants to look for new ports. Strange currencies abounded. Large profits were made even as official government receipts diminished. Fishing, whaling, and slaving expanded, as did the coastal and West Indies trade. Shipbuilding in New England boomed. Between 1674 and 1714, 1,332 ocean-going vessels were constructed in New England shipyards; from 1712 to 1720 alone, 700 were built (Samuel W. Bryant, *The Sea and the States*). A century of discovery and exploration had ended. The age of sea-borne commerce had begun.

American literary attitudes toward the sea changed, partly as a result of the growth of maritime commerce and technology, but also in response to aesthetic changes in the eighteenth century. Renaissance and seventeenth-century sensibilities were replaced by neoclassicism and the Age of Reason in Europe and America. The decades following the Restoration saw a shift away from the often internecine theological concerns of the seventeenth century toward what was styled the Grand Alternative: Nature (and in Jefferson's phrase, "Nature's God"). Nature in the eighteenth century meant not wildness but its opposites: order, unity, law, proportion. In literature (as in religion), decorum became a universal ideal.

Behind these changes lay a changed perception of the universe. Its scientific basis became clear and Newtonian laws replaced Providence as the guiding force of the cosmos. The rationality of science allowed Americans to understand events and experiences according to the laws of nature. Natural phenomena were no longer to be explained by the supernatural or the mysterious. Nature and its effects were the result of material causation, not Divine Intervention. Providence, as it was used in the eighteenth century, came to mean the Divine Order, not the intercession of God in specific events; storm and shipwreck were explained as disorders within the orderly frame of the universe. Providence provided only the assurance of a regulated universe held in the Divine Mind by the laws of nature, not interference in specific events.

The changed sense of Providence presented a new cultural framework. No longer did storm, shipwreck, battle, or capture by pirates symbolize favor or mission. What made sense of these phenomena was not the intervention of God to make His purposes known. Therefore, no longer was it possible for a community to see in the suffering and deliverance of the storm-tossed individual its own special call to greatness; providential terms once used to explain (and even celebrate) the vicissitudes of life on the sea lost their credibility. Indeed, many writers were actively antagonistic to such explanations, describing them as remnants of an age of appalling ignorance, fear, and superstition.

Aesthetically, this new perception of the world led to a certain distance between artist and subject. Thinking they understood the laws that regulate the universe, human beings were able to stand back and observe the whole appreciatively. The concept of the Sublime provided a way to apprehend and describe the mysterious, explosive power of nature contained within an ordered whole. In literature, the aesthetic distancing was reflected in the neoclassicism of the age and its concern with order, proportion, and decorum, and the conscious reliance on classical models and modes of expression. Imagery moved away from the detailed and sensuous in nature toward the abstract and heroic. This increasing rationalism and classicism in the eighteenth century greatly influenced the literature of America, including its sea literature.

As the seventeenth century ended, Americans tended to view the sea less as an avenue for exploration than as a highway of commerce, and this changed perception altered sea literature. The expanding commercial might of the colonies and a growing sense of American nationalism were reflected in the narratives, poems, and emerging fiction of the sea in the 1700s. The sense of communal uniqueness and mission found in the providential literature of the seventeenth century found new expression in the nationalistic fervor of the Revolution and the War of 1812. The special mission of America came not from providential experience so much as from the blessings of nature. In the rational view of the eighteenth century, the New World offered again, as in the dreams of the Renaissance explorers, a pristine land preserved by nature and protected by the vastness of the surrounding ocean from the corruption of Europe—a fit home for Liberty and the New Man, the American.

Writers of sea literature in the eighteenth century saw the sea as a shield and protector on the one hand, and, on the other, a dark reminder of the portentous power of Nature unleashed and uncontrolled. The transatlantic crossing remained a pivotal experience, a symbol of evolving (and enlarged) cultural expectations. The American colonies emerged as a nation expected to fulfill the promise of Nature. The providential wilderness became the New Eden. Where once men and women took courage and saw their faith rewarded by the Lord's wonders in the great deep, they were now made to see the sublime order of the mechanical universe from the perspective of what Philip Freneau called "Dread Neptune's wild, unsocial sea."

Freneau's work represents the fully developed attitudes of American sea literature in the eighteenth century. No clear line identifies the beginning of those attitudes: like indicators of an unknown coast before land is sighted, the literary evidences appear in such hints as the scent of land, or scattered boughs on the surface of the sea. The coast turns out to be irregular. Older aesthetic attitudes endured long into the new century, as we have seen in the narratives of Peirce, Bailey, and Hammon. Other writers early indicate a changing tone and perspective, and signal the coming transition.

In this regard it is useful to look at a writer whose work reflects something of the change, Richard Steere (1643–1721). Steere's *Monumental Memorial of Marine Mercy*, published in Boston in 1684, is thoroughly reflective of Puritan theological and aesthetic attitudes. In 1713, Steere published a second volume of poetry. The title poem, "The Daniel-Catcher," written in the 1680s in England, is Puritan and anti-Catholic in tone. Other poems in the volume, written later, reveal a changing aesthetic. Among these is "On a Sea-Storm Nigh the Coast" (89).

> All round the Horizon black Clouds appear;
> A Storm is near:
> Darkness Eclipseth the Sereener Sky,
> The Winds are high,
> Making the Surface of the Ocean Show
> Like mountains Lofty, and like Vallies Low.
>
> The weighty Seas are rowled from the Deeps
> In mighty heaps,
> And from the Rocks Foundations do arise
> To Kiss the Skies

Wave after Wave in Hills each other Crowds,
As if the Deeps resolv'd to Storm the Clouds.

How did the Surging Billows Fome and Rore
    Against the Shore
Threatning to bring the Land under their power
    And it Devour:
Those Liquid Mountains on the Clifts were hurld
As to a Chaos they would shake the World.

The Earth did Interpose the Prince of Light,
    Twas Sable night:
All Darkness was but when the Lightnings fly
    And Light the Sky,
Night, Thunder, Lightning, Rain, and *raging* Wind,
To make a storm had all their forces joyn'd.

No providential design informs this poem. It is instead an aesthetic contemplation of natural phenomena. Steere's subject is not, as in one of his earlier poems, "The Excellence of PRESERVATION," from a storm at sea but the composition of the storm itself. The imagery of the poem is more abstract than that of the 1684 poem and lacks the latter's accumulation of concrete detail. The point of view is different, too. Gone is the immediacy of the narrator enmeshed in the action of the event. Here Steere is consciously constructing a seascape drawn from the vantage point of distanced observer. Despite the violence of the action, the tone of the poem suggests a dispassionate quality of observation, an angle of mind essentially reasonable rather than passionate, having about it a quality of scientific interest or curiosity. The narrative attitude is one of compelled wonder and admiration at the fury of the storm, of its vast primeval power, its threat of a universe barely held in check. In short, Steere is looking at what other writers define as Sublime.

    Steere's interest in the aesthetic composition before him is a scientific one. He is curious about the mechanics of the storm, its constituent parts, and their relationship one to another. He focuses on the storm's testing of the limits of the natural order as earth blots out the sun ("the Prince of Light") and threatens to reduce all to chaos. This aesthetic contemplation marks one pole of the eighteenth-century attitude toward nature and the sea. More commonly, such contemplation is accompanied by matter from the other pole—pure didacticism of one kind or another: moral, patriotic, or even scientific. Rarely does an aesthetic response occur alone. The Sublime offered a framework from which the mystery and power of disorder could be safely viewed, but usually with a caveat against too much of a good thing. Outside of social control, disorder is harmful and destructive, and eighteenth-century writers for the most part treat it warily.

Steere's poem thus marks a frontier. The poet and minister Mather Byles represents a more typical attitude, combining sentiment, a changed sense of providence, and a conventional expression of the sublime. Byles's "Hymn at Sea" (1732) invokes the providence of the Master Builder of the universe—"Thy Pow'r produc'd this mighty frame / Aloud to thee the Tempests rore"—and for Byles the variety and disorder of nature compel wonder not in and of themselves (in their art), but as they argue for the existence of a distant eighteenth-century providence:

> Each various Scene, or Day or Night,
> LORD, points to thee our ravish'd Soul'
> Thy Glories fix our whole Delight:
> So the touched Needle courts the Pole.
>
> (17–20)

The deficiencies of the "Hymn at Sea" are readily apparent—it was parodied by fellow poet Joseph Green not long after it appeared—but its point of view and aesthetic attitude are typical of sea literature before the Revolution. Byles represents the vitiated Puritan intellectual climate of the eighteenth century. In its accommodation with the Age of Reason, the force of Puritanism was diverted into sentimentality and emotionalism. The result was a set of stock responses characteristic of the American literature of the period.

Perhaps the most excessive of such productions is the ludicrous and nearly incoherent verse account of nine Yale undergraduates caught in a squall on Long Island Sound in 1727 ("Rolling and Tossed on the Foaming Brine / Recall the Day our fatal Canvas spread"). The narrative is never allowed to interfere with the extravagant heroic diction and incongruous imagery. The gratitude attempted by the author, John Hubbard, is as unfocused as the verse, developing no conscious views either of providence or of art. However, the tendency in Hubbard's verse to sensationalize events is an increasingly common characteristic of sea literature in the eighteenth century.

The narrative of Captain David Harrison's ill-fated voyage in the *Peggy*, bound to New York from the Azores in 1765, exhibits literary influences and purposes not found in seventeenth-century sea literature. Harrison makes clear what in Steere's "On a Sea-Storm Nigh the Coast" was only implicit, that the work was for entertainment as well as for instruction. Harrison's references to Providence are perfunctory and mannered. He admits what no Puritan could bring himself to say, that imagination and entertainment stand equal to didacticism in the purposes of sea literature. The language of Harrison's narrative reflects the sentiment and sensationalism of the age. The earlier narratives of Henry Norwood (1649) and John Dean (1711), for example, contain instances of cannibalism among starving sailors and ship passengers. Norwood's description is very much understated, and Dean's, while graphic, is straightforward. But Harrison describes similar circumstances aboard the *Peggy* in the elevated and effusive style of the eighteenth-century English novel:

> The miserable Black, however, well-knowing his fate was at hand, and seeing one of the fellows loading a pistol to dispatch him, ran to me begging I would endeavour to save his life.—Unfortunately for him I was totally without power. They therefore dragged him into the steerage, where, in less than two minutes, they shot him through the head.— They suffered him to lye but a very little time before they ripped him open, intending to fry his entrails for supper, there being a large fire made ready for the purpose;—but one of the foremast-men whose name was James Campbell, being ravenously impatient for food, tore the liver from the body, and devoured it raw as it was, not withstanding the fire at his hand where it could be immediately dressed. The unhappy man paid dear for such an extravagant impatience, for in three days after he died raving mad, and was, the morning of his death, thrown overboard,—the survivors, greatly as they wished to preserve his body, being fearful of sharing his fate, if they ventured to make as free with him, as with the unfortunate negro. (23–25)

In the nineteenth century, Edgar Allan Poe read a shortened version of Harrison's narrative and drew on some of its events for *The Narrative of Arthur Gordon Pym.*

By the outbreak of the Revolution, the providential sea-deliverance narrative of a century earlier no longer existed. Providence as an aesthetic influence of consequence was gone, its reference reduced, as in Harrison, to mere convention, its original symbolic force lost in expressions of piety and moralizing. In the seventeenth century, survivors of the sea deliverance (and their audience) could lay claim to insight enlarged and corrected by explicit doctrinal interpretations of the experience. In the eighteenth century, sea-deliverance survivors are simply restored to a harmonious social order.

Conventional piety in the manner of Harrison continues in narratives and poems written after the Revolution. Barnabas Downs, Jr., was a young New England sailor whose privateer, *General Arnold,* was caught in a severe winter storm and driven aground on a shoal in Plymouth harbor in December

1779. Seventy-two of the one hundred and five crewmen froze to death, and many of the survivors, including Downs, lost limbs afterward. Stylistically, Downs's narrative (published in 1786) resembles Harrison's: "It is not possible to describe sensations at this period; death appeared inevitable, and we waited every moment for its approach! Even now, when I recollect my feelings, it is difficult to steady my pen!" His piety is entirely conventional both in his narrative and attached poem.

Two New England broadsides published in 1792 further illustrate the conventional quality of providential sea literature in the eighteenth century. Indeed, the sentimentalism and moralism of these works is mannered even by eighteenth-century standards. The coffin-emblazoned *True Account of the Loss of the Ship Columbia* consists of fairly reportorial prose and twenty lines of poetry ("Here! see! the dread effects of Heav'ns decrees, / The Ship COLUMBIA tost by boist'rous seas"). The final lines illustrate the contradicting values of Puritanism and sentimentalism in eighteenth-century American literature:

> From hence this lesson learn, that all our joys
> Are but delusive, vain and empty toys,
> Yet 'tis allow'd the sympathetic tear,
> Must drop, and yield to natures tender care.

A True and Particular Narrative of the Late Tremendous Tornado (also with coffins) is a broadside describing in prose and verse the storm that struck New York harbor on July 1, 1792, "When several PLEASURE-BOATS were lost . . . and THIRTY Men, Women and Children, (*taking their* Pleasure *on that* Sacred Day) were unhappily *drowned* in NEPTUNE's raging and tempestuous *Element*!!!!!!" A stern warning to Sabbath-breakers from Massachusetts to the Middle States, both prose and poem mix neoclassic diction and imagery with Puritan moralizing and commercial vision:

> Our *Commerce* shall increased be,
> He'll prosper those who skim the sea
> And bring our ships with wealth from far
> Because our *GOD* is always near:
> Our *Arts* will thrive thro' our domain,
> Nor *Science* rear her head in vain.

The declining aesthetic influence of Providence in sea-deliverance poems and narratives was not without influence in another genre. Ministers could no longer depend on a belief in Providence to control the aesthetic of the lay writer of sea deliverances. Instead, they adapted the narrative's dramatic context to sermons. A single example: On December 17, 1769, William Whitwell addressed his congregation on the occasion of a series of maritime disasters that had befallen Salem in recent weeks. A generation earlier, from the same pulpit, John Barnard, secure in the knowledge that his audience would understand and respond to the typology of the sea deliverance, preached not on Philip Ashton among the pirates, but on the three Hebrews in the fiery furnace. Whitwell, on the other hand, brought the sea-deliverance experiences directly into his sermon, adapting its drama for purely theological purposes. In the middle of his sermon, Whitwell, like Thacher, Steere, Ashton, Bailey, and others, took his audience into the deep:

The sky, which looked serene and fair, gathers blackness: The Winds, which were moderate, rise and blow with violence: The sea, which was as smooth as glass, tosses its waves mountains high: The night comes on: The storm increases: The winds rage: The rain beats upon you, and the sea is ready to swallow you up: "you reel to and fro, and stagger like a drunken man"; "are at your wit's end," and "your soul melts within you because of trouble." Oh! who can describe how great the trouble is? your hearts can paint it in more lively colours than I am capable of. (8–9)

Whitwell's experiment with narrative technique reflects his compromise with the age. Tales of sea adventure reflected the reading public's taste for sensationalism and the exotic, and collections of shipwreck and maritime disaster narratives, in addition to numerous printings of individual accounts, appeared in England and America throughout the eighteenth century. *A Journal of the Travels and Sufferings of Daniel Saunders, Jun.* (1794) and Benjamin Stout's *Narrative of the Loss of the Ship Hercules* (1798) are representative of this period.

Saunders's is a straightforward account of the wreck of the *Commerce*, of Boston, on the coast of Oman and the crew's month-long trek to Muscat. During that time they were robbed and beaten, and suffered terribly from heat, cold, thirst, and hunger. "Some of the company," Saunders observed, "were driven to the sad necessity of making use of a sustenance, to save themselves from perishing, too disagreeable to be named." Only eight of seventeen Americans from the *Commerce* survived the journey. At Muscat, "snatched from the very jaws of death, thanks to the Supreme Disposer of all events, we were once more placed in a situation to seek a living in this variegated, troublesome world." Before he got back to his native Salem, Saunders was impressed on a king's ship, where he was kept several weeks, until he made his escape, a phenomenon over which the United States and Great Britain would go to war within a decade.

Saunder's narrative is complemented by Benjamin Stout's more lively account of the wreck of the *Hercules* on the southeast coast of Africa. The narrative is prefaced by a long-winded dedication to President John Adams inviting the president's aid in assisting the African natives against the oppression of the Dutch. It also calls for commercial development of the region by the establishment of an American colony. Stout was an eighteenth-century man of reason (and sentiment), who wrote sympathetically and accurately of the natives of South Africa, whom he describes as compassionate, generous, and hospitable—children of nature, emphatically not savages. He recommends that the natives be taught the reasonableness of Christianity, but no doctrine, which he regards as litigious and unreasonable. His stated motives for publishing are educational, not theological. The work itself falls into two parts, the storm at sea that crippled the *Hercules* and forced Stout to run her ashore, and his description of the South African countryside and her peoples. The storm scene replicates in graceful prose the natural Sublime described by Steere nearly a century earlier:

> Although bred to the sea from my earliest life, yet all I had ever seen before, all I had ever heard of or read, gave me no adequate idea of those sublime effects which the violence and raging of the elements produce, and which, at this tremendous hour, seemed to threaten nature itself with dissolution. The ship raised on mountains of water, was in a moment precipitated into an abyss, where she appeared to wait until the coming sea raised her again into the clouds. (3)

The threat of dissolution, central to the theory and emotion of the Sublime, is redeemed by the social order of the African natives. Stout's narrative (unconsciously, one suspects) contrasts the two controlling symbols in the iconography of eighteenth-century sea literature: the disordered and disorienting force of the "wild, unsocial sea," and the controlling Newtonian universe reflected in the social order. The violent and disorienting experience of the storm and shipwreck is contrasted with the march to Cape Town, an endeavor assisted by organized human, and humane, solicitude. In their natural state, the native Caffirs embody more purely the organizing principles of the neoclassic cosmos than do the corrupt, "savage" Dutch, a comparison that recalls the Renaissance myth of the Golden Age.

The eighteenth century also saw an increase of American sea experience in fiction. Its antecedents were in the providential tradition of sea deliverance, in promotional and discovery literature, and in the English fictional models of Bunyan and Defoe. Such antecedents ought not to be thought of, though, as some primal literary protoplasm out of which the novel may be said to "rise." Rather, each of these forms had its own aesthetic constraints and opportunities. All of them were influenced by changing cultural assumptions. Locke's epistemology, for instance, says Roger Stein, may well have influenced the development of fictional form as much as the presence of related narrative forms.

There are cultural and aesthetic similarities, for example, between the sea-deliverance narrative of Benjamin Stout and that of William Williams's *Mr. Penrose: The Journal of Penrose, Seaman.* Written sometime in the 1770s, *Penrose* details the life of Llewellyn Penrose, who as a young man determines to go to sea against his mother's wishes, a matter he later regrets, finding that "the ocean seldom softens the passions." Later he is unintentionally marooned on the coast of Nicaragua. Penrose survives rather in the manner of Crusoe, though with less material aid. He establishes relations with the local Indian population and marries two successive wives. The heart of Penrose's story is the development of family and community in a primitive setting. Penrose, cut off from civilized society by an unsocial sea, is redeemed, morally as well as physically, by children of Nature in a New World Eden. In the course of his exile, Penrose relinquishes the corruption of civilization and internalizes the values Benjamin Stout associates with native South Africans. Penrose's community prevails through crisis, adventure, and tragedy. When offered the opportunity, he refuses to return to civilized society, and dies in Nicaragua twenty-seven years after his arrival there. His last wish is that his son bring his journal to the English-speaking world. Hoping his experience will serve as a moral exemplum to his countrymen, Penrose, like Columbus, seeks to preserve his story for posterity. Williams's novel is convincingly told, with a wealth of effective detail and realistic characterization that raise it far above its peers in early American fiction.

Charlotte Temple (1791) and *The History of Constantius and Pulchera* (1794) are rather different matters. Susanna Rowson's *Charlotte* is not a sea novel at all, but is worth mentioning here because of the ocean's powerful, albeit brief, symbolic role. Charlotte's "voyage" is toward moral corruption. The physical ocean is at once a symbol of the boundary between virtue and vice, and a reminder of the unsocial, destructive force of unrestrained nature. Rowson's didactic novel uses the sea as a negative symbol and a foil for the place of reason in social relations.

Constantius and Pulchera, on the other hand, makes more extensive use of the sea in a plot whose peregrinations include separation of the lovers, forced betrothal, impressment, escape, disguise, battles, shipwrecks, and near-cannibalism. The disorder of the unsocial sea is reflected in the vicissitudes of the lovers. Misfortune threatens to destroy their relationship even as the storm at sea threatens nature with dissolution: "Their ship, by the flashes of lightning, seemed to be wrapped in sheets of fire, which were accompanied by such dreadful peals of thunder as indicated that the final dissolution of universal nature was fast approaching" (18). The pair are ultimately reunited and make their escape from the disorder of the sea, redeemed by their experience and made worthy not only of each other but also of the New World. Constantius pledges that together "we will cross the Atlantic, we will revisit the land of freedom," there to be married and live as the new man and woman in a world removed from the moral and political disorder of Europe.

While many of the adventures of Dr. Updike Underhill, the narrator of Royall Tyler's *Algerine Captive* (1797), are initiated by water, the sea experience in the novel is largely circumstantial. Underhill's real voyages are moral and intellectual. Like Penrose, Underhill's travels lead him to reconsider his ethical stance, particularly toward slavery. Made a slave himself, Underhill learns through painful experience the moral limitations of humanity. The wry detachment and easy satire of the first volume gives way in the second to a process of disillusionment and insight. The sea is the moral boundary that Underhill crosses and from which he returns, chastened.

Perhaps the oddest of early sea novels in America is *The Female Marine* (1816), a work that includes a number of genres in its picaresque structure. Like the heroine of *Constantius and Pulchera*, Lucy Brewer, the female marine, spends considerable time disguised as a man. Unlike Pulchera, however, Lucy goes to sea to escape the disgrace of having lost her virtue ashore. Her redemption at sea comes in fighting aboard the *Constitution* in the War of 1812. For the patriotic and nationalistic Lucy, the sea is a field of action on which redemption is won through heroic effort.

Charles Lennox Sargent's *Life of Alexander Smith* (1819) uses many of the external events of the mutiny on the *Bounty*. Smith, an American, goes to sea early, learning his craft on the Grand Banks. Later, he ships to Europe and India and is abandoned on an island off Madagascar. There, in the manner of Crusoe and Penrose, he survives and later escapes. Smith then finds himself again abandoned, this time in the Falklands. He makes his way to England and joins the *Bounty*. From that point the book follows the general pattern of historical events and ends in 1814 as the first generation of Pitcairn children marry. Thematically, *The Life of Alexander Smith* has close ties to *Penrose*, for in the island survivals of Smith we find a familiar pattern of redemptive experience in the midst of an exotic environment. Pitcairn is the most significant locale, for there, like Penrose in Nicaragua, Smith learns to abandon the decayed moral accoutrements of civilization. Isolated from Europe and, interestingly, America, he builds the utopian community made possible by nature and nature's children. In this way, the novels of Williams and Sargent and the narratives of Saunders and Stout mark the beginning of a shift in the focus of writers seeking exotic, unspoiled nature. America, so long the New Eden of sea dreamers, gives way to the more mysterious shores of Africa and the Middle East, South America and the South Pacific.

Eighteenth-century sea novelists such as Williams and Smith had their counterparts among American dramatists. Treatment of the sea experience by American drama, which began only at the very end of the eighteenth century, was given added impetus by the naval conflict with the Barbary pirates and by the War of 1812. Most of the theater products of the period were occasions for patriotic effusions and demonstrations of martial enthusiasm. Many were musicals with a few lines of dialogue inserted between the songs. Dramatic realism and rounded characterization were not major concerns of the authors, and stock characters and stage types such as the "Yankee" and "Jack Tar" were common. For the most part, depictions of sailors were limited to such types. Nautical language was used primarily as an aspect of the stereotypical tar—simple, bumbling, coarse.

James Ellison's *American Captive* (1812), William Dunlap's *Yankee Chronology* (1812), and Susanna Rowson's *Slaves of Algiers* (1794) all have Barbary settings. *The American Captive*'s Jack Binnacle is a bombastic cardboard sailor, all patriotism and sentiment, his language liberally spiced with stage nauticalisms. The nationalist muscle-flexing of the play concludes with the American bombardment of Tripoli. Dunlap's *Yankee Chronology* was written around a song of the same title, whose first eight verses celebrate the Declaration of Independence and the Revolution. Two verses on the War of 1812 were added for the play. The characters, Ben Bundle, his father, and the father's friend O'Blunder, exist primarily to sing the songs. In between, Ben relates his impressment into the British navy, his escape, and his service on board the USS *Constitution* during her victory over the *Guerriere*. Songs such as "Freedom of the Seas" and "The Yankee Tar" give some indication of the sentimental and patriotic nature of the productions. Another play, A. B. Lindsley's earlier *Love and Friendship* (1809), a sentimental comedy set on the waterfront in Charleston, South Carolina, differs in some respects from the others in that life aboard ship is described with some accuracy and attention to detail.

The experience of the sea as portrayed in these early productions is limited to martial triumphs and celebrations of American naval power. More than the other genres of the period, the theater reflects Americans' growing sense of themselves as a maritime force and as a nation whose coming dominance of the sea mirrors a parallel subjugation of the land.

The chorus of *Yankee Chronology* praises those sons of Columbia who "are lords of the soil—they'll be lords of the sea." This vision of land and sea compressed into a single Jeffersonian concept receives its most eloquent treatment in a work not usually considered sea literature. However, J. Hector St. John de Crevecoeur's *Letters from an American Farmer,* justly celebrated for its vision of the American land and character, contains as well a comprehensive conception of America as seascape. Five of Crevecoeur's middle chapters are devoted to Nantucket and Martha's Vineyard, whose people are profoundly influenced by the sea, their American-ness as fully shaped by sea experience as is the farmer's by the continent. Coming to America, Crevecoeur tells us in his story of Andrew the Hibernian, "is like going to sea"; and the Nantucket seamen form a community of tranquility, freedom, industry, and prosperity: "the greatest part of these are always at sea," and "the sea which surrounds these is equally open to all, and presents to all an equal title to the chance of good fortune." The isolation of Nantucket and the breadth of the ocean protect the inhabitants from the oppression of Europe: "After all, is it not better to be possessed of a single whale-boat, or a few sheep pastures; to live free and independent under the mildest of governments, in a healthy climate, in a land of charity and benevolence; than to be wretched as so many are in Europe, possessing nothing but their industry: tossed from one rough wave to another?" (135–36). Nantucket in Crevecoeur's vision is the epicenter of a vast American sea venture in which the New Man, the American, asserts himself on the oceans of the world as fully as on the prairies. "They navigate to all parts of the world"—every ocean has become an American lake.

The Martha's Vineyard whalers symbolize this vision. Crevecoeur's extended description of the taking of a whale is both a metaphor for the dynamism of the American enterprise and a catalog of the risks that accompany it. In the Nantucket Shoals, Crevecoeur finds another vivid metaphor for the danger and isolation of the American national situation, a danger transformed into opportunity by human energy in combination with nature. At the same time that the ocean threatens destruction, it provides a fertile field of action for American enterprise.

However, the sea is not only a metaphor for the American experience Crevecoeur envisions. It is also itself the source of mysterious knowledge. Crevecoeur, like the archetypal returning sea hero awestruck by the sublime beauty, mystery, and meaning of the sea, is compelled to recreate the experience for his audience. The ultimate role of the sea in American experience is, for Crevecoeur, imaginative and philosophical, as a rich source of transcendent ideals and meanings for which the New Man plows the sea:

> I had never before seen a spot better calculated to cherish contemplative ideas; perfectly unconnected with the great world, and far removed from its perturbations. The ever raging ocean was all that presented itself to a view of this family; it irresistibly attracted my whole attention. . . . My mind suggested a thousand vague reflections, pleasing in the hour of their spontaneous birth, but now half forgot, and all indistinct: and who is the landman that can behold without affright so singular an element, which by its impetuosity seems to be the destroyer of this poor planet, yet at particular times accumulates the scattered fragments and produces islands and continents fit for men to dwell on! Who can observe the regular vicissitudes of its waters without astonishment; . . . Who can see the storms of wind, blowing sometimes with an impetuosity sufficiently strong even to move the earth, without feeling himself affected beyond the sphere of common ideas? . . . How diminutive does a man appear to himself when filled with these thoughts, and standing as I did on the verge of the ocean! (158–59)

The comprehensiveness of Crevecoeur's vision of the sea is equalled only by that of Philip Freneau, the touchstone for American sea literature before Cooper.

Freneau shared with Crevecoeur a view of the ocean as a broad field of action on which to assert the coming glory of America. The sublime mystery and terror of the ocean is for Freneau balanced by the power of American ships such as the frigate *Alliance:* "where other ships find a grave, / Majestic, aweful, and serene, / she sails the ocean, like its queen." The ability, ingenuity, and courage of American seamen are a match for nature's power. The *Alliance* is "Undaunted by the fiercest gales, / In dreadful pomp she ploughs the main, / While adverse tempests rage in vain."

Unlike the dramatists of his era, Freneau was able to utilize sea imagery convincingly in works whose theme is essentially patriotic and nationalistic. "The British Prison Ship" (1780) successfully combines nautical language and heroic verse, and uses ships as integral parts of the work, not merely as background. Freneau's fierce indignation at the conditions in which the prisoners were held is integrated with a fast-paced sea narrative to provide a fully rendered work.

Yet Freneau's poetic stance toward the sea was ambivalent: his public poems describe the ocean as an American sphere of influence, while his more contemplative works evoke the transience of life at sea and contrast the vulnerability of human experience with the power and timelessness of the ocean. His personal attitude appears to be ambivalent as well. His "Lines by H. Salem" balance the landsman's "supposing the sailor a wretch, / That his life is a round of vexation and woe," against the seaman's knowledge that "If the sea has its storms, it also has its calms," and that with a good ship "From tempests and storms I'll extract some delight." In the end, however, Old Salem is moved principally by the opportunity for profit: should Neptune "pay [him] with farthings instead of a pound," he will bid farewell to the sea.

In poems such as "Pictures of Columbus" and "Discovery," the ocean is evoked as either a great desert, needing man's improvement, or at its best, a shield protecting the New World against European tyranny and vice. And the structure of "The Sea Voyage" mirrors the eighteenth-century preference for society over the unsocial sea. In smooth neoclassic couplets, Freneau evokes the voyage from the social tranquility of an "island green and fair" into the sea and storm ("Away the fluttering canvas bore, / And vow'd destruction to the mast"). In a movement as regular as the meter, the tempest rises, strikes, and subsides, and the next morning "gentle breezes warm and fair, / Convey'd us o'er the wat'ry road" to land and the narrator's "Charming Caelia." The structure is as organized as Richard Steere's a century earlier, yet instead of a voyage that illustrates the will of Puritan Providence, we see a progress from social order to natural disorder to social order again.

"The Sea Voyage" represents Freneau's aesthetic attitude toward the sea at its most facile and conventional. In a number of other poems, we find him evoking a human situation far more vulnerable and transient, a sea more mysterious and powerful. In "The Hurricane," for example, nature again threatens dissolution, a threat against which all of man's laws and hypotheses fail. Society itself is exposed as frail, insignificant, helpless.

> The barque, accustomed to obey,
> No more the trembling pilots guide:
> Alone she gropes her trackless way,
> While mountains burst on either side—
> Thus, skill and science both must fall;
> And ruin is the lot of all.

Such poems as "Captain J. P. Jones's Invitation" and "The Bermuda Islands" underscore the vulnerability of man's place in nature. Only the thinnest of membranes keeps the sailor from death: "A watery tomb of ocean-green / And only one frail plank between!" The seeming calm of the ocean itself is all deception: "Be not deceived—'tis but a show, / For man a corpse is laid below." The sea may be a broad highway to green and golden isles, but "here many a merchant his lost freight bemoans, / And many a gallant ship has laid her bones."

"The Wanderer," "The Argonaut," and "Hatteras" likewise evoke the poignance of sea experience, its essential loneliness and lack of connection with human society ("Dread Neptune's wild, unsocial sea" again). Each of these poems contrasts the unity and emotional connectedness of life ashore with their opposites in a life at sea. The Wanderer, like the lost bird that lights upon his ship, is a "Sad pilgrim on a watery waste" driven by fate from his native home ashore. Man at sea is forced to endure an alien element that affords no rest, offers no end except "To perish on this liquid field." The sailor's only home is the empty ocean "Upon whose ancient angry surge / No traveller finds repose!" Even the beauty of the sea is deceptive: "With masts so trim, and sails as white as snow, / The painted barque deceived me from the land." The ancient argonaut's hard-won wisdom is for himself learned too late:

> 'Tis folly all—and who can truly tell
> What storms disturb the bosom of that main,
> What ravenous fish in those dark climates dwell
> That feast on men—then say, my gentle swain!
> Bred in yon' happy shades, be happy there,
> And let these quiet groves claim all your care.

Likewise, in "Lines Written at Sea" in 1809, Freneau found nature itself disconnected in the alien world of the sea. Everything is uncertain, all familiar indicators changed: "The glow of stars, and the breath of the wind / Are lost!—for they bring not the scent of the land!"

In 1819, Washington Irving's "The Voyage" appeared in the first installment of *The Sketch Book*, Freneau's last published poems were printed in 1824, and the same year James Fenimore Cooper published *The Pilot*. The new age in American sea literature had begun. Increasingly the sea experience became the focus of romantic iconography and symbolism, its power to excite, awe, delight, and terrify increasingly the principal object of the writer's imagination. The concept of the Sublime as a means of morally and aesthetically controlling the disordered elements of nature no longer held together. The unsocial aspects of the sea now became a crucible in which individual rather than communal experience was tested and realized. The subjectivity of experience and the multiplicity of perception fashioned a new world to be explored.

The sea literature of the American eighteenth century, however, had its own cultural and aesthetic constraints, assumptions, and forces. The period is one of cultural crosscurrents and eddies, of changes often subtle, sometimes contradictory. The providential design in sea literature of the seventeenth century became in the eighteenth increasingly mannered and sentimental as the concept of providence itself changed from one that saw God as an active, primary force in human affairs to one in which the Divinity was rather a grand and distant governor of a mechanical universe. In literature, neoclassicism replaced the plain style of the early Puritans as well as the measured cadences of the Elizabethans in America. The sea became a vast canvas on which human experience was drawn in the foreground. The sea's natural disorder and unsocial character were safely, if tenuously, approached through the concept of the Sublime. The commercial expansion and nationalistic prowess of the new American states produced a sea literature that boasted of both. The Garden of Liberty replaced the Renaissance discoverers' vision of the New Eden.

But always in sea literature there are continuities with earlier and later times. The archetypal sea journey remains. The ocean crossing is in every age a transforming experience that brings new knowledge and insight, redeems the initiate and transforms secular space into sacred. The burden of telling his tale continues to weigh on the returned voyager, compelling his voice as he remembers the sail, the wave, the dark spars touched with fire. In one of his early works, "Pictures of Columbus," Philip Freneau remembered the storm that threatened Columbus's return to Europe in 1493 and recreated the admiral's preparation of the cask bearing the account of discovery that he hoped would preserve his name even if he himself perished. As a writer of sea literature, Freneau recognized intuitively his debt and his continuing communion with the great Admiral of the Ocean Sea, celebrating in the language of revolutionary America the primal act of American sea literature:

> Let's obviate what we can this horrid sentence,
> And, lost ourselves, perhaps, preserve our name.
> 'Tis easy to contrive this painted casket,
> (Caulk'd, pitch'd, secur'd with canvas round and round)
> That it may float for months upon the main,
> Bearing the freight within secure and dry:
> In this will I an abstract of our voyage,

And islands found, in little space enclose:
The western winds in time may bear it home
To Europe's coast: or some wide wandering ship
By accident may meet it toss'd about,
Charg'd with the story of another world.

DONALD P. WHARTON

Chapter Three

COOPER AND
HIS CONTEMPORARIES

From the Puritans' earliest soundings off an unexplored continent to Herman Melville's minute mappings inside the body of a whale, American independence from England—religious, political, economic, literary—was inextricably linked with the sea. In the first half of the nineteenth century particularly, America's growing military and commercial strength established the sea as a powerful source of national sentiment. In his message to Congress in June of 1812, President Madison outlined five reasons to declare war on the British, four of which—the impressment of American sailors, the seizure of cargo on merchant vessels, the raids on American coastlines, the interdiction of trade with America—involved illegal acts at sea. In the war that followed, the *Chesapeake*, the *Constitution*, and the *United States* became vehicles of patriotic furor and pride. In the South and West (where cotton and tobacco industries were dependent on the efficiency of export shipping) as well as in New England, the nation came to equate its independence with guarding its rights at sea.

In the 1820s, 1830s, and 1840s, American oceanic commerce began to rival that of Great Britain as well. The Atlantic Ocean, theater of Anglo-American hostility in both the American Revolution and the War of 1812, became a thoroughfare for the flourishing "packet" trade. Keen competition between companies encouraged constant innovations in ship design to balance speed with cargo capacity. Eventually, a packet ship of the Black Ball Line or Dramatic Line could, with luck, make the New York–Liverpool run in sixteen days. And beyond the Atlantic, the sealers and whalers of New Bedford, Nantucket, New London, and Sag Harbor crisscrossed the South Pacific, Indian, and Arctic oceans. As Jeanne-Marie Santraud outlines in her study, *La Mer et le Roman Américain Dans la Première Moitié du Dix-Neuvième*

Siècle, American commercial vessels filled up foreign as well as native harbors, thriving on the competition from without and within. Even as steam was being harnessed in such a way as to mark the end of the sailing era, the desire for more speed on longer voyages gave rise to the famous "clipper ship" design. Perhaps best exemplified by designer John W. Griffiths's *Sea Witch,* the clipper, symbol of a Yankee willingness to push the limits of wood and canvas, was an exotic bloom late in the season of American sail.

The sea served, too, as the appropriate arena for American *literary* nationalism, largely because of the efforts of James Fenimore Cooper, the nation's preeminent author of the sea in the first half of the nineteenth century. Exploiting fictional possibilities along the full sweep of American maritime history (indeed, dating back to Columbus), and sensing the deep and complementary relationship between ocean and continent in the American psyche, Cooper established the sea novel as a viable genre in American literature. There were many other prose writers of his era who explored the ocean for its literary possibilities, and who thus deserve mention in this chapter—among them Washington Irving, Nathaniel Hawthorne, Nathaniel Ames, William Leggett, Edgar Allan Poe, Henry David Thoreau—but for the sheer volume of his prose, for its nautical specificity as well as its spacious suggestiveness, Cooper had no peer. In the course of a thirty-year literary career, Cooper wrote thirty-one novels, eleven of which might be considered sea novels. In addition, he wrote volumes of nonfictional prose connected with his nautical interests, including review essays (1821–22), a history of the U.S. Navy (1839), an autobiography of a former shipmate (*Ned Myers,* 1843), a lengthy review of the *Somers* mutiny trial (1844), biographical sketches of American naval heroes (1842–45), and constant marine observations in his travel writings.

Cooper's literary interest in the sea was earned before the mast. After being expelled from Yale following a series of college pranks, he spent a year as a common sailor aboard the merchantman *Stirling,* which sailed for England and Spain in 1806. Chased by pirates, boarded by British press gangs, and battered by storms, the *Stirling* sold the seventeen-year-old Cooper on a career at sea. He received his naval midshipman's warrant in 1808 but was eventually assigned to Oswego, New York, to oversee the construction of warships, a tedious and frustrating substitute for one who longed to be attached to a saltwater vessel. He quit the navy and married in 1811, though his interest in the sea persisted. When, in time, he turned to authorship to support himself and his family, Cooper found in the sea novel an imaginative fulfillment of his earlier career aspirations.

Cooper died in 1851, the same year *Moby-Dick* appeared. The coincidence is strangely fitting, for Cooper can be seen as unmooring the novel of the sea from the essentially landed concerns of British authors Tobias Smollett and Sir Walter Scott, and steering it toward Melville in time for his masterpiece. This is an oversimplification, surely, but all of Cooper's sea novels betray the grand visions and technical awkwardnesses of just such a transition. As Cooper began his literary career, the sea had yet to play a major role in prose fiction. Smollett was a surgeon in the Royal Navy who used the sea from time to time in his rambling, picaresque plots. His most memorable nautical creation is probably Dr. Mackshane of *Roderick Random* (1748), the ship's surgeon who glories in amputation (and who served as a model for Melville's Cadwallader Cuticle in *White-Jacket*). Sir Walter Scott's *Pirate* (1821), by contrast, contains lyrical evocations of the sea while placing almost none of its scenes there. It was after talking with a friend about this work, in fact, that Cooper decided to write his own version of the sea novel, one that "might present truer pictures of the ocean and ships," as he wrote in his preface to the 1849 edition of *The Pilot.* First published in 1824, *The Pilot* succeeded in doing this while imitating Scott in other ways; indeed, Cooper was known early in his career as "the American Scott." That this epithet typed him unfairly is perhaps best suggested by Cooper's exploitation of another literary source, more revealing than Scott or Smollett: Byron's poems, particularly "The Corsair" and "Lara." As Thomas Philbrick points out in his landmark study, *James Fenimore Cooper and the Development of American Sea Fiction* (to which this discussion is indebted throughout), Cooper's portraits of the moody John Paul Jones in *The Pilot* and Captain Heidegger in *The Red Rover* (1828) owe much to Byron, as do other scattered hints of exoticism in his fiction.

Cooper, of course, is best known as the author of the Leatherstocking Tales, five novels of the American forest that, put in sequential order, follow their frontier hero Natty Bumppo westward until he dies beyond the Mississippi on the great American prairie. It is worth noting that just as *The Pilot* immediately followed *The Pioneers* (1823), nearly all of Cooper's Leatherstocking Tales give way to sea novels. When *The Red Rover* appeared just after *The Prairie* (1827), the *North American Review* expressed relief that Cooper was back in his own element: "The quarterdeck is his home." It would seem that Cooper's meditation on the continental frontier inevitably led him out to that even vaster and less stable expanse by which it was surrounded. Moreover, images of the sea flow *into* his landbound works. Clichéd saltwater tars always find their way into Cooper's forest, but the sea encroached on his conception of the continent in other ways as well. *The Prairie,* for instance, contains a host of famous oceanic metaphors for the flat landscape west of the Mississippi. In *The Last of the Mohicans* (1826), Natty Bumppo considers how, despite its apparent willfulness, the cascading waterfall of Glens Falls, New York, flows "steadily towards the sea, as was foreordained from the first foundation of the 'arth!" (ch. 6). The subtitle of *The Pathfinder* is *The Inland Sea*, referring to Lake Ontario, on and around which much of the action takes place. Even in his celebrated novels of the forest, Cooper was always aware that the ground on which his characters dwelt was "but dry land," as Thoreau later said.

Cooper's vision of the sea was extraordinarily rich, and so too was his influence. His sea-writing career divides itself loosely into two decades, the 1820s and the 1840s; in the discussion that follows, the early Cooper novels and the later ones serve as a frame for the treatment of his contemporaries. In the twenties, Cooper composed *The Pilot, The Red Rover,* and *The Water-Witch* (1830), a popular romantic trilogy that attains something of the symbolic coherence of the Leatherstocking Tales. Heavily Byronic and patriotic, these novels are dominated by themes of smuggling and captivity, and thus too by a shoreline—rather than oceanic—consciousness. The 1830s was a decade of great personal turmoil for Cooper, during which his popularity waned and his sea fictions nearly ceased altogether. But cued by Cooper's early success, contemporaries took up the task, producing an impressive body of short sea tales. Cooper returned to the sea with a vengeance in the 1840s, and his eight novels from this era both reexamine the romantic assumptions of his earlier works and demonstrate the symbiosis of American nautical fiction.

I.

The "sea novel" can be seen as a contradiction in terms, for those very elements that make a novel seaworthy work against its being a novel at all—at least one in the British tradition with which Cooper was familiar. Male-dominated, removed from the tissue of society, the world of the sea will inevitably impress the domestic reader as exceptional, artificial, or fantastic. But in the 1820s, with the War of 1812 still fresh in the national memory, these very qualities suited the sea novel to a young American literature that was quite consciously trying to divorce itself from British models. Many of the qualities Richard Chase assigns to the American "romance" in the opening pages of *The American Novel and Its Tradition*—the emphasis on the strangeness rather than the familiarity of experience, the need to create a world rather than fit into one—aptly describe the sea novel.

Cooper's early sea novels embrace the radical oppositions implicit in the form. The passages of these works actually set at sea—especially in their lyrical depictions of water, vessel, and sailor—promote both a maritime and literary nationalism. Poised against sea plots, however, are substantial domestic plots that muddy the waters of patriotism considerably. As the women marry and take their sailors home, as the men discover themselves wealthy, Cooper softens the stark geometry of ocean, ship, and mariner, to accord with the inevitable "corruptions" of femininity and civilization. *The Pilot, The Red Rover,* and *The Water-Witch* establish the terms of this conflict and probe its sources and ambiguities.

Like the forest, Cooper's sea is that trackless area of possibility on which a nation traces its own history; capable of optical illusion, the sea undulates and has hidden recesses. Like the forest, too, it is a region of spontaneous violence—but with a difference: forest violence, noble or ignoble, is man-made, and has behind it the assumption that the environment can be mastered to outsmart an enemy. While Cooper portrayed plenty of nautical battles between political opponents, *the sea* is often the enemy itself, frothing in unpredictable anger, sometimes killing off its most appreciative denizens, as in the case of *The Pilot*'s Long Tom Coffin. Though obviously romantic, Cooper's descriptions of the sea—its interior animations, its hidden life—answer to both sailor and author, and constitute a marked improvement over the "troubled foam" of Byron:

> The surface of the immense waste was perfectly unruffled . . . but the body of the element was heaving and settling heavily, in a manner to resemble the sleeping respiration of some being of huge physical frame. (*The Water-Witch,* ch. 7)

And unlike his forest, Cooper's sea cannot be "improved" or transformed into towns and cities; the conditions it presents to ship and sailor are at once spontaneous and eternal.

To survive in such a protean environment, the vessels of Cooper's early trilogy, as Philbrick has suggested, tend to be small, light, fast, and maneuverable: they are metaphorical sea fowl capable of floating and gliding. In addition to their speed and grace, these vessels possess a fairylike, supernatural ability to appear and disappear in mysterious fashion. Cooper often writes of the "delicate tracery" of spars and cordage as seen from a distance, their "exquisite symmetry" that is comparable to the "delicate work of a spider." To a distant observer, these ships seem born of the very mists through which they float and often mysteriously fade into them altogether. When a military vessel attempts to raid the *Water-Witch* in the mist, her boarding irons fall directly into the sea. Cooper invokes myths of specter ships such as the *Flying Dutchman* to explain the uneasiness such a disembodied vessel causes in the mind of the perceiver.

In nearer focus, the instant when vessels *yield* to the pressure of the wind or water is a mystical one in Cooper's mythology. Whether just getting under way or going down for good in a high sea, Cooper's vessels are only reluctantly tamed by the elements:

> There was an instant when the result was doubtful; the tremendous threshing of the heavy sail, seeming to bid defiance to all restraint, shaking the ship to her centre; but art and strength prevailed, and gradually the canvass was distended, and bellying as it filled, was drawn down to its usual place, by the power of a hundred men. The vessel yielded to this immense addition of force, and bowed before it, like a reed bending to a breeze. (*The Pilot*, ch. 5)

Cooper's ships, which so much *seem* to be one with the environment when viewed from a distance, literally are so when sped by wind and sea.

Combined with his conceptions of sea and vessel, Cooper's sailor completes a triangle of heightened experience which the author viewed as distinctly American. *The Pilot* is the first fictional treatment of the American Navy, and the common sailor, typified by Long Tom Coffin, is wise, loyal, and courageous. Coffin demands "plenty of sea-room, and good canvass," and like his vessel not only rides the sea but is *of* the sea. At a crucial juncture in a battle with the British, Long Tom is thrown overboard and presumably drowned, but "issuing from the sea," reemerges with his harpoon "like Neptune with his trident" and promptly impales the British commander. Later in the book as the *Ariel* is shipwrecked on the dangerous reefs of the North Sea, Tom faces the choice of abandoning his vessel or certain death. "I saw the first timber of the Ariel laid, and shall live just long enough to see it torn out of her bottom; after which I wish to live no longer" (ch. 24). In less dramatic fashion, Cooper similarly dignifies Dick Fid of *The Red Rover* and Trysail of *The Water-Witch*.

Cooper's early cast of sea characters also includes a "middle hero" (a term used by H. Daniel Peck), a young man divided at heart between the love of a woman and a love of the sea, as well as a Byronic presence such as John Paul Jones or Captain Heidegger. Cooper's portraits of Jones (based largely on Nathaniel Fanning's *Narrative* [1806]) and Heidegger (based loosely on Byron's "The Corsair") both emphasize their moody inaccessibility, their aversion to domesticity, their adopted—and therefore heightened—sense of American patriotism. Both men are driven by private hatreds and secret wounds, and while Jones is a shadowy figure in *The Pilot*, Heidegger can be seen as the prototype for Jack London's willful and loquacious Wolf Larsen.

As a world by itself, Cooper's sea is the male domain of American history, peopled with brave colonists aboard game vessels, testing their courage against the enemy and the elements. But the poetry of the sea chapters acquires its peculiar pitch only when sounded against the physical and psychological dangers of the shore. These books are not set at sea so much as in the *margin* between land and sea, and thus they are, as the subtitle of *The Spy* (1821) first suggests, more tales of "neutral ground." One of the greatest dangers to the vessels comes from hidden rocks and shoals, and it takes a brilliant presence such as John Paul Jones or Tom Tiller to navigate in this world. Doing so successfully is a matter of feminine delicacy as much as masculine strength. Psychologically, the "off shore breezes" bring thoughts of women to the middle heroes of Cooper's novels—creating a conflict between their affections and their duties as seamen. A subtheme of *The Pilot* involves John Paul Jones's criticism of Barnstable for his vulnerability to sentiment, but given the fact that the young American captain ends up safely married at the end of the tale while Jones's sea mission fails, it is difficult to assess Cooper's own view of the matter.

Thus the terms of opposition become confused. Looked at in a certain way, the early trilogy of novels constitutes a retreat from (rather than a reinforcement of) the issues of American independence, even as they were being updated in Cooper's time by the Monroe Doctrine and the election of Andrew Jackson. *The Pilot* takes place during the Revolution, but the two subsequent novels push backward in time—*The Red Rover* to 1759, *The Water-Witch* to even earlier in the century—seeking, as Peck has argued, to find a world protected from the forces of history. In this sequence, Cooper progressively blurs his oppositions, and in the end deliberately obscures the line between sea and land, male and female, America and England.

The Water-Witch, in particular, examines how the manly world of the sea and the domestic world of the shore interpenetrate. The story takes place at a time, for instance, when being a good American meant being a good Englishman. The maleness usually associated with ship and sailor is undercut by the very names of the vessels, *Coquette* and *Water-Witch*, and by the fact that master Seadrift turns out to be a woman in disguise. The *Water-Witch*, a "hermaphrodite brig" (a vessel both square-rigged and schooner-rigged), is commanded by Tom Tiller, an elusive character who combines a variety of roles: Byronic presence, common sailor, and middle hero. Most significant of all, Cooper concludes the novel with a woman choosing to run off to sea with her man rather than with a man settling down into a mansion with his wife. As Thomas Philbrick suggests in his essay on this work, the forces of history eventually do destroy the protective fabric of illusions Cooper has created throughout the tale; the *Coquette* is finally dismasted and burned. It is a tribute to Cooper's complexity as an artist that he examined and undermined his own fictional assumptions as he wrote; this very self-consciousness makes his first three romantic novels of the sea worth reading and rereading.

II.

While Cooper was establishing the sea as an area of American experience capable of challenging and transforming his own landed fiction, Washington Irving, America's most celebrated author of the era, was probing the relationship between continent and ocean in quite different terms. In Irving's "histories" of the 1820s and 1830s, land and sea meld in the romantic consciousness of a discoverer or adventurer. His Columbus volumes (*The Life and Voyages of Christopher Columbus* [1828] and *Voyages and Discoveries of the Companions of Columbus* [1831]) begin with the discovery of a new continent and end with the discovery of a new ocean. Earth and water thus complement one another, taking turns as that resistant element through which the explorer must pass to confront the capaciousness of his own imagination.

Columbus, by clinging to the belief that he had discovered a new route to the Indies, is a victim of the "riot" of his own anticipations. Similarly, what drives Balboa across the mountains of Mexico (and eventually gets him beheaded) is his obsessive vision of a paradisal Pacific. As Balboa first spreads sail on the Pacific, "the range of the unknown world before him," Irving brings his Spanish narratives of discovery full circle: an ocean has become pathway to a continent and a continent has become pathway to an ocean. Much the same kind of substitution is invoked in *Astoria* (1836), Irving's history of John Jacob Astor's failed attempt to extend his fur-trading empire to the Northwest Coast in the early 1800s, and a book used by both Poe and Cooper. From New York, Astor sends out simultaneous land and sea expeditions to the mouth of the Columbia River, and as the back-and-forth narrative follows the halting progress of the parties, continent and ocean become interchangeable strategies in a fantastic board game.

Despite the fact that they end inevitably with petty politics and betrayal, Irving's histories thus possess a grand circular spaciousness. His brief but suggestive fictional treatments of the sea are tied even more intimately to the psychology of an observer, specifically a genteel first-person narrator. Geoffrey Crayon of *The Sketch Book* (1819–20) is not an adventurer but an antiquarian, seeking out the old land rather than the new. In the sketch "The Voyage," Crayon climbs to the maintop and, like Melville's Ishmael, finds this "giddy height" conducive to reverie. In one sense, the ship and sea provide the male society, the monastic stillness, and above all the access to the imagination that Crayon will repeatedly seek in the libraries and museums of

England. And much as he does in those preserves, Crayon at sea betrays his own morbidity in the guise of a genial good nature. On the "gentle undulating billows" Crayon projects "shapeless monsters"; a floating piece of mast becomes an occasion for melodramatic speculation; a storm at sea brings with it "funereal wailings." Crayon is titillated by tales of horror, and when the captain tells his story of shipwreck, he fixes on the "drowning cry" of those who perished. The ocean, which Crayon muses on as a "vacancy" or a "blank page of existence," becomes in fact a record of his own psychological preoccupations.

Compared to Cooper, the literary attention Irving paid to the sea was minute, but his *style* was widely imitated by sea authors and was particularly influential in fostering a school of short sea fiction that flourished in the 1830s, one in which Nathaniel Hawthorne, Nathaniel Ames, William Leggett, and Edgar Allan Poe all participated. Sentimental, first-person, gothic, this fiction can be seen as the sea-borne counterpart to the tales from the Old Southwest that were also popular at the time with the civilized reading audiences of Boston and New York. The exaggerated sea "yarn" was widely read, in the *Knickerbocker*, in *Burton's*, *Graham's*, and *New-England* magazines, and even made its way into women's magazines such as the *Ladies Companion*. Irving's two-page sketch "The Haunted Ship" (1835) is typical of this movement at its best in the way the tale reflects ironically on the teller. In the guise of searching out an eyewitness account of a ghost story at sea, the narrator actually displays his own genteel isolation from the world of his story.

Nathaniel Hawthorne, who was raised in the seaport of Salem, son and grandson of ship captains, who fought against the abuse of sailors as Consul in Liverpool, and to whom *Moby-Dick* is dedicated, wrote little on the sea itself, though its margins clearly impinged on his literary imagination. Hawthorne edited two full-length works, memoirs by friends Horatio Bridge and Benjamin F. Browne, which largely concern life at sea. Further, he was nearly named historian of the famous South Seas exploring expedition eventually led by Lieutenant Charles Wilkes, a trip that might have sent his career in a different direction entirely. But even in *The Scarlet Letter*, "the broad pathway of the sea" looms as an immense and impossible avenue of escape. While Hester Prynne fantasizes about "drowning" her red letter in "mid-ocean" and starting a new life with Dimmesdale, she confines herself eventually to the "moist margin of the sea," a border that teases out the warring elements in her conflicted nature. Hester's cottage, located on a peninsular verge, paradoxically faces both the sea and the West.

Hawthorne's Irvingesque first-person accounts of the sea are even more revealing. The Salem Custom House, described in the introductory chapter of *The Scarlet Letter*, filled with retired sea captains, ship's agents, owners, sailors, is the very meeting ground of sea and shore. The sea, for some such an active life, a culture in which a "boy of fourteen took the hereditary place before the mast, confronting the salt spray and the gale, which had blustered against his sire and grandsire," becomes for Hawthorne's narrators a source of self-consciousness. Like Irving's narrators, they must justify their positions as "idlers" and observers. Such is the case in "The Village Uncle" (1835), related by a weary old fisherman who invokes the sea when speaking of the richness of his past life: his wife Susan is a "mermaid"; his fellow fishermen "have all been christened in salt water," etc. But the tale is one of impotence and sterility, of a man trying to speak a life of adventure now that he has not lived one. The sea, which provided the original "village uncle" with a wealth of manly and global experience, has provided the narrator with only local fishing tales, which he repeats tediously in an attempt to give his life definition. Similarly, "Footprints on the Sea-Shore" (1838), a story Melville underlined repeatedly in his own copy of *Twice-Told Tales*, is told by a man who lingers on the shoreline and lyrically declaims his sense of solitary independence—all the while showing evidence of a voyeuristic loneliness. The sea, as in Irving, is a mirror of interior consciousness, a sea of self-justification.

Perhaps the two authors who came closest to perfecting the sentimental sea tale were Nathaniel Ames and William Leggett. Ames, anticipating Richard Henry Dana, Jr., was a Harvard man who went to sea, both in the navy and in the merchant service. He is most famous for his nonfiction, particularly for his description of a fall overboard in *A Mariner's Sketches* (1830), which lies behind a passage in Melville's *White-Jacket*. But he wrote fiction as well (*An Old Sailor's Yarns* [1835]), loose, sentimental, and much of it concerning the love of a sailor for a beautiful woman. Most fascinating is the attitude of the narrators toward their stories. Clearly influenced by Irving, Ames's tellers are genteel and whimsical, and focus on those small moments when sea and shore intersect, such as when women are "whipped" aboard (physically hoisted in a sling), or when the sailor must act the courtly lover. They are capable of displaying an almost unbearable self-consciousness, but his fiction rises above the play of manners when Ames gives the sense of a narrator being taken over by his subject, as happens in the short stories "Old Cuff" and "The Rivals." Each contains a more or less omniscient account of the remarkable adventures of a sailor, followed by an eyewitness verification; the gentility of the tales is thus overlapped by the psychology of the observer. (Another sea author, John W. Gould, gains something of the same effect in his *Forecastle Yarns* [1845] through the use of interior narration.) Ames, it should be noted, was a bitter critic of Cooper; he claims, in *A Mariner's Sketches*, to have read *The Pilot* out loud to the crew of a naval vessel and records their disgust with it.

William Leggett's fiction, sparer and more controlled than Ames's, is an example of the sentimental sea tale at its best, and again the Irvingesque narration plays a vital role. Leggett was a naval midshipman in the 1820s who, in the course of his stint, contracted yellow fever in the West Indies and suffered the persecutions of a commander in the Mediterranean for his ability to quote Shakespeare. Following a duel, he was court-martialed and resigned his commission to take up journalism and authorship. His sea stories, collected in two volumes (*Tales and Sketches by a Country Schoolmaster* [1829] and *Naval Stories* [1834]), bring a literary sophistication to the idea of the sea as romantic projection.

Working from surface to depth, many of Leggett's stories begin in a dead calm or tranquil evening and end with some gothic image of violence. The "perfect calm" of "The Encounter" ends with the quartermaster impaled on the bowsprit of another vessel. "The Main-Truck, or a Leap for Life" takes place in the port of Mahon, "one of the safest and most tranquil places of anchorage in the world." The story ends with a father threatening to shoot his son unless he leaps from the top of the mainmast into the sea. Leggett's sea is a pool of primal energies capable of welling up at any minute. The men who observe or relate these horrors (many of his tales are in first person, and in all there is a mediating observer) experience a kind of sympathetic wounding. The narrator of "The Main-Truck" says, "I myself had the sensations of one about to fall from a great height." The imagination has its own horrors. Two of the stories actually end with the hackneyed device of a man having discovered he was only dreaming the incidents of the tale, but at his best, Leggett captures the "shadowy and inverted image" the sea records on the mind of the observer. His narrators, at once observant and naive, strike a note of modernity and seem to anticipate the rhythms of a Sherwood Anderson.

Leggett is a transitional figure between Irving and Edgar Allan Poe, in terms of the gothic content of his tales as well as the sophistication of his technique. In the short stories "MS. Found in a Bottle" (1833), "A Descent into the Maelstrom" (1841), and the short novel *The Narrative of Arthur Gordon Pym* (1838), Poe creates first-person accounts from which the narrators never fully return, thus pushing the sea yarn to its epistemological limits. His fictions explore geographical limits as well: "MS." and *Pym* are both tales of the South Pole, a region that fascinated Poe. He was interested, for example, in the "Symmes Hole" theory (which posited that the earth was hollow at the poles and habitable within) and followed the career of its colorful proponent Jeremiah N. Reynolds, whose name, legend has it, Poe was calling out on his deathbed. He kept track, too, of the expedition of Lieutenant Charles Wilkes (1838–42), which confirmed the existence of a southern continent. "A Descent," by contrast, takes place near the *North* Pole (Poe's plots gravitated inevitably to the extreme), but as with the other stories, its setting is a sea of nightmarish interiority, a warped mirror of consciousness, equally threatening in fact and in memory.

The narrator of "MS. Found in a Bottle" is an odd combination of Irving's Crayon and Melville's Ishmael. Like Crayon, he is an effete passenger rather than a sailor, and a "dealer in antiques." Yet his abrupt tone—"Of my country and my family I have little to say"—and his admissions that he is "haunted like a fiend" by a nervous restlessness, foreshadow the desperation of Melville's narrator in "Loomings." Thrown off course in a storm and eventually pitched into the rigging of a ghost ship in a high sea, the narrator realizes his voyage has become one of primal "DISCOVERY" (a word that appears on a studding sail), but has surpassed his powers of comprehension. "[H]urrying onward to some exciting knowledge—some never-to-be-imparted secret, whose attainment is destruction," Poe's ghost ship, like Melville's *Pequod*, swirls downward into a great vortex. But while Ishmael emerges on Queequeg's coffin, all that pops up of Poe's narrator is his preserved manuscript. ("The Oblong Box" [1844] provides Poe's version of the coffin-in-the-sea theme.)

The storyteller in "A Descent into the Maelstrom" claims to have undergone a similar experience off the coast of Norway, surviving (Ishmael-like) by lashing himself to a barrel. He describes the slow, circling descent of his vessel in great detail, if with a questionable grasp of hydrophysics. (Poe used the *Encyclopedia Britannica* for his account.) But if the interior storyteller survives the sea, the exterior narrator (to whom the story is told) does not survive the tale: after introducing his subject and expressing his distress at what he is hearing, the narrator mysteriously disappears. In its use of a narrative frame, the story is similar to the tall tales of the Old Southwest—in which a gentleman-narrator is typically hoodwinked by a local yarn spinner—except that in Poe's story the narrator never returns to close the tale. His identity is effectively erased, not by the sea itself, but by its haunting presence in the imagination.

By and large, the 1830s were distinguished more for their short sea fiction than for their sea novels—Joseph C. Hart's *Miriam Coffin* (1834) and Robert Montgomery Bird's *Adventures of Robin Day* (1839) are notable exceptions—but Poe's *Arthur Gordon Pym* caps off many of the assumptions of the gothic sea tale. Like Hart's book, *Pym* contains more anticipations of *Moby-Dick*, such as its meditation on whiteness and its digressive chapters. Moreover, as the tale proceeds through a catalog of horrors (including starvation and cannibalism) and finally ends with the narrator blending with the polar shroud, it can be seen, like *Moby-Dick,* as vexing the problem of the uninterpretability of experience. Invariably, what Pym sees or feels is beyond his ability to articulate: "We saw nothing with which we had been formerly conversant" (ch. 18). The continual images of doubling, self-consumption, and polar opposition take us out of a rational world of mediation and into a surreal world of categorical extremity. *Pym* is awash in a dream ocean of terror, a pre-Freudian sea of the id. The painstaking attempts at technical, historical, and geographical accuracy (there is a disquisition on "lying to" in chapter 7 that seems a parody of Cooper) become a grand metaphor for Pym's progress toward his own inscrutable source and center. The book ends with a series of gnomic linguistic symbols, another kind of preserved manuscript. As John Irwin has argued, Pym's journey inward is also toward the source of language. As if the *tale* is more threatening than the experience it relates, Pym survives the polar shroud but dies in the course of writing about it. Poe's subjective sea impinges on the act of creation itself and challenges the assumed superiority of the teller over his memory.

In the same year that *Pym* was published, Henry David Thoreau and his brother John took a "fluvial excursion" down the Concord and the Merrimac, a voyage reconstituted in *A Week on the Concord and Merrimac Rivers* (1849). Though not truly a contemporary of Cooper (he was born twenty-eight years later), Thoreau deserves special mention in this chapter for the way he invested his prose with apparently incongruous but illuminating nautical metaphors, ones that hint at spacious, spiritual alternatives to man's mentally landlocked existence.

Raised in Concord, seventeen miles from the ocean, Thoreau, like Cooper, sensed the porousness of the continent (which he punned was "not continent" at all): "Our town, after all, lies but farther up a creek of the universal sea" (*Journal,* April 15, 1852). He had seafaring ancestors; his grandfather owned a business on Boston's Long Wharf; he read widely about the ocean—including geodetic surveys—and developed in his childhood a lifelong fascination with canal boats and boatmen, local shipwrecks and far-flung nautical expeditions. In *A Week* and his later *Cape Cod* (1865)—the one traveling up an artery of the ocean into the continent and the other up an arm of land into the sea—Thoreau examined the literal exchanges between water and shore as figures for inward meditation and spiritual discovery. While the leisurely and whimsical current of *A Week* contrasts sharply with the naturalistic descriptions of shipwreck in *Cape Cod* (Thoreau casts off Irving's sentimentality in this regard), both demonstrate the importance of Thoreau's shoreline consciousness: "equally mariners and terreners," he writes in his essay "Paradise (To Be) Regained," thinkers must forever position themselves on the periphery of a new element, a new world.

Not connected directly to the sea at all, Walden Pond was nonetheless, as Emerson said in his essay on Thoreau, "a small ocean; the Atlantic, a large Walden Pond." Thoreau used nautical metaphor in *Walden* (1854) as a way of viewing life freshly and paradoxically, as a way of grasping the tiller while holding the plow. The sea voyage, particularly, provided Thoreau with a design that rounded Poe's sense of horizonless psychological westering to a more seasonal and organic sense of personal renewal. In "The Village" chapter, Thoreau "sets sail" on foot from Concord for the "snug harbor" of his cabin, commenting along the way on the "well-known beacons and headlands" of one's usual course in life. At the start of the book, Walden Pond is described as a "good port and a good foundation," but as the work moves toward winter, Thoreau's journey is intellectually "out-bound." In winter, "We are islanded in Atlantic and Pacific and Indian *Oceans of thought*" (*Journal*, January 30, 1854). Spring is the season of rebirth and, in terms of the nautical metaphor, return to port. As he says in "Autumnal Tints," "The ideal course of a thinking man is a voyage and a return to harbor—a round trip symbolizing his life." The importance of his interior exploration is underscored by the repeated use of marine metaphors in the "Conclusion" of *Walden*. Here, for instance, is his famous exhortation: "Nay, be a Columbus to whole new continents and worlds within you, opening new channels, not of trade, but of thought." And further on, Thoreau makes reference to Charles Wilkes's "South-Sea Exploring Expedition" as a way of urging his readers to "explore the private sea, the Atlantic and Pacific Ocean of one's being alone."

Like Edgar Allan Poe, Thoreau had an interior voyaging impulse, but while Poe penetrated the sea of unconsciousness, taking his narrators to the very edges of form and sanity, Thoreau's voyage, equally perilous, took him more deeply into the conscious motives and methods behind his everyday existence, into the "Symmes Hole" of the self.

III.

James Fenimore Cooper wrote no novels between 1834 and 1838, a period during which he felt wounded by newspaper reviewers, but kept his nautical interests alive by collecting material for his monumental *History of the Navy* (1839). The work that marks Cooper's return to fiction writing, *Homeward Bound* (1838), was originally intended to be a novel of manners examining

Jacksonian America through the eyes of a Cooper-like gentleman, Edward Effingham, who returns to the states with his daughter, Eve, after an extended stay abroad. But as Cooper writes in the preface, "the cry was for 'more ship,' until the book has become 'all ship.'" Thus, though not intended to be such, *Homeward Bound* can be seen as Cooper's "purest" sea novel, for nearly all the action takes place aboard the *Montauk*, a London–to–New York packet. (Cooper completed the design of his original plan with the companion piece *Home as Found* [1838].)

That the domestic novel *becomes* a sea novel is significant, since it unites the two opposing strains of Cooper's earlier fiction, and in so doing establishes a tendency that will recur throughout the second half of Cooper's career: the sea will not be considered a romantic world by itself so much as a lens through which to view a troubled contemporary world, a lens capable of isolating and focusing essentially continental dilemmas. His sea fictions, in short, become novels of critical principle rather than of patriotism. This is clear when we examine how the geometry of sea, ship, and sailor changes in his last eight novels, which, in addition to *Homeward Bound*, include *Mercedes of Castile* (1840), *The Two Admirals* (1842), *The Wing-and-Wing* (1842), *Afloat and Ashore* and *Miles Wallingford* (double novel, 1844), *Jack Tier* (1848), and *The Sea-Lions* (1849).

In *Mercedes of Castile*, Cooper's retelling of the discovery of America (much of it based on Irving), Columbus's three small vessels move through the "pathless" and "shoreless" Atlantic "like human beings silently impelled by their destinies toward fates they can neither foresee, control, nor avoid" (ch. 14). More and more in the later sea novels, Cooper liberates his plots from the shoreline, and in the "mysteries of the open oceans" evokes a sense of Melvillean uncertainty and hidden fate. In *Homeward Bound*, the *Montauk* sails absurdly off course to avoid being boarded and searched by the British vessel *Foam*. Wrecked on the coast of Africa, threatened by Arabs, the *Montauk* is eventually boarded by the *Foam* anyway. The sea, which still hisses, froths, glitters, and breathes, is not an escape from civilization (from questions of law, religion, love, etc.), not a world apart, but serves instead to heighten and define these civilized dilemmas in their most starkly symmetrical form. In *Afloat and Ashore* and *Miles Wallingford*, Cooper's double novel delivered in first-person voice, Miles repeatedly finds himself reduced to a speck of existence in a shoreless sea. In each instance his thoughts turn toward God and Clawbonny, New York (his upstate residence). The sea's depredations test and refine one's principles, as in *The Sea-Lions*, where, against the bleak whiteness of the polar landscape, Roswell Gardiner carries on a heated internal debate concerning Unitarianism and Trinitarianism.

To be sure, Cooper borrows some of his new sense of the sea from the gothic tale and from Poe—starvation and cannibalism are prospects raised in a couple of the novels; a race to the South Pole is the subject of *The Sea-Lions,* and in that novel as well as in *The Two Admirals* appears the theme of the double—but, unlike in Poe, Cooper's characters never fully merge with the horrors of the ocean and are most often delivered back to a safe, landed existence. The terrors of the sea, the close calls, now usually lead to a greater love of God, woman, or ancestry, as opposed to country. Less pictorial and resonant than in his early romantic sea novels, the interplay of continent and ocean becomes increasingly symbolic and allegorical in his later ones. From the vastness of the Atlantic, the cultured passengers of the *Montauk* are rudely deposited on the coast of Africa, thus establishing the pure poles of opposition (ocean/desert) that begin W. H. Auden's study, *The Enchafèd Flood.* There is still plenty of action in the margin between land and sea, still the dangers from reefs and shoals, but gone is that poetic sense of insinuation from one state of existence into another. Now sea and land stand against each other in tense and symbolic opposition.

Against an open sea, Cooper created his ships in more technical and realistic detail than in his early works. The reasons for this are varied and provide a glimpse into the sometimes incestuous borrowing and lending patterns associated with nautical fiction in the 1830s and 1840s. First, Cooper's nationalistic trilogy of sea romances had spawned a school of successful popular novelists who codified his themes and characters in outlandish terms. Maturin Murray Ballou, Benjamin Barker, Robert Burts, Sylvanus Cobb, "Harry Halyard," Joseph Holt Ingraham, and the prolific Edward Zane Carrol Judson ("Ned Buntline") sensationalized Cooper's sea tales, all but overwhelming those more interesting and legitimate imitations such as Charles J. Peterson's *Cruising in the Last War* (1850). After seeing his first novels reduced to the formulas of piracy, temperance, shipwreck, sea serpents, and specter ships, Cooper no doubt wanted to reestablish his own authority and expertise as an author of the sea. Second, Cooper himself had just finished his *History of the Navy,* a work filled with statistical information about naval battles and strategies. But finally, Cooper's new realism is due in large part to the influence of Richard Henry Dana's *Two Years Before the Mast* (1840), an enormously popular record of life in a merchantman. Dana, too, was widely imitated, most notably but not exclusively by nonfiction writers. J. Ross Browne's *Etchings of a Whaling Cruise* (1846), reviewed by Melville, gives an inside view of life on a whaler much as Dana had done for merchant vessels; similarly, Charles F. Briggs's *Working a Passage* (1844) is an unvarnished account of a sailor's experience on a transatlantic packet. The short story writers John

[79]

Codman and John Sherburne Sleeper ("Hawser Martingale") managed to combine the old romance of Cooper and the new realism of Dana in their blend of fiction and essay.

Cooper and Dana read each other carefully, seeking both inspiration and audience. In the opening pages of *Two Years Before the Mast,* Dana cites Cooper's unintelligible—but captivating—descriptions of nautical maneuvers in *The Red Rover* as a rationale for his own use of technical vocabulary. Cooper, in turn, responded to Dana's popularity by heightening the attention *he* paid to nautical detail, digressing freely, for instance, on scudding, the construction of the *Santa Maria,* the positioning of the fleets in the British naval battles of 1745, and so forth. Describing specific shipboard exercises, Cooper is apt to bewilder a reader with knight-heads, fore-topsails, capstan-bars, cat-falls, head-yards, spankers, and brails, all of which are to be found in a short section from chapter 13 of *Jack Tier.*

In Cooper's later novels, the ship thus becomes the sum of all its working parts, a massive, technical phenomenon quite the contrary of the sprightly water fowls of his earlier works. One still finds vessels described in terms of their "symmetry" or "delicate tracery," and occasionally vessels "skim" over the seas, but more typical is this description of the *Caesar* in *The Two Admirals:* "the huge hull became visible, heaving and setting, as if the ocean groaned with the labor of lifting such a pile of wood and iron" (ch. 19).

The human effort involved in getting and keeping a vessel in motion is another distinct revision of an earlier conception. While in his early novels Cooper emphasized the grace or terror evoked by a ship ceding itself to the elements, he now tends to emphasize just how difficult it is to budge. In *The Wing-and-Wing,* the crew of the lugger *Feu-Follet* try to heave it off the rocks of a Mediterranean islet "until it seemed as if the hemp of the cable were extending its minutest fibres," but to no avail. Shipwreck is again common, though now most often in deserted or uninhabited regions where the process of getting a vessel under way is a matter of life and death. The heaviness of the ship, its inevitable entanglement in its own spars, its refusal to "give" to collective human effort, become matters of crucial concern to the novelist. Similarly, the smaller boats carried in or alongside the larger vessel receive a great deal of attention in the way they must be hoisted or dropped with the right combination of muscle and skill.

Completing Cooper's later conceptions of sea and ship is his sailor who, unlike the Byronic presences and ideal patriots of his earlier fiction, tends to partake of the greater continuum of human experience. The gentleman-captain becomes a predominant figure, at ease both afloat and ashore and capable also of seeing the metaphorical connections between. "Life is like a passage at sea," explains Captain Truck in *Homeward Bound* (ch. 18), introducing an extended conceit, and indeed, a number of gentleman-captains are called on to make grand moral choices brought on and isolated by the circumstances of the sea. In *Afloat and Ashore* and *Miles Wallingford*, which take place against the background of American neutrality during the great European war of 1797–1803, Miles suffers countless reversals at sea, to the point where he himself becomes reduced to a state of metaphysical neutrality. As he floats on the wreckage of his own vessel, Miles says: "I knelt . . . and prayed to that dread Being, with whom, it now appeared to me, I stood alone, in the center of the universe" (*Miles Wallingford*, ch. 22). The sea is that element which quickens moral dilemma, and here the sailor is an existential pioneer, repeatedly brought to the abyss only to be saved. Recognizing that at sea he is forever vulnerable to the wash of history, Miles retreats to his ancestral residence with his new wife to build a life that history and noisy patriotism cannot touch. Other Cooper captains undergo similar conversion experiences, religious or otherwise. In *The Two Admirals*, Rear Admiral Richard Bluewater must decide, at a crucial moment in battle, whether his loyalty to the Young Pretender's claim to the British throne in 1745 is greater than his personal loyalty to his own fleet commander. And as suggested earlier, Roswell Gardiner of *The Sea-Lions* ponders the divinity of Christ while weathering the horrors of an Antarctic winter.

Like his earlier Byronic heroes, Cooper's new captains tend to be troubled and intelligent, undergoing crises of conscience while making brilliant nautical maneuvers, but the source of psychological disturbance, hidden and resonant in his early trilogy, becomes itself the issue of the fiction in his later novels. Those that most clearly resemble his earlier work are *The Wing-and-Wing* (an updating of *The Water-Witch*) and *Jack Tier* (a darker version of *The Red Rover*). In each case, Cooper begins with the assumptions of the original and works them into a more complicated moral design.

Because Cooper had not himself sailed many of the seas he wrote of in his later tales, he relied increasingly on his sources—on Irving's *Astoria*, for instance, in *Afloat and Ashore*, and in *The Sea-Lions* on William Scoresby's *Account of the Arctic Regions* (1820), which Cooper reviewed when it first appeared. He borrowed freely from historical narrative and pieced it into his own imaginative vision. Melville, we know, worked the same way, eventually piecing Cooper's vision into his own. When Melville reviewed Cooper's *Sea-Lions* in 1849, he praised the "grandeur" of its "lonely and terrible scenes." When *Moby-Dick* appeared two years later, *The Pilot* was quoted among the

"Extracts." Melville knew early Cooper and late Cooper, and pushed beyond. The circles of influence in American literature of the sea expand outward, and Cooper can be seen as navigating the distance between Scott's shoreline sea of domestic sentiment and Melville's open sea of metaphysical speculation.

HUGH EGAN

Chapter Four

PERSONAL NARRATIVES, JOURNALS, AND DIARIES

After Christopher Columbus sailed literally and metaphorically beyond the Old World and into the New, he narrated his experiences in a way that, as Chapter 1 shows, reveals the struggle between the archetypal myths underlying American sea literature and the literary style of the age. Yet Columbus's narration of his voyages also reveals an ideological struggle that would become similarly typical (*Four Voyages to the New World*). In telling of his encounters with the "Indians," his classifying them sheds light on his own culture's values; in his discussion of colonization, he reveals the political and economic substructure of Western society; and in his fervent desire to christianize the natives and to find the Earthly Paradise, he expresses the religious foundations of the West's political and cultural beliefs. In short, his narration of the events he experienced, as it indicates his ideological assumptions, shows a profound sense of what would later be manifested as American specialness: he maps out what is, if not the Earthly Paradise itself, then the surest providential path toward it. Yet Columbus also records the effects of his voyages on himself, for to him as later to William Bradford and Benjamin Franklin, sea voyages mark critical stages in the individual development of the sailor-narrator.

These issues, ideological and psychological as well as literary, later become the focus of many nonfictional American sea narratives, journals, and diaries* during the formative period of American culture, from the American Revolution to 1900. Although most such documents of this period are flat, unadorned records of events—and are therefore of limited literary (though

---

*For practical reasons the words "account" and "narrative" are used in this chapter to refer to all three of these different but closely related forms of writing.

of considerable historical) interest—many contain significant reflections on the events and interpretations of their meanings. That is, just as the actions of Americans at sea expand the boundaries of the nation and the self, so their accounts attempt to shore up American values, both historical and individual. In his 1856 preface to Elkanah Watson's narrative, Winslow C. Watson perceived such a connection during the first half-century of nationhood. Noting that it was a period that "embraced the epoch of the War of Independence, and of those amazing mutations which have marked the transformation of dependent colonies into a mighty nation," with Whitmanesque optimism Watson goes on to say that the narrative provides "data by which the vast progress of the Republic, in its prosperity and power, may be best realized and most adequately appreciated." Although nonfictional sea narratives may not provide the "best" measure of the history of the Republic, they nevertheless offer a sound gauge of the nation's and its citizens' concerns and values.

Not to be forgotten is that those concerns and values sometimes diverged according to race and gender. Although many black slaves suffered the horrors of the Middle Passage to America, few narrated their experiences; however, Olaudah Equiano's stirring account (1789) ably details these experiences—inhumanely crowded and filthy conditions, brutal treatment, disease, starvation, and death. Later, other slaves took to the sea in attempts to escape their situation, as did, for example, Henry "Box" Brown, who had himself shipped as cargo. Many black men also served as sailors on merchant and naval vessels, although again relatively few produced narratives or journals. As though illustrating stereotypical male-female differences, women writing personal accounts of maritime experiences (very few wrote formal narratives, and fewer still were published) rarely commented on politics, issues of the day, and larger human questions. Almost all of the surviving women's journals of the period, manuscript and in print, were written by the wives of sea captains accompanying their husbands on whaling or commercial voyages. Typically, these women wrote about their emotions and those of others, about religious concerns, about their husbands—in their roles as both captain and spouse—sometimes about the community aboard ship, and occasionally about the scenes encountered ashore. Another frequent topic is their sense of isolation, of exclusion and inessentialness because of being the lone woman aboard, the lone person not officially engaged in the work of the ship. In the genre of the public narrative, though, Abby Jane Morrell's account of the long exploring voyage commanded by her husband Benjamin Morrell (whose own narrative exists for comparison) reads like those written by many of the men, but the gender-genre issue is made problematic by the fact that her narrative was ghost-written by a man.

The men who went to sea tended to be more independent, more restless and impulsive, than their fellow countrymen; and when "the independence of the U.S. was not to be forever annihilated," as the man-of-war's-man Samuel Holbrook later characterized the situation, many Americans shipped out to establish and protect that independence (ch. 2). And to assert it— for "independence" and "freedom" are watchwords in most sea narratives of the period. Andrew Sherburne, a common sailor on the *Ranger*, explicitly wishes that his own narrative might help Americans "properly appreciate the freedom they enjoy," but both navy men and privateers, though usually ascribing little literary value to their narratives, often claim that they are narrating experiences typical of the Revolution, and therefore reemphasize the values common to Americans. Rare exceptions, such as the Loyalist Louisa Susannah Wells's *Journal*, may express joy at sailing "out of the dominion of Congress," but such accounts serve mainly to cast contrasting light on the more dominant praise for freedom, Congress, and America. To highlight these values further, many narrators attempt to reverse the cultural stereotype (mainly held by those in the Old World) of America as an uncivilized backwater. In a revolutionary turning of the concepts of civilized/barbarian, American sailors like Josiah Cobb, for example, feel it their "duty" to counter the impression of their enemies' so-called "high-souled gentlemanly conduct" by claiming that it is the British who are "barbarous" and "inhuman" (ch. 11). That is to say, the Americans' narratives seek to establish the legitimacy of American civilization by placing it on a plane above and certainly separate from the mother country's.

Nowhere is this separation more clearly in evidence than in the numerous accounts of sailors held captive by the British. Narratives by Cobb, Benjamin Waterhouse, Charles Herbert, and Thomas Dring chronicle the most extreme insults and the most "unnatural" treatment by the British, whose infamous imprisonment of the captives—in prisons such as Dartmoor and prison ships such as the *Jersey*—was characterized by severe deprivation, squalor, brutality. As Dring says (in Albert Green's *Recollections of the Jersey Prison Ship* [1829]), "Among the varied events of the war of the American Revolution, there are few circumstances which have left a deeper impression on the public mind, than those connected with the cruel and vindictive treatment which was experienced by those of our unfortunate countrymen whom the fortune of war had placed on board the Prison-Ships of the enemy" (ch. 1). Yet such experiences only increase the patriotic fervor of both the narrators and their audience. During their absence, for instance, the captive Americans cling even more tightly to their nascent nation and its customs: they rejoice in any positive news that has evaded British censorship, and they raucously celebrate in prison the anniversaries of the Declaration of Independence.

Like the Puritan captivity narratives from which they descended, these narratives from the Revolution emphasize personal and religious dimensions beyond the political. Both Waterhouse, a ship's surgeon, and Dring, a captain, conclude by stressing how the captive comes to feel strengthened morally and spiritually by his ordeal. Dring explicitly acknowledges the hand of Providence in his deliverance, as do most of the captives. Sherburne's *Memoirs* is in fact a conversion narrative, which shows his turning to God when in danger, his recognizing that "God spared me" in prison and shipwreck, and his consequent devoting himself to religious activities (which the latter half of his narrative describes). He ends the volume with a sermon, "Appendix. Address to sailors, especially those of the American Navy"—not the equal of Father Mapple's sermon, but an eloquent precursor nonetheless. Thus crucial to these captivity narratives, as to the genre as a whole, is their intention, often literally their mission, to affect their audience.

The characteristic concerns of these revolutionary narratives are clearly evident in the two most important firsthand accounts of the period, those written by John Paul Jones and a sailor under his command, Nathaniel Fanning. Brash yet calculating, devoted to the cause yet self-promoting, Jones typifies in heightened terms the flexibility necessary for the revolutionary sailor to prevail. He is equally adept at narrating vividly the battles with the *Serapis* and at Whitehaven, and the negotiations at the French court—not to mention his battles and negotiations with the ladies of the court. Like most other sailor-narrators, Jones stresses his commitment to America and freedom and his enmity for the "barbarous and unmanly" British. In a typically impassioned passage, he states, "I was, indeed, born in Britain; but I do not inherit the degenerate spirit of that fallen nation, which I at once lament and despise. . . . They are strangers to that inward approbation that greatly animates and rewards the man who draws his sword only in support of the dignity of freedom" (Sands, *Life and Correspondence*, 211). Also evident in his remarks is a strong sense that his war experiences, like those of the captives, are crucial to his personal development; indeed Jones is determined to make them so. To him the war offers not only a way of serving his country and the cause of freedom, but a way of "acquiring honor," of elevating his rank, of making a name for himself. In Robert C. Sands's edition of Jones's *Life and Correspondence* (1830), this ambiguity of character is deeply at issue, for

Sands devotes considerable space to defending Jones and glossing over his flaws. Just as American attacks on the British show a desire for legitimacy, so Jones's and his editor's defensive comments seek to legitimize him as an American hero.

Fanning has no such desire. Like many common sailors who described their experiences, Fanning rebels as much against officers in general—and Jones in particular—as against the British. His narrative, as lively or even mercurial as Jones's, describes his captivity by the British, whom he excoriates frequently and at length, and his later service under Jones. With the voice of a true democrat, Fanning censures—also frequently and at length—his commander's insensitivity and brutality toward the common sailors as well as his egoistic ambition. Nevertheless, Fanning does acknowledge Jones's heroism and defends him against criticisms leveled by the British. Fanning is, in short, a dedicated American who means his narrative to serve as exemplum and sermon on the cause of egalitarianism and freedom: "O my country!" he exclaims, "how happy a lot has Providence placed her in. Thank God, there are no royal leeches there, and I sincerely pray to him that we may never have any" (170).

Such paeans to liberty and complaints about tyranny continue, with America's own struggle, through the War of 1812. John Hoxse, self-proclaimed *Yankee Tar* (1848), tells of his considerable suffering at the hands of the British, but he rejoices in his patriotic sacrifices and trusts to Providence. Captivities by the British and the Algerians provide the occasion for James Durand, Samuel Holbrook, and George Coggeshall to reiterate their belief in Providence and in patriotism. Common sailors continue to rail against their barbarous enemies and against their own officers' inhuman practice of flogging, the most vicious and often-cited example of the anti-democratic government on board ship. Like their country, though, the sailor-narrators emerge from these experiences stronger, more independent, more confident.

Its nationhood to some degree secured, the United States continued to develop its power and influence throughout the nineteenth century, a process reflected vividly in the sea narratives of the period. Basically, these narratives demonstrate American concern in five major areas, which hearken back to Columbus: political, to extend the official American presence into such areas as China or Africa; scientific, to collect data (geographical, geological, botanical, etc.) about little-known regions; ethnological, to observe and record the customs of other peoples; religious, to convert those peoples to Christianity; and commercial, to extend and deepen American trade in whale products, seal skins, and manufactured goods. In his address to the House of Representatives (the first section of his *Pacific and Indian Oceans*, 1841), Jeremiah N. Reynolds, a relentless advocate of nautical exploration and development, stresses the importance of each of these aspects for further development of America, since "the maritime enterprise of our ancestors was an important element of our subsequent power."

During the course of the century, American recognition of this importance included sponsoring and encouraging a number of expeditions and voyages to virtually every geographical area. David Porter's *Journal* of 1812–14 serves as a bridge between the war narratives and these voyages of exploration, as he voices some militant criticism of the imperialistic practices of the European powers and celebrates, for example, the Chilean revolt against Spain. Porter's patriotic fervor also allows him to see, less accurately, a "republican" form of government akin to his own country's even in a tribe of Marquesans, with whom he makes a treaty and whom he enlists in the war against Britain. Edmund Roberts's account of his *Embassy to the Eastern Courts* during the 1830s provides interesting details of his relations with China, Siam, and other nations, but his introductory wholesale condemnation of "the abject condition of morals among the inhabitants of the Indian ocean" exposes a cultural chauvinism slightly more overt than Porter's. Matthew C. Perry's official report (compiled by Francis L. Hawks) of his midcentury expedition to Asia describes in even greater detail—but with little deeper appreciation— his experiences in the Asian courts and cities, information eagerly awaited by an American public whose interest had been heightened by the discovery of gold in California and the consequent possibility of extending trade with Asia. Admiral D. G. Farragut's cruise to Europe, Africa, and Asia, narrated by James Eglinton Montgomery, met with "honor," "respect," and "admiration" not only for its commander but also for "the country," which had just survived the Civil War. Montgomery's dropping of a vast number of foreign dignitaries' names further attempts to demonstrate that America's status abroad has not been diminished by the conflict; as Montgomery ends his narrative, "the brilliant cruise of the *Franklin* passed at once into history, and on its pages illustrates our unchecked prosperity at home and our national power abroad" (ch. 42).

Little of the pomp and circumstance of these accounts is evident in the narratives of the common sailors who served under these and other officers. Rather, the sailors, such as Nathaniel Ames, Charles Nordhoff, Samuel Leech, and Henry Mercier, detail life on board: the typical crewmen, their yarns and complaints, their duties and pastimes, the organization of the ship—plus a few passages describing the beauty of Rio or the corruption of Callao. Here are evident few great personages or exotic cultures, but many of

[88]

the everyday, down-to-sea adventures and hardships of the average American tar. In keeping with the typical nineteenth-century concern for reform and proletarian conditions, the most common theme of the sailors' narratives is not so much patriotism as egalitarianism: most complain at length about abuses—especially flogging—by the officers. Leech gives a particularly stirring account of flogging and indicts the mechanistic system of order on board men-of-war, by which discipline leads all too quickly to tyranny. "Treat him as a man," Leech says, "and the sailor will respond in kind; treat him as a cog in the nautical machine, and the sailor's baser nature will prevail" (ch. 2). Nordhoff voices little hope for better life in the navy and opts for the merchant marine, for "I had had a surfeit of bondage" (ch. 16). Ames's is one of the strongest voices for sailors' rights, so frequently abused by "the generally overbearing, tyrannical and inhuman conduct of Yankee skippers" (102), even in the merchant marine. Others, including Leech and Mercier, see the sailors themselves as partly responsible for their plight, particularly to the degree that they overindulge in grog. They are part of the sea-going temperance movement that Melville parodies in his fiction.

Some officers' narratives show an effort to respond to these problems. Charles S. Stewart's *Brazil and La Plata* describes the successful results of his experiment in maintaining discipline without flogging. Samuel Samuels, who worked his way up to captain and through virtually every possible hardship, gains his mutinous sailors' respect and obedience through what he, following Disraeli, calls "muscular Christianity." And Rev. Walter Colton's religion heightens his sympathetic recognition of the almost hopeless hardships of man-of-war life. Christianity, temperance, and social reform go hand in hand at sea as in America during the midcentury; however, the restrictions of shipboard life heighten and sharpen these microcosmic correspondences.

Most expeditions also made some effort to advance scientific knowledge along with political influence. Flora and fauna of the exotic regions occupy page upon page in the narratives; Porter, for example, acknowledges that many previous writers have described the Polynesian breadfruit tree, yet he proceeds with his own description. By far the most important scientific descriptions, however, stem from the U.S. Exploring Expedition of 1838–42, commanded by Charles Wilkes. Wilkes's own voluminous *Narrative*, but one of about forty published and unpublished accounts of the expedition, certainly cannot be faulted for lack of detail, although his many critics savaged its repetitious and difficult prose and its author's condescension and pettiness. Yet Wilkes's *Narrative* was very influential: not only for the general public but also for Cooper and Melville, Wilkes provided a wealth of information about the Antarctic and the Pacific.

In the latter half of the century, the polar regions were a major focus of scientific study, and nowhere are the hardships of nautical life more clearly drawn than in the narratives of the sailors to these regions. Elisha Kent Kane, Frederick Schwatka, and Adolphus Greely describe terrifying cold, suffering, sickness, and death in the Arctic. Nevertheless, in his "abject misery and extreme wretchedness," in Greely's prefatory words, is also evidence of the "extraordinary spirit of loyalty, patience, charity, and self-denial" that allows them to survive. In short, in the Arctic narratives the sailors attempt to demonstrate—to quote Greely again—"that indomitable American spirit": in the context of the midcentury turmoil that threatened to destroy America, such statements, both the exploration itself and the narration, reveal the individual American's anxious reassertion of American specialness. Equally illustrative yet perhaps more vivid and accomplished is Frederick A. Cook's *Through the First Antarctic Night, 1898–1899,* in whose introduction Cook, the only American in a Belgian expedition, characterizes the voyage as "scientific exploration along the edge of the unknown." Caught in the Antarctic ice for over a year, Cook and his shipmates experience what he calls the worst of sea life, the lack of anything sympathetic to man in the ocean, now frozen, surrounding them. As all deteriorate physically and mentally, Cook comes to know what is elemental in man: "In these dreadful wastes of perennial ice and snow, man feels the force of the superstitions of past ages" (ch. 25); "animal sentiments take the place of the finer, but less realistic human passions" (ch. 27). Even in this literal heart of darkness, though, Cook sees an equally elemental democracy in which "there simply exists no longer a Commandant, no captain, no officers. We are all ordinary workmen" (ch. 29). This basic commonality reflects not only the American egalitarianism and the proletarian revolutionism of the period but also the literary realism at its height at this time.

In the early chapters of his narrative, Cook presents a relativistic, even sympathetic portrait of the natives of Patagonia and Tierra del Fuego, and such ethnographic descriptions pervade the sea narratives into and throughout the nineteenth century, even though the authors may be primarily concerned with science, politics, or commerce. These observations of native peoples generally fall into one of three categories: romantic, chauvinistic, or relativistic. Writers from the early part of the period, like Porter and John Ledyard, are especially prone to overly romanticized portraits of the natives they encounter. Ledyard presents a lush and exotic view of the life and appearance of the nobly savage Hawaiians, who are superior in his eyes to Captain Cook and the civilization he is advancing. Porter is "inclined to believe that a more honest, or friendly and better disposed people [than the Marquesans] does not exist"; he denies their savagery, lauds the "elegance" of their buildings, and concludes that morally and physically "they rank high in the scale of human beings" (ch. 14).

In direct contrast to such views are those of sailors who see natives as decidedly *ig*noble savages, those whose cultural chauvinism reduces the cultures they encounter to a level beneath their own. Fitch Taylor glories in the changes wrought on the Hawaiians, who "before were a heathen and a savage nation"; religions other than Christianity are to him "the superstitions of heathen nations" (vol. 2, sec. 8). A. W. Habersham, visiting the Japanese, condemns their culture in no uncertain terms: "Natural depravity and impurity of taste is perceptible at almost every turn . . . but then it must be remembered that they are half-civilized Orientals, and heathens at that" (ch. 14). As the recurrence of the term "heathen" implies, the basis of much of the chauvinism is religious: these sailor-narrators, like most of the women, share the common belief that "Christianity" and "civilization" are synonymous and are the standard by which all humanity is to be evaluated. Nowhere is this attitude more prevalent than in the volumes of Stewart, whose nautical mission to the South Seas gave him ample occasion to demonstrate his absolute intolerance of any native customs at variance with his Christianity.

Some visitors do perceive other cultures in considerably more relativistic terms. Cook considers—with little romanticization—the whites' supplantation of the Fuegians "a crying sin," for "they do not need a new set of morals as badly as we do ourselves" (ch. 8). Even though his *Narrative* is infused with his Christianity, Captain Amasa Delano does not allow it to preclude a respect for other religions and cultures. He refuses to censure the religion of the Malays, for example, because it has elements as rational as Christianity if one thinks of them with the "catholic spirit" of the "reflecting traveller" (ch. 7). Later, the ever-tolerant Delano concludes, "National prejudices, to a certain extent, may be very useful, and possibly necessary; but they are always attended by considerable evils in the narrow and intolerant spirit which they perpetuate, and in the contentions which they produce. The more enlarged a mind becomes in its views of men and the world, the less it will be disposed to denounce the varieties of opinion and pursuit" (ch. 14). Each of these attitudes, moreover, reveals as much about its author's conception of America as about ethnography. Ledyard and Porter expose their romantic optimism for their republican nation; Habersham and Stewart their equally idealistic faith in the redeemer nation; and Cook and Delano the tolerance that, in ideal conception, assimilates diverse peoples into the melting-pot nation.

An interesting, feminist note on this issue is sounded by Harriet Allen of the whaler *Merlin* in 1871. Having dined by invitation on the African island of Johanna with its prince, she writes:

> I should have enjoyed the dinner very much if the ladies could have sat with us at table. The Prince said their religion will not allow it. Being a woman I felt insulted and degraded for their sakes. How I long for the power to take them away, to give them their birth right—freedom. And not only them, but all miserable women slaves the world over. (May 21)

Allen was unlike her seagoing sister-writers in her active concern about women's inequality, at home as well as in exotic cultures; but other comments suggest that even she was intellectually unaware of her legal kinship to the "women slaves" with whom she sympathizes.

As Allen's narrative demonstrates, some of the most powerful voices for tolerance and equitable treatment were those of the women and blacks aboard. Many women who experienced and recorded life at sea tried to do something about the state of the sailors' minds, bodies, and souls, including distributing Bibles and temperance pamphlets. While the dominant nineteenth-century attitude of ships' officers seems to have been that sternness, rigorous discipline, and the occasional use of force (including flogging) were necessary adjuncts of command, the journals and diaries of their wives sometimes express a belief in the efficacy of softer words, better physical treatment, and the salubrious effects of applied Christianity. Margaret Fraser, aboard the *Sea Witch* in 1852, tried to modify her husband's profanity and violence (he was a notorious "driver" of his men) with timid, wishful statements in her manuscript journal. Captain Fraser was unmoved. But this gender gap did not exist in all cases: though these women were sometimes naive about the malleability and responsiveness of the men under their husbands' command, it is clear from journals written by both men and women that at least some captains were able to make policy of such principles. In slaves' narratives, such as those by Henry Bibb, Lunsford Lane, Solomon Northup, James W. C. Pennington, and John Thompson, the sea frequently represents the freedom that they seek and occasions a number of eloquent statements on equality and opportunity for all people.

Religion, closely tied to these humanistic issues, is a fourth area of concern to both men and women. For a missionary and naval chaplain such as Stewart, Christianity obviously informs every observation, description, and narration of events. Similarly, that of Ames and W. S. Ruschenberger aids in their overcoming the hazards and hardships of shipboard life and provides a standard by which they can evaluate the morality and spirituality of their shipmates and the others they encounter. Many women recorded sincerely pious prayers, affirmations of belief, and statements of spiritual striving. Particularly on whale ships, where their captain-husbands conducted their Sundays at least as actively as advised by the tortuous logic of Melville's Captain Bildad ("Don't whale it too much a' Lord's days, men; but don't miss a fair chance either, that's rejecting Heaven's good gifts" [ch. 22]), women took the sanctity of the Sabbath and its nonobservance at sea as a common topic. They were usually more religious themselves than their husbands and pointedly noted the rare occasions when they met captains who permitted no more Sabbath work than what was essential for safety and well-being. Eliza Fraser accompanied her husband, Captain Samuel Fraser, on a voyage and follows in her narrative the traditional captivity narrative form: shipwrecked on an island, captured by "savages," she endured the trials by deepening her faith, until "the God of mercy interposed" and rescued her.

In the accounts of many American men, though not often in those by women, advancements in religion, society, and science were at most secondary to the advancement of commerce. The prospect of material gain led many to sea—and to narrate their experiences on their return. A multitude of sailors tell of gaining little more than enough to stave off abject poverty, often due to the kind of parsimony on the part of owners and captains that Ames castigates—or that Melville fictionalizes in his Bildad and Peleg. Some merchant captains such as Edmund Fanning and Richard J. Cleveland do narrate travels that gain them fortunes. Rarely is individual gain a theme in these narratives, however; Cleveland, for example, states that "the fortune I had gained was amply sufficient to enable me to live independently," but he is "too long accustomed to a life of activity" and, like Melville's Bulkington (though apparently for different reasons), remains ashore as little as possible (ch. 11). For him, and for others like Delano, the merchant marine is an outlet for their independence, restlessness, and adventurism. George Coggeshall recognizes that his experiences typify "the progress of our growing commercial marine" and so narrates his early voyages "that the present generation may be able to contrast our then infant commerce, and the inefficiency of our merchant marine, with the rapid strides it has since made" (ch. 1).

The independence and adventure—and the hazard—of commercial sailing are epitomized by the narratives of whaling. Though F. D. Bennett devotes more space to the ethnography of the South Sea Islands than to whaling, and passengers such as Mary Chipman Lawrence tend to focus on the more pleasant aspects, the more vivid and gripping reports emphasize the dark side of whaling life. The accounts of whalers such as J. Ross Browne and Ben-Ezra Stiles Ely detail drudgery and horrors akin to those suffered by the polar explorers. Tyrannical captains and villainous owners, long voyages fraught with danger, and hard-earned but meager rewards—such are the conditions of life described by Browne and Ely. "It is painful beyond description," Ely says, worse than slavery (ch. 4). Both accounts are tinged with bitterness as well, for both authors recognize that whaling has helped "enrich and aggrandize a nation," in Ely's words, but has grossly abused the sailors themselves. Mere abuse, however, was the least of Owen Chase's problems. His short *Narrative of the Most Extraordinary and Distressing Shipwreck of the Whale-Ship Essex* (1821) turns "Extraordinary" and "Distressing" into understatement, for Chase describes being stove by a whale, forced to endure weeks in a small, open boat, and reduced to despair, starvation, and even cannibalism. His suffering is an almost unimaginable extreme of that shown in many whaling narratives, themselves documenting an extreme of the hardships inherent in nineteenth-century nautical life. Such shipwreck narratives, collected and reprinted throughout the century, were very popular and provided not only insight into life at sea but also the lurid excitement of Gothic fiction. In short, even as the sea narratives chronicle the growth of America and the values that guide it, they make equally clear the price many individual Americans paid to secure that growth.

The price of American development is also visible in the relatively few narratives concerning the wars of the last half of the century: the Mexican War, the Civil War, and the Spanish-American War. The first included few naval engagements—the siege of Vera Cruz is the notable exception—but did give the opportunity for naval narrators such as Raphael Semmes to muse about manifest destiny: "The next generations of Mexicans may have cause to look back with satisfaction upon the struggle of their country with the United States, as the starting point whence a new impetus was given her, in that great race of civilization, which is to fit her for her ultimate incorporation into the Anglo-American family" (*Mexican War*, ch. 22). Actually, the events of the next generation led Semmes to look back with much dissatisfaction on the Civil War. A Confederate admiral and an accomplished if vitriolic writer, Semmes describes at great length in his *Memoirs of Service Afloat* (1868) a destiny manifestly changed, for to him South and North are *"two nations"* whose Puritan and Cavalier heritages are, and will continue to be, incompatible; "God has created men in different moulds," Semmes says in a peculiar twist of American providentialism (ch. 22). Ever the impassioned if not embittered spokesman for the South, Semmes sermonizes about states' rights, about Yankee avarice and brutality and hypocrisy, and about the glorious mission of the raider *Alabama.*

Narratives by the Northern sailors tend to be much less excited and exciting, less analytical about the cause and practices of the war; most, such as those by Samuel Pellman Boyer and Robley D. Evans, simply repeat the dull routine of life in the blockade of the South. Winfield Scott Schley's *Forty-Five Years Under the Flag* (1904), however, evinces the youthful enthusiasm of a midshipman under the illustrious Farragut. Schley and Evans go on to describe events in their careers after the war, both taking active part in the Spanish-American War. Their narratives reinforce their conviction, or hope, that America itself has weathered the Civil War in fine trim, and the navy they describe in their final pages is the impressive right arm of an imperial power of worldwide scope and influence. Evans ends his *Sailor's Log* (1901) appropriately on July Fourth, with the triumph of America over Spain, and in a rousing, patriotic paean to America. National legitimacy, which the sea narratives of the Revolution had sought to establish and which the nautical narratives of the nineteenth century had developed, is in full evidence at the century's end.

Equally in evidence is the theme of individual development, which in many if not most of the narratives is the microcosmic accompaniment to the development of the nation and the national ideology. The growth of America—as Emerson and Whitman state—is commensurate with the growth of the American, and both are fostered by nautical experiences. Most narrators who discuss their motivations for going to sea recognize that the desire for personal change or betterment is a large factor. Women express other motivations too: duty to husband is one; others make simple expressions of love. For still others, previous loneliness ashore during their spouses' long voyages made the vicissitudes of the unknown preferable to the certain pain of the known. Keeping the family together—almost an article of faith with some—led them to consider seafaring virtually a necessity. In their time and place, it was almost a matter of course that some wives accompany their captain-husbands, while for others, staying home was economically unfeasible. And a few, very few, say they felt more at home at sea than they did ashore. Restricted, more than any of the men, simply by their being female, they did choose to go to sea; but their reasons generally contrast with those of men, implying their conditional, dependent status.

The further reasons expressed by men come in statements of romantic longings nurtured by their youth or their reading (Cobb), of desire to escape from social or familial disappointments (Samuels), of a need to become physically healthy (Nordhoff, Taylor, Ely, F. A. Olmsted), or of a desperate financial necessity (Browne, Durand). Horatio Bridge put it succinctly (though perhaps more audible here is the voice of his editor, Nathaniel Hawthorne): "Many go to sea with the old Robinson Crusoe spirit, seeking adventure for

JOHN SAMSON

its own sake; many, to escape the punishment of guilt, which has made them outlaws of the land; some, to drown the misery of slighted love, while others flee from the wreck of their broken fortunes ashore, to hazard another shipwreck on the deep" (ch. 1). As he sets out, Bennett echoes Milton in saying, "the world was before us—and our destination was involved in an agreeable uncertainty" (preface). For many, the land and society left behind are indeed a sort of paradise lost, and the uncertainty may be less than agreeable; yet the possibility of a new life and an eventual salvation (whether religious or secular) opens their minds for possible changes.

The various trials at sea, then, act as a catalyst and crucible to free the sailor for a confrontation with what is essential or basic to himself, his values, his capabilities. This is fundamentally a process of definition, the individual coming to define himself much as Crevecoeur attempts to explain "what is an American," contextually. That is, the sea, Melville's "dark side of this earth," is radically alien to the individual but is also the context in which the self is defined. This alien yet formative element may be the absolutist shipboard government that reduces the common sailor to insignificance, or the exotic locale and culture he visits, or the nature that seems hostile to humanity. As Ruschenberger notes, initial seasickness is a sort of mythological initiation, just as weathering Cape Horn is to many an almost archetypal descent into the underworld. From either experience, and generally from the harsh and alien conditions of sea life, the sailor may realize what he values in himself and his society, and thus may emerge a new man.

His return, with the attendant reacculturation, completes this process, particularly when the sailor returns to construct a narrative of his experiences or to publish his journal—which women hardly ever did, largely, perhaps, because they consciously wrote for a small, familial audience. The prefaces to sailors' books frequently reveal the psychological impetus to and benefit of writing: Cobb's narrative attempts to "rescue" his memories from "oblivion," Delano's is a providential tool to save him from depression, and Hoxse's may "tend in some measure to soothe [his] downward passage to the tomb." Moreover, many intend their narratives to benefit other individuals as well. Both Leech and Samuels, for example, wish their narratives to serve as a caution to youths contemplating shipping out, that they might learn in a school of softer knocks. These sailor-narrators, in perceiving their experiences as a pattern of personal growth or as a model for social emulation (or caution), thus finally emphasize a reintegration—of the self and of the self with its society—that is the individual equivalent of the nationalism prevalent throughout the narratives of this period.

The classic description of the psychological development of a sailor—

perhaps the classic nonfictional American sea narrative—is Richard Henry Dana's *Two Years Before the Mast* (1840). In smooth and vibrant prose admired by Melville, Dana describes his shipping out to strengthen himself both physically and psychologically, which he accomplishes through the sailorly experience he ably details. The effete and ailing bookworm grows stronger in the many demanding tasks of the common sailor; the Boston Brahmin gains a wider perspective on humanity as he lives with a typical range of nautical types, from the "handsome sailor" Bill Jackson to "the most remarkable man I have ever seen—Tom Harris" (chs. 13, 23). He returns with a "Hurrah for Yankee Land" and with a sense of security, "a calmness, almost an indifference" that marks his mature self-reliance (ch. 26). He returns, also, as "a voice from the forecastle," narrating his firm identification with and sympathy for the plight of common sailors, for whom his writing and actions continue to lobby. Dana's effectiveness is attested to not only by his continuing popularity but also by the number of imitations: Nicholas Isaacs's *Twenty Years Before the Mast* (1845), Jacob Hazen's *Five Years Before the Mast* (1854), Charles Erskine's *Twenty Years Before the Mast* (1890), and so forth.

The self-assured independence Dana learned is at the forefront of two works that—along with Evans and Cook—mark a triumphant end to the century's sea narratives. Joshua Slocum narrates, with typical Yankee understatement, *Sailing Alone Around the World* (1900), a declaration and achievement of independence on the individual level that parallels on a smaller scale the cruise of Teddy Roosevelt's Great White Fleet during the first decade of the new century. If Slocum's independence is literal, Mark Twain's, demonstrated in his *Following the Equator* (1897), is intellectual and ironic. Stating at the outset that "to be tied in any way naturally irks an otherwise free person and makes him chafe in his bonds and want to get his liberty" (vol. 1, ch. 1), Twain uses the context of an around-the-world cruise to give free rein to his iconoclastic wit. Twain's is an irony that liberates—from the perceived absurdities of cultural practices, habits, and beliefs—and it seems most easily achieved in the comfortable irresponsibility of sea travel. "If I had my way," he says, "I would sail on forever and never go to live on the solid ground again" (vol. 2, ch. 26). Twain probably intends a metaphorical significance here: an avoidance of solidity that—as with Melville's Bulkington—is philosophical more than geographical and that he echoes at the end of *The Mysterious Stranger*.

If published sea narratives of this period moved from a call for independence to a demonstration of it, they also moved stylistically from romanticism to realism—paralleling literary as well as cultural history. Early personal narratives, such as Porter's, showed a heightened, emotional sensibility and prose style not far removed from Poe's, but in the generic demand for factuality are elements of a literary realism that will only later develop in fiction. As numerous prefaces show, the narrators' reading in sea fiction and nonfiction serves further to define generic parameters, but the allusions also occasion arguments against the romantic perspective and thus further promote the realistic cause. Not merely a sub-sub-genre, sea narratives from the Revolution to 1900, like the nautical experiences they describe, have a significant impact on American culture in their own right and in their influence on more mainstream genres: on the poetry of Whitman, the drama of O'Neill, and the fiction of Cooper, Melville, London. To use Melville's metaphor in *Redburn*, they are the junk, "odds and ends of old rigging . . . the yarns of which are picked to pieces, and then twisted into new combinations, something as most books are manufactured" (ch. 24). From the junk are woven the ropes that support the masts, control the sails, and in general keep the ship sailing; so these narratives, journals, and diaries, at times seemingly composed of insignificant or incoherent nautical "yarns" merely strung together, in fact provide revealing insight into the basic values and concerns that hold American literature and culture together as they sail onward.

JOHN SAMSON

Chapter Five

HYMNS, CHANTEYS, AND
SEA SONGS

They that goe downe to th' sea in ships:
Their busines there to doo
In waters great. The Lords work see,
I'th deep his wonders too.

"I hear America singing," Walt Whitman wrote, "The boatman singing what belongs to him in his boat, the deckhand singing on the steamboat deck." America has always been a singing nation, and her maritime concerns have always been reflected in the lyrics of her songs. The Founding Fathers had their hymns of the "waters deep" that spoke of the symbolic quality of the natural world and their spiritual relation to it. The early republic found in the popular ballad the perfect medium for expressing the enthusiastic patriotism surrounding the exploits of its young navy. The rise of chanteying provided an opportunity for the celebration of every maritime endeavor of the nineteenth century from slaving to whaling to the creation of a vast American maritime empire, while the decline of the age of sail saw lyrical interest in the sea shift to the world of the passenger liner and ultimately fizzle out as America began to lose touch with her maritime heritage. Twentieth-century resurgence of interest in things maritime has been accompanied by new directions in lyrical interest, as the sea has become viewed more and more as a place for recreation, as well as an ecological resource deserving protection.

The "varied carols" of Whitman's poem might refer as well to the many forms through which American songwriters have found expression. Not surprisingly in a maritime nation such as ours, writers have throughout three centuries consistently turned almost without deliberation to nautical subjects, whether their inspiration has come through experience or through the

ROBERT D. MADISON

imagination. It would be difficult to find any musical trend in our culture that was not represented by nautical examples—thus suggesting the pervasiveness of maritime imagery and awareness in a nation that does, after all, boast one of the largest *interiors* of any country. As a restless people, our culture is indelibly marked with those most ancient symbols of restlessness and change, the ship and the sea.

The appreciation of American nautical lyrics might well begin with America's first published work, the Bay Psalm Book. More properly *The Whole Booke of Psalmes Faithfully Translated into English Metre* (Cambridge, 1640), it clearly demonstrates several characteristics that describe American sea songs in general. First, the individual lyrics are largely anonymous: we know that the work was prepared by perhaps a dozen clergymen, but we are largely unable to assign particular individuals to particular psalms. It happens that, as we survey the range of American sea song, the best-known examples tend to be anonymous.

Second, the psalms underwent alteration as edition after edition was printed, in a manner analogous to the variation found in lyrics transmitted orally. Again, the best-known American sea songs are those that have been recorded from the oral tradition by folklorists and that vary significantly from version to version. Thus the study of any particular song is usually based on one "archetypal" version—or, at the other extreme, becomes a study of varieties.

Third, the examples in the Bay Psalm Book are not particularly "American." Although touted as the first American published verse, the psalms, of course, are biblical in origin and were translated by displaced Englishmen. Although other chapters in this book show an unmistakable "American attitude" toward the sea, the very difficulty of assigning the origins or reducing the subject matter of the sea songs almost precludes isolating a particularly "American" contribution to the genre. There are, of course, notable exceptions, which are discussed later in this chapter.

Finally, the Bay Psalm Book presents a set of lyrics that, although designed as a "plaine translation" intended solely for "prayse," are unfortunately and perhaps ironically totally dominated by form, as every reader of them has noticed. The naiveté of the versification of these psalms is typical of almost all of the significant nautical lyrics in the American tradition. However lofty nautical themes were to become in other genres, in the sea song the humble accentual verse of the popular ballad remained almost a constant. Whether adopted for hymns or chanteys or even the parlor, the ballad stanza, with all its irregularities and permutations, has dominated American sea song.

If we could identify particular authors and particular versions of sea songs for study, we would still have a great problem with national boundaries. All those sea songs whose origins are connected with the sea itself, or whose transmissions took place on board sea-going vessels, have been tossed together in one great intercultural bin. If such an item as "American" sea songs could be isolated, it would still have to trace its development not simply from the English ballad tradition but also from literary and musical traditions from Ireland, Africa, the West Indies, and the South Seas—as well as, ultimately, that great "Lingua-Franca of the forecastle," as Herman Melville called it.

Actually, the most popular psalm book throughout much of American literary history was not the Bay Psalm Book or its progeny, but Isaac Watts's *The Psalms of David, Imitated in the Language of the New-Testament.* There can be no doubt of the tremendous presence of Watts in the development of American hymnody—his influence was perhaps greater than that of Shakespeare on the rest of our literature—and yet Watts is as indisputably British as Shakespeare himself. If for a hundred and fifty years every American knew Watts as he knew his gospel, then the version of Psalm 107 permeating the American mind is not the Bay Psalm Book translation that stands as a headnote to this chapter but the following one:

> Would you behold the works of God,
> His wonders in the world abroad,
> Go with the mariners and trace
> The unknown regions of the seas.

Watts shared the enthusiasm of eighteenth-century England for maritime endeavors and seems here to be steering away from the literalness of the Bay Psalm Book translation and toward a preromantic or even gothic notion of the sea. It was this direction that American sea literature would come to share.

It is no surprise that a nation of extraordinary maritime development should in time develop a ballad literature topically its own. Nautical songs of the American Revolution gave the first evidence of political emancipation, while the form and general theme of the songs showed that, as in most other ways, we were still indebted culturally to mainstream British songwriting. But such songs as the anonymous "Bold Hathorne" and Francis Hopkinson's "The Battle of the Kegs" showed that with particular vessels and events the American naval ballad had a subject matter and a slant of its own, namely, the celebration of Yankee ingenuity and wit at the expense of British stolidity, coupled with a mockery of the vast British naval establishment as confronted by a greatly "inferior" American force. While patriotism would mark the naval ballad for the next eighty years, Hopkinson's work shows the adaptation of homespun Yankee humor to the sea song, one that would manifest itself not only in topical songs but also in the more "highbrow" poetry of James Russell Lowell and later in the prose of the Southwest humorists and

Mark Twain. The first significant American writer of naval ballads was Philip Freneau, whose patriotic verse does not partake of his romantic tendencies so much as it follows the established pattern for such topical songs: that is, an "improved" ballad with many traces of eighteenth-century poetic diction. If there is a simplification of language, it is toward humorous Yankee dialect. There is no trace of an author behind his work, no romantic sensibility, no conspicuous sea imagery. Neptune's altar, apparently, needed no further polishing.

Freneau's lyrics cover the period from the American Revolution to the War of 1812, and it is interesting that he, along with his anonymous contemporaries, seemed to avoid the influence of Charles Dibdin, whose sentimental sea songs overwhelmingly dictated the major trend of the sea song in nineteenth-century England. Several of Dibdin's songs became enormously popular in America, but "Poor Jack" never was able to replace "Brother Jonathan" as the archetypal seaman in the American naval song. One might speculate that the rejection of the completely sentimental song of the Dibdin type was related to the Yankee scorn of the British naval establishment, as expressed by Freneau in his "Battle of Stonington":

> The old razee, with hot ball,
> Soon made a farmer's barrack fall:
> A codfish flake did sadly maul,
>> About a mile from Stonington.
>> The bombs were thrown, the rockets flew,
>> But not a man of all their crew,
>> (Though every one was in full view,)
>> Could kill a man of Stonington.

The close of the second war with Britain and the subsequent "sentimentalizing" of America did establish an outlet for the sentimental sea song, and that was in the parlor. If the naval ballads of the school of Freneau were intended to inspire the patriotism of the sailors themselves, the new class of "nautical" lyric, freer in form and larded with sentimental pseudonautical imagery, was directed usually toward the feelings of those on shore who were expected to extend their "sympathy" to the "plight" of those at sea—or were to tremble in vicarious fear or admiration of the wonders of the deep. James G. Percival, one of the ranking American poets before Bryant, was guilty of such a piece in his "Naval Ode" ("Our walls are on the sea / And they ride along the wave"), while James Fenimore Cooper included an example of such a song in *The Water-Witch; or, The Skimmer of the Seas* (1830), the first verse of which reads:

My brigantine!
Just in thy mold, and beauteous in thy form,
Gentle in roll, and buoyant on the surge,
Light as the sea-fowl rocking in the storm,
In breeze and gale thy onward course we urge—
My water queen!

"He often sings thus," comments a character after the performance. Cooper's lyric was almost immediately set to music and published in sheet music form, presumably for distribution to parlor pianofortes rather than to sea chests.

If there was a sentimental form that did reach the sailors themselves, it was the elegy. Such songs were usually purveyed to the sailor in one of the many songsters published between the War of 1812 and the Civil War, whose motives could be patriotic, religious, or sentimental. Probably the most widely known elegy was Dibdin's "Tom Bowling," but that song is by no means the most moving. Given a sympathetic reading, William D. Gallagher's "Wreck of the Hornet" demonstrates the power of sentiment:

Soon the rough tar's prophetic eye
Saw many a floating shroud on high,
And many a coffin drifting by—
And on the driving gale
Beheld the spirits of the deep,
Above—around—in fury sweep—
And heard the dead's low wail,
And the demon's muttered curse.
And on the fierce and troubled wind,
Rode Death—and, following close behind,
A dark and sombre hearse.
And soon the barque a wreck was driven,
Before the free, wild winds of Heaven!

The very title of another elegy—published in Phineas Stowe's popular *Ocean Melodies, and Seamen's Companion* (1849)—gives a clue to the content: "Affection's Tribute, to Lieutenant Henry Eld, Jr." begins, "A mother's yearning heart of tenderness / Watched for the coming of a home bound sail," and ends with "The Lord hath given—sorrowing mourner say! / 'Shall He not take away?'"

Still, the period between the wars may be fairly called the golden age of sea song, just as it was America's golden age of sail. Larger vessels and smaller crews encouraged the use of songs to regulate the work: these songs, or "chanteys," as they came to be called, remain today probably the best known of all sea songs, although they probably were neither as widespread as modern writers seem to think, nor did they ever die out as completely as many writers since the 1880s have claimed.

Chanteys, like the early psalms, are dominated by form. With very few exceptions, they existed solely to provide a rhythm for hauling on lines or heaving on the bars of a capstan or windlass. They take their form, as might be expected, from the ballad, but they emphasize the refrain or chorus to the point that verse content becomes almost meaningless. Because the songs were sung only as long as there was work to be done, the verses are largely episodic where there is any action at all, and seldom is there developed a continuous story line. Frequently the verses (as well as the choruses) are nonsense. Nevertheless, since a large number of chanteys seem to have been developed from other songs or ballads, there are many identifiable subjects.

Chanteys fall into three main groups, depending on their use: long drag, short drag, and capstan chanteys. In form, the first two tend to be similar, with alternating lines sung by the lead singer and the crew, but with a varying number of pulls in each chorus. Thus the verse,

> Boney was a warrior,
> Ch. Away hey-yah!
> A Warrior and a tarrier,
> Ch. John Fran-swor!

could be a short drag chantey if pulls were given only on "yah" and "swor" in the two choruses, but would become a long drag chantey if two pulls were given per chorus—on "way" and "yah" and on "John" and "swor." A short drag chantey was used for work that demanded relatively few pulls of enormous strength, while the long drag chantey was used for work that was not so immediately exhausting but which extended over a longer time—a matter of one or two minutes instead of a few seconds. Nothing in the form of the verse, however, pointed out the difference, and thus from ship to ship the hauling chanteys were found to be largely interchangeable, so long as the men at work knew how many pulls they were expected to give.

In a capstan chantey, the form might be quite different. Here the song provided a marching rhythm rather than a hauling one, it might or might not have a chorus, and it tended to be longer, although seldom is a single capstan chantey long enough to get a job done—especially when heaving up an anchor. Here there *was* an opportunity for plot, but the themes of capstan chanteys are far from being exclusively nautical. One might hear both the nonnautical "John Brown's Body" and the briny "Homeward Bound" used to raise an anchor. The "sea song" nature of a chantey thus remains largely in its use, not its subject.

When a chorus is present in a capstan chantey, it sometimes appears after an uninterrupted four-line verse sung by the lead singer, who is largely not participating in the work—hard work at the capstan is literally too exhausting to allow both continuous heaving and continuous singing:

> From Liverpool to Frisco a-rovin I went,
> For to stay in that country it was my intent,
> But gals and strong whiskey, Like other damn fools,
> I soon was transported back to Liverpool,
> Singin' row—Row, bullies, row!
> Those Liverpool gals they have got us in tow!

The longer rest provided by the continuous verse in many cases might therefore be said to be a particular aspect of the form of the capstan chantey, but frequently the pull in a short drag chantey is also led up to by a fairly long line of verse and line of chorus. And despite the exceptions, most capstan chanteys collected still follow the responsorial form of "Boney" (above).

The other major type of song that flourished during the golden age of sail has been designated the "forecastle song" or "forebitter." This category comes also from the use—or nonuse—of the song, for forecastle (pronounced "foke-s'l") songs are defined as songs that sailors sang at sea while they were not working, usually while in the forecastle (their living area in the bow of the vessel) or near the fore-bits (sturdy posts in the deck, which incidentally make good seats). Again, the category is a very imprecise one, but it is the one in use among folksong collectors and we are pretty much stuck with it, although it fails to help us in a literary evaluation of the lyrics. The forecastle repertoire, as represented in sea-song collections, consists almost entirely of traditional ballads and topical songs. Transmission of such songs was almost entirely oral, so there is no such thing as a definitive version of any of them, barring first printings of original broadside songs such as "Captain Kid," or traceable (or debatably traceable) songs such as "The Sailor's Grave" or "Sacramento." Most scholars admit that the repertoire included all the songs currently popular ashore as well, regardless of subject matter or origin; they modify their definition of "forecastle song" to "leisure songs sung at sea about the sea or sailors" when they come to choose what they actually will include or discuss in their books.

Ultimately, the most successful of the chanteys and forecastle songs as lyrical poetry—apart from traditional ballads of long and usually British pedigree—seem to have been those about ships, with the rather late songs such as

"Alabama" and "Dreadnaught" heading the list. It is curious that no chantey appears to be purely about the sea itself—such sentimentalized Byron was reserved for the musical settings of the likes of the British Felicia Hemans, whose popularity seems to have been greater in the parlors of America than in her own country.

The study of sea songs in relation to other works of literature has proved as interesting as the study of the songs per se. Many American nautical writers have included sea songs in their own works, including Cooper, Melville, and Dana. While most mention or include specimens, a few incorporate them thematically in their own fiction—notably Herman Melville. Joan Tyler Mead's model study of such thematic use finds that Melville uses sea songs to transform chapter 40 of *Moby-Dick*, "Midnight, Forecastle," into a kind of "soliloquy of the crowd" in which the crew members, like the speakers in the preceding soliloquies, are self-revelatory, and, unknowingly, self-damning. For the songs represented by the fragments offer some insight into the men's motivations for their enthusiastic support of their captain. Simultaneously, the songs initiate a series of images that indicate the enormity of Ahab's quest—a concept the men cannot understand. These perceptions follow from two preliminary considerations: first, Melville's general use of sea songs in his fiction; and second, the thematic relationship between the song fragments and the men's wild activity in "Midnight, Forecastle."

Even without reference to larger contexts, these sea songs offer interesting areas for study. As outlets for frustration aboard ship, several chanteys were quite vocal in their disapproval of persons or policies on board vessels. As the chanteys were sung from ship to ship, the chantey became considered a legitimate forum for such grousing, under the seaman's adage "Growl you may but go you must." Likewise, as a particular chantey gained popularity, it was likely that it was sung as much for humor as for any direct conditions aboard the current vessel. Most notable among chanteys of this type is the pumping chantey, "Leave Her, Johnny, Leave Her"—where the reference is to the ship, not a woman.

While frequently the *dramatis personae* in the songs seem to be real officers of actual vessels, other, "mythic" characters also occur: Stormalong, the old shellback who is mourned when he dies, Reuben Ranzo, the lubberly sailor who succeeds with the help of the captain's daughter, and, of course, the Dead Horse, hero of the ceremonial chantey used to celebrate the working-off of the first month's advance.

The development of marine technology during and after the Civil War lessened the demand for chanteys—*Harpers Monthly* in 1882 referred to the sailor himself as a vanishing species and his songs as a topic of archeological science. That this was an overstatement for the time is proved by the dozens of the "last of the chanteymen" who have appeared as informants or sometimes scholars themselves in the folksong anthologies. Nevertheless, *Harpers* more or less accurately referred to these exceptions as "rare specimens preserved in such marine museums as the 'Sailor's Snug Harbor,' and like places." One interesting eddy on the more literary side of sea song was Young E. Allison's continuation of Robert Louis Stevenson's "Fifteen Men on the Dead Man's Chest" in his "Derelict: A Reminiscence of 'Treasure Island'" (first published in a version called "A Piratical Ballad" in 1891). But no vitality seems to have leaked into the sea song until the sinking of the *Titanic*. And even then, after an incident that inspired a literary outburst, the lasting lyric is of music-hall or playroom variety, its false sentiment inspired by humor rather than pathos. ("It was sad, so sad.") The redemptive feature of such songs is that their music is finally unmistakably American, the product of ragtime rather than religion. Such seems true also of that most influential of songs, widely dispersed through sheet music, "The Ship that Never Returned"—with its analogues in the railroading song "The Wreck of the 97" and in the "folk revival" of the sixties ("Let's get Charlie off the MTA"). Tone and tune certainly dominated the sea song at the beginning of the century: the tune was American; the tone was no longer a sincere or even sentimental reaction to the sea, vessels, and seamen.

The American sea song, as it existed in its heyday, no longer permeated American popular culture. With the collapse of sentimentality toward the end of the nineteenth century and with the end of the age of sail, nautical imagery was no longer fused with religious sentiment, and the American sea songs might be said to have lain dormant. Two unlikely movements in music have resurrected the sea song in the last twenty years, however—one backward-looking in its attention to traditional forms and subject matter, the other forward-looking, if not in lyrical form at least in music, and certainly in subject matter.

After World War II a few imaginative investors attempted to bring to life the moribund schooner trade of the northeast coast. People, rather than lumber, stone, or lime, became the chief article of commerce, not as slaves but as paying passengers. As the schooner industry grew into a year-round trade, wintering in the Caribbean and summering in Maine, it actually became possible for young men and women to scratch out a livelihood on these old boats—and some new ones—which were sailed entirely the old way—"With our hands," as Irving Johnson of the brigantine *Yankee* would say. The revival in maritime interest coincided with the great folk revival of the sixties, and yielded up—not surprisingly—a tremendous interest in sea songs and chanteys. Although the folk revival is certainly dead, the interest in sea songs remains high among the schooner people and has produced several noteworthy sea-song writers. Chief among these is Gordon Bok, whose "Fundy" and "Herring Croon" indicated in the 1960s the direction that twenty more years of songwriting were to take. Influenced largely by the poetry of Ruth Moore as well as by traditional forms, Bok's lyrics frequently fuse Maine mysticism and mythology with realistic or even naturalistic descriptions of New England maritime industries. His best lyrics may be compared favorably with the best of Maine's local-color writing. Another notable American sea-song writer is Larry Kaplan, whose "Ballad of Old Zeb" is based on the life of schoonerman Zeb Tilton. The schooner writers in general have felt the influence of their Canadian counterparts Gordon Lightfoot, whose "Ballad of Yarmouth Castle" (1966) and "Wreck of the Edmund Fitzgerald" (1976) are internationally appreciated, and Stan Rogers, whose "Barrett's Privateers" emphasized singing, rather than working, in its elaborate choruses. Slightly outside this tradition is John Denver's "Calypso," based on the celebrated career of Jacques Cousteau's research vessel.

It might be said that one particular development in marine technology is responsible for the other main drift of the American sea song in the last twenty years. As the traditionalists still look reverently to Maine, so the other group looked westward to California—and beyond—to celebrate the culture of the surfboard. With music as unmistakably American as America has ever produced, the Beach Boys and their imitators helped to define as well as celebrate southern California beach life. They extolled the virtues of a splendid tan, a giant wave, and two girls for every boy. Although it is doubtful that any of Brian Wilson's lyrics will appear in anthologies of serious literature, his music will remain one of the most vital and imaginative examples of the sea song in America.

Although American literature began with a book of psalms, lyrics—whether hymns, chanteys, or sea songs—have played a fairly inconspicuous part in the development of our literature. Their popularity with a large, nonliterary segment of our population, however, may ultimately make them especially valuable in determining the response of the *average* American to the "wonders of the deep." The growing emphasis on noncanonical writing in America, with a burgeoning interest in popular cultural forms, once more is focusing scholarly attention on the subgenre of traditional song. Sea-song materials, whether by themselves or as aspects of more ambitious literature, promise to be the locus of newly perceived meanings. Finally, traditional sea songs again offer viable forms for the "ocean reveries" of today's "crowds of water-gazers."

ROBERT D. MADISON

Chapter Six

POETRY IN THE MAINSTREAM

The publication in 1853 of a small anthology of poetry called *Thalatta* was hardly a historical occasion. It is not numbered among the works we enumerate (*The Scarlet Letter, Moby-Dick, Walden, Leaves of Grass*) when we point to the first half of that decade as the initial rich flowering of our young national literary tradition. Nor could it compete for popularity with *Fern Leaves from Fanny's Portfolio*, which sold more than seventy thousand copies in the same year. But from the standpoint of American publication history, it is noteworthy as the first American anthology of poetry devoted entirely to the subject of the sea. Edited by Henry Wadsworth Longfellow's younger brother, Samuel, in collaboration with Thomas Wentworth Higginson (later to become Emily Dickinson's "safest friend"), the volume contains 127 pieces by close to one hundred poets of the period. The collection's subtitle, *A Book for the Sea-Side*, would suggest that the younger Longfellow was attempting to capitalize on the recent success of his more eminent brother's volume called *The Seaside and the Fireside* (1849). Its publisher—Ticknor, Reed, and Fields—made sure that its entire stable of prominent poets (Emerson, Longfellow, Lowell, Whittier, and Holmes) was represented.

Poetry anthologies of the 1840s and 1850s tended to be comprehensive in scope (*The Poets of America*) or, when more restricted (*The Female Poets of America*), rarely focused on a single subject. The gift annuals called "flower" books (*The Amaranth, The Dahlia, The Violet*) were miscellaneous collections of prose and verse embellished by engravings. Even the numerous collections called *A Sailor's Companion* or *The Mariner's Library*, designed for seamen's use, represented a variety of subjects or gathered together anonymous naval songs. Longfellow and Higginson appear to have been the first to discover one of the few subjects that could muster selections by virtually every nineteenth-century British and American poet who achieved prominence during the period. Every generation since has rediscovered that fact and has produced its own collections of nautical verse, drawing heavily from its own writers. For despite the passing of the romantic age of sail and the more prosaic age of steam, poets—even landlocked ones—continue to explore their feelings and ideas, their culture, and their relationship to the past through the inexhaustible resources of the sea. American sea poetry for the period in question is substantial in bulk and richly varied in subject and theme, to a point where it almost defies classification. Nevertheless, some generalizations are possible, not the least of which is that most of it was written in response to what we might term "the call of the sea."

That call is a call to the voyaging life. And while many of the poems we will examine deal with real voyages, the majority of them are metaphorical. The dominant literary legacy of this period is its symbolic mode of perception. Two of its prose masterpieces, *Moby-Dick* and *Walden,* explore the symbolism of water and the journey. But sea poetry in any period tends to be symbolic because the subject of the sea tempts one to representational expression. The topography of the poems becomes, in effect, their typology. Currents, tides, the depth and vastness of the sea, calms and storms, wind and wave, the sea's relation to the land and sky, its changeability, its coloration, all have figurative associations. Out of the configuration of sea, ship, journey, and voyager comes a multiplicity of associations and experiences, all bearing on the journey-of-life archetype.

The call of the sea, therefore, is the summons to explore life's conditions and challenges, its tragedies and triumphs. The sea itself may be viewed as a chasm of catastrophe or as an avenue to salvation. The voyage may be fraught with peril or filled with expectation. All the elements of the life-voyage are in the paradigm of the sea, for the sea is always the same and yet always changing. From one vantage point, it seems the epitome of stagnation, from another, the image of vitality. The voyage may seem to be one of constant strife or of unbounded freedom, the voyager, uncertain of his course or resolutely visionary. The ceaseless rhythms of life and death, with all of their mysterious mutations, are reflected in the motion of the ebb and flow of the tides themselves.

In Longfellow's "The Secret of the Sea," the poem's speaker recalls the experience of a landsman, the Spanish Count Arnaldos, who asks an ancient helmsman to teach him the "wondrous song" he sings. The helmsman, whose song is about "the secret of the sea," replies: "Only those who brave its dangers / Comprehend its mystery!" As the poem concludes, the speaker gazes out to sea, with the same unfulfilled longings as Count Arnaldos. The scene is reminiscent of the "thousands upon thousands of mortal men" in the first chapter of *Moby-Dick* "fixed in ocean reveries," "water-gazers" drawn to the sea by "the image of the ungraspable phantom of life," what Melville then calls "the key to it all."

The sea as an alluring repository of secrets or mystery, as a phantomlike environment that possesses the key to life, can be understood in a variety of ways, and was by the poets who answered its call. At one end of the emotional spectrum are the escapist fantasies in which the music of mermaids is an aerial narcotic that "soothed each tumult of the mind," in the words of James Gates Percival's "The Mermaid." William Gilmore Simms creates a similar dream in "The Evening Breeze," a blissfully languid underworld where the turbulence of wind and wave is unknown and where "the fairest of mermaids will lull thee to sleep." In a substantially longer version of the fantasy, Simms traces the journey to this ocean paradise.

The opening scene of *Atalantis* (1832), a blank verse dramatic poem in three acts, consists of an argument between the work's principal antagonists on the nature and source of their power. The sea demon, Onesimarchus, holds the princess of sea nymphs, Atalantis, captive because of his superior physical strength. Atalantis is able to resist his sexual advances because of her spiritual power. The standoff ends when Atalantis comes upon a prostrate, handsome Spanish knight, Leon, the only survivor of a shipwreck caused by Onesimarchus. Instantly struck by an erotic love she has never felt before, she kisses him back to consciousness and eventually escapes with him to her kingdom beneath the sea where, she says, "in coral groves, / Thy head well pillow'd on my happy breast, / I'll sit and watch thy slumbers, blest to soothe / Thy ever beating pulse" (III.iv). But before that happens, we learn that Leon had been on a voyage with his sister, Isabel, to an island "whose bowers of beauty and eternal spring / Recall the first sweet garden of our race, / Before it knew the serpent" (II.i). There are strong suggestions that this Eden, without the strictures of civilization, would have been an amatory bower of bliss for the siblings. Despite the death of Isabel, Leon gains his Eden in the relationship with Atalantis, and although we are led to believe that the union is an ideal mating of the physical and the spiritual, we also know that Atalantis wins Leon's heart by telling him she will strive to act and look like Isabel. Leon has the best of all worlds. Not only is he "free from the jar, / The heat, the noise, the dust of human care" (III.iv), he has also found a way of avoiding the sexual taboos of the world he has left behind:

In William Cullen Bryant's "Sella" (1862), the title character, a beautiful young woman with a compelling curiosity about all bodies of water, gains the power to "walk at will the ocean floor" when she puts on magical slippers given to her by a sea nymph. Initially torn between her mountain home and the "glorious realm" of the sea, she gradually becomes a kind of sea goddess on earth. When her ability to return to the sea is terminated by her brothers who cast her slippers into a stream, Sella channels her power toward practical activity—teaching humans to dig wells, and to build irrigation systems, dams, and reservoirs. Unlike Leon, who finds a permanent escape from our world, Sella is obliged to return to it but, with the sea's secrets, to return to and help to create a more tolerable dwelling place.

In the hands of a Poe, of course, the enchanting sea kingdom becomes a nightmare. The colorful groves and bowers of both Sella's and Atalantis's sea worlds are replaced by the "hideously serene" impression of shimmering towers and turrets dominated by the figure of death in "The City in the Sea." Not a world to escape to, it is, nevertheless, the bleak fate of all, Poe seems to suggest, to inhabit this grotesque kingdom. And although the city is said to be "in the sea," the poem's imagery is essentially of a landscape entombed in a viscous, gelatinous atmosphere with an eerily stagnant kind of motion. It is the city of the living dead from which one never escapes.

If Poe's ghastly fantasy is antithetical to the ocean idyll, it jars us to the realization that fantasies of the kind written by Bryant and Simms are distinctly in the minority in nineteenth-century American sea poetry, that the "ungraspable phantom" some of Melville's "water-gazers" are peering at is the phantom of death. In an age of pleasure boats and love boats where major catastrophes of travel usually occur in the air, it may be difficult for us to understand the perspective of those for whom the sea was, proportionately speaking, considerably more important and more hazardous for travel and mercantile activity than it is today. Without the benefit of radar, satellite weather forecasts, ship-to-shore radio networks, powerful diesel engines, or computer-assisted navigation—to name a few of the conveniences of contemporary seagoing—vessels were vulnerable to storms at sea and even more vulnerable to the treacherous conditions of coastal navigation. In the New England fishing fleet alone, three major disasters wiped out many of the vessels of Cape Cod and of Marblehead in the ten-year period from 1837 to 1846, taking a large toll in lives. Americans would have understood William Bradford's desire to bless the God who delivered the Plymouth pilgrims two hundred years earlier from the "perils and miseries" of "the vast and furious ocean," for their experience was not far removed from Bradford's. Indeed, for the millions who came to America by tolerating the foul conditions of overpacked immigrant ships, the experience may have been worse. On slave ships it was infinitely worse.

To be sure, there were poetic treatments that made light of such perils. Epes Sargent's "Life on the Ocean Wave," published as a poem in 1847 but popular as a song long after, presents the portrait of a seaman who is bored by the "dull, unchanging shore" and who finds stimulation in the uncertainty, even the danger, of "a home on the rolling sea." The sailor of this poem is not unlike the title character of Poe's *Narrative of Arthur Gordon Pym*, who continues to feel "a passion for the sea" despite a near-catastrophic boating mishap off the coast of Nantucket—whose longing for the suffering and disasters of the sea may even have been intensified by the wreck. The difference is that Sargent's poem sounds like a jingle for the recruitment of seamen, while Poe's opening chapters present the more complicated psychological phenomenon of an individual who lives on the edge, who comes alive by facing death.

The casual disregard, in Sargent's poem, of the sea's potential destructiveness, when found in poems where the sea appears to respond to that human challenge, becomes a harbinger of death itself. The best-known example is Longfellow's "Wreck of the Hesperus" (1841), a ballad in the tradition of "Sir Patrick Spens" (and similarly based on history) in which the captain, in a gesture of defiant omnipotence, ignores the advice of an experienced crewman and his own instincts, and plunges headlong into a hurricane. The disaster that results is not unexpected, and seems justified, except that others innocent of the captain's arrogance are victims of it. In the end, we are left with the gruesome image of the captain's daughter, who went on the voyage "to bear him company," floating ashore as a corpse amid the storm's debris—lashed to a broken mast and sheeted in ice. A variation of that incident appears as one of the musician's tales in Longfellow's *Tales of a Wayside Inn* (1863). In "The Ballad of Carmilhan," a ship's captain listens to another skipper's yarn about the ghost ship *Carmilhan* and responds scornfully to the warning to avoid the area of the mid-sea where it roams, especially during a storm. On his next voyage, he comes upon the *Carmilhan* in a gale. When he tries to ram it, he plunges through the specter and into rocks. The same fate awaits a group of mariners who ignore the warnings of fishermen in Mary E. Hewitt's fanciful "Scena" and swim toward rocks where they are being called by sirens—who eventually devour them.

In all three cases, the tragedies could have been avoided had the principals been more respectful of what Auden called the sea's "primitive potential power." Time and again, however, under different circumstances and in the works of numerous writers, the idea seems to persist that the sea, more than any other domain in nature, is that environment where the human ego suffers most in the face of an implacable power against which human skill and will are insignificant. On a considerably larger scale, this is the challenge Melville assigns to Ahab. And with a different conclusion, it is the realization that the religious skeptic Roswell Gardiner eventually comes to in Cooper's treatment of the theme in *The Sea-Lions.* The scope of the sea's power, often associated on an elemental level with divinity, is such that the challenge is doomed to failure. Thus, to cite only one more example of this theme in the sea poetry of the period, Elizabeth Oakes Smith's "The Drowned Mariner" presents the spectacle of a sailor rocking wildly "in the shrouds" high up the mainmast, oblivious to the danger and enjoying the ride on this "uncurbed steed." As the ship's lurch dips the mast toward the water the sailor suddenly finds himself staring into the face of a drowned shipmate. Moments later, he joins him. Death comes too suddenly in all these poems for remorse or humility. There is just enough time to acknowledge the might of the "Great Sea-God."

The call of the sea as a voice of death takes many other manifestations. The "voice" is often that of a dead or dying sailor calling his lover to him. Examples range from the brief but powerful "Ballad of Nantucket" (1859) by Thomas Bailey Aldrich to the tediously long (more than one thousand lines) "The Wreck: A Tale" (1843) by James Gates Percival. In Aldrich's poem, Maggie goes down to the docks whenever a ship arrives, hoping her Willie will be on it. Despite the gentle reproach of friends that "Your Willie's in the ocean, / A hundred fathoms down!" Maggie insists that "It is so sad to have him / A-sailing on the sea!" One day, she walks into the sea, "gone to Willie / Across the Spanish Main." In Percival's version, a young woman looks out from a vantage point several times each day, searching for a white pennon with a red heart on an incoming vessel—the sign of the return of her lover. After several years, she watches helplessly as a ship bearing the pennon sinks in a powerful, fast-moving storm before it can make it to port. When she finds his lifeless body floating in the shallows, she embraces it and whispers, "I come, my love." They are finally reunited in death. In Simms's "Cape Hatteras," the roar of a storm subsides momentarily as a drowning sailor calls to his wife. The wife "sprang to join him, and the sullen seas / Closed over them forever." In one of the best versions of this kind of sea poem, Whittier's "Sisters," a young woman hears at night during a hurricane the voice of her older sister's lover tell her he is "in peril from swamping." Because the older sister does not hear the voice, the younger one suddenly realizes that as she had loved him secretly, so he had also loved her. The voice from the sea reveals that "Life was a lie, but true is death."

Whitman's "Out of the Cradle Endlessly Rocking" (1859) partially reflects the conventions of these poems in which lovers are united in death. The mournful chant of the "he-bird" whose mate is dead is a powerful message of love and an inspiration to the future songs of the boy who hears it. But it ends with an image of separation, "We two together no more." If the boy as a poet is to be a "uniter of here and hereafter," he must see beyond the song of death. It is then that the voice of the sea is heard, rhythmically chanting the "delicious word death." In the eternal rhythms of the "old crone" endlessly rocking the newly born, the boy finds the answer to separation. In the paradox of the gestation of life through the substance of death, the birds are united and the boy can now become a "chanter of pains and joys." The sea gives the future poet not only the unifying vision but also the total freedom for "peering, absorbing, translating" that comes in having transcended the fear of death. The gentle rhythms of the sea, as in Longfellow's "The Tide Rises, the Tide Falls," while they insist on recognition of the reality of individual death (the she-bird in Whitman, the traveler in Longfellow), also develop awareness of a larger pattern of continuing life. In Whitman, death is, in fact, the beginning of the ultimate transcendental trip. Certainly, that is suggested in the pattern of "Passage to India" (1871), as we move in the early sections of the poem through the achievements of civilization in space and time but always toward "passage to more than India," to a "primal thought" that comes outside the dimensions of this world. In the cluster of brief embarkation poems Whitman wrote between 1871 and 1891 ("Joy, Shipmate, Joy," "Now Finale to the Shore," "Old Age's Ship and Crafty Death's," "Sail Out for God, Eidolon Yacht," "Old Salt Kossabone"), the message is always the same—the voyage of life being over, the truer, longer, bolder, freer voyage is about to begin. In the voyage of life, one must always proceed cautiously, study the charts, and return to port. But in the voyage after life, all sails are raised for an "endless cruise" on "really deep waters."

Another "voice" from the sea in which the cruise seems endless but in a less visionary way is the journey of the ghost ship—already alluded to in the legend of the *Carmilhan*. In most cases, the legend can be traced to a historical event. Longfellow's "Phantom Ship" is based on Cotton Mather's account in *Magnalia Christi Americana* of the disappearance at sea of a New Haven vessel, *Fellowship*, on a 1646 voyage to England. When an image of the ship sinking at sea appears in a cloud over New Haven a year and a half later, the inhabitants thank God who has taken this means, they say, "to quiet their troubled spirits." In this particular legend, the ship makes one appearance and then disappears forever. Most, however, in the tradition of the Flying Dutchman, sail on endlessly and return to certain locations regularly. In Whittier's "Dead Ship of Harpswell," for example, the grieving families of the dead sailors are spared the ghastly sight of the corpses by a merciful angel who repeatedly sails the ship of death up Maine's Harpswell Sound only to push back to the open sea just before coming ashore.

Conflicting versions of the same event in poems by different authors indicate the extent to which ghost-ship accounts were derived from unreliable oral tradition. Whittier's "Palatine" is based on a letter from an acquaintance on Block Island who grew up listening to accounts of the event. In Whittier's version, a ship of German emigrants is wrecked in a gale on the coast of Block Island with apparently no survivors. Voracious islanders swoop down on the vessel and pick it clean before setting it on fire. Simms sets his version of the event, called "The Ship of the Palatines," on the Carolina coast and describes the massacre of all the emigrants by the ship's master and crew, who then torch and abandon the vessel after appropriating the emigrants' treasures. In actual fact, the ship came ashore on Block Island in late 1752. The crew had plundered the emigrants' possessions and abandoned the ship but had not massacred its passengers. The islanders rescued all of the passengers but one. The common element in both accounts is the return of the flaming ship each year, presumably as a kind of spectral monument to its victims. The ghost ship is most often depicted as a recurrent apparition that serves as a memorial to victims of crime. The victims are usually passengers and range from the African slaves of Celia Thaxter's "Cruise of the Mystery" to the wealthy young Spanish widow of Richard Henry Dana, Sr.'s "Buccaneer."

Dana's poem, popular in the 1820s and 1830s, reminds us that ghostly apparitions need not be ships. The buccaneer, Matthew Lee, massacres his passengers for their possessions. The Spanish widow leaps from the ship before she can be killed, and shortly thereafter her white horse is thrown overboard by the crew, who burn the ship before going ashore on an island (Simms acknowledged the influence of this poem on his Palatine account). A year later, the burning ship makes its appearance off the coast of the island Lee now reigns over. But at the same moment the "Spirit-Steed" rises from the water and comes ashore. Lee is compelled by some mysterious force to mount the horse who brings him to the edge of a cliff where they watch the ship burn throughout the night before both the ship and the horse fade into air in the morning. This event is repeated the following year. On the third anniversary, the horse returns and carries Lee to his death in the sea. On this occasion, the animal is referred to as "that pale Horse," an obvious allusion to the figure of death in the Book of Revelation. (The best-known use of Revelation's pale horse in modern sea literature, of course, appears in John Millington Synge's *Riders to the Sea* [1904].) Frances Osgood's "Lutin-Steed" is another account of a phantom horse that carries, in this case, three brothers to their deaths in the sea.

There is also a large body of poetry in the period that focuses on past and recent sea disasters and tragedies but without the spectral element. Longfellow's account of the death at sea of the sixteenth-century British navigator in "Sir Humphrey Gilbert" may have been inspired by the disappearance of the 1845 expedition of Sir John Franklin in search of a Northwest Passage, one of Gilbert's dreams. The ten years that passed before anything was heard of the Franklin party produced a massive amount of speculation in the press and a small number of poems, including Thomas Buchanan Read's "Passing the Iceberg" and "Ballad of Sir John Franklin" by Nathaniel P. Willis. Bloody naval engagements are commemorated in Longfellow's "Cumberland," on the sinking of that ship by a rebel ironside during the Civil War, and in section 35 of Whitman's "Song of Myself," on John Paul Jones's 1779 encounter with the British *Serapis*. Whittier recalls (inaccurately) the aftermath of the sinking of a Marblehead fishing vessel in 1806 in "Skipper Ireson's Ride" and describes in "The Swan Song of Parson Avery" the disaster that struck the Avery and Thatcher families on a 1635 trip from Newbury to Marblehead during a hurricane. The list of similar poems by lesser poets, most of which were first published in the newspapers and journals, seems endless. Joanna Colcord collected many of them in an unpublished anthology now housed at the Peabody Museum of Salem. Understandably, then, part of the perception of the sea in the poetry of the period reflects an anxiety that goes beyond gothic or romantic preoccupation with death. As Longfellow emphasizes in "The Building of the Ship," "The tales of that awful, pitiless sea, / With all its terror and mystery, / The dim, dark sea, so like unto Death" came from sobering experience with wind and rock. Sargent's "Life on the Ocean Wave" is an amusing lark but Longfellow's "Twilight" is closer to the truth. The poem portrays the extreme anxiety of a woman and a child in a fisherman's cottage as a storm rages outside, and concludes with the dreadfully ironic question: "And why do the roaring ocean, / And the night wind, wild and bleak, / As they beat at the heart of the mother / Drive the color from her cheek?"

The turbulent sea is an awesome manifestation of potentially destructive power. Bryant's "Hymn of the Sea" describes the devastating effects of that power on ships that are "whirled like chaff upon the waves." But the poem also acknowledges another kind of power—the power to create—in images of the birth, fertilization, and colonization of islands from coral and volcanic activity. For Bryant, both kinds of activity are expressions of divine will, but the creative force is the stronger and prevails ultimately. While the sea's association with life in this poem is somewhat strained by its didactic intention, the life force of the sea is something we all recognize on at least an elemental level and acknowledge in our own way. Immersion in the water may mean death by drowning. But for Whittier's Parson Avery, it is also a "baptism" that opens for him "the sea-gate of . . . heaven." In Emma Willard's

"Rocked in the Cradle of the Deep," "the germ of immortality" grows from "wreck and death." Frances Osgood's "Spirit's Voyage" describes the Polynesian ritual, after burial of the dead on land, of sending the deceased's spirit to sea in a canoe laden with fruit for the voyage. All of these poems recognize two seas—Whitman's "Sea of the brine of life and of unshovell'd yet always-ready graves."

The ebb and flow of the tides is the most persistent manifestation in sea poetry of the sea's vitality and fertility, a constant reminder of the immutable rhythms of birth and death and of other similar patterns of human experience. Longfellow's "Tides" suggests that deep sorrow is a kind of emotional death that seems permanent to the sufferer but, like the "insurgent waters" of the tide, "thought and feeling and desire" inevitably return. In Lowell's "Seaweed," the death is a spiritual one brought on by "the burden of the day," but the revival of faith is said to be as certain as the return of water to the drying seaweed on the shore, a somewhat different conclusion from the one drawn by Arnold on a similar subject in "Dover Beach." Both poets insist that the return of the inner condition is as predictable as the recurrence of a natural phenomenon governed by physical laws. In "The Tide Rises, the Tide Falls," Longfellow sees that ineluctable pattern not only in the inevitable ebb and flow of the tide but also in the equally telling imagery of the day's cycle from light to darkness and back to dawn—the time of reawakening. The final stage for the poem's traveler is darkness and the ebb of life. But the final stanza emphasizes the renewed flow of life in a continuing journey.

Whitman comes to a similar conclusion in the cluster of eight sea poems in the annex to *Leaves of Grass* called "Fancies at Navesink," his most sustained portrait of the mysteries of the tides. In "And Yet Not You Alone," the least desirable experiences—failure, despair, death—are associated with "twilight and burying ebb," but without the subdued sense of melancholy of Longfellow's version. Rather, as in "Out of the Cradle Endlessly Rocking," those experiences are embraced as part of a "boundless aggregate" of time and motion, matter and spirit, prevailed over by the "fluid, vast identity" Whitman mentions in "You Tides with Ceaseless Swell." The "clue," another echo of "Out of the Cradle," is in that "unseen force" from which "the tide and light again" return, weaving "from Sleep, Night, Death itself, / The rhythmus of Birth Eternal." In the final two poems of the group, Whitman places his own life in the context of this cosmic dance of birth and death. The panoramic perspective of "By That Long Scan of Waves" provides him with the opportunity to place his achievement in the context of "God's scheme's ensemble," to sum up his life as "some wave, or part of wave" in the endless pattern. "Then Last of All" associates consciousness, "the brain that shapes, the voice that chants this song," with the same law that governs tidal motion.

This near-identification of self with surf, of being with a life-rhythm, explains the pervasiveness of surging motion—sometimes softly undulant, sometimes more vibrantly pulsating—in the very form of much of Whitman's poetry, including many poems without sea images. For if Whitman's vision of cosmic wholeness is best expressed in the symbol by which each blade of grass, unique and complete in itself, is part of the whole field (each poem, part of the whole work), his insistence on the pursuit of freedom, power, and vitality within that cosmic scheme is embedded in the incantatory effluence of the poems' rhythm, propelling us toward new experiences and discoveries. "The mystic human meaning" flows through those rhythms.

Although not as insistent on tidal imagery, the "Sea-Drift" section of *Leaves of Grass* is an even more suggestive affirmation of the paradox of a life force in the rhythms of death. "Drift" has the sense here of both matter and motion—the stuff that comes ashore and the currents that bring it in. The "stuff" may be a sound—the song of a shorebird or the "delicious word" that comes out of the sea. Or it may be bits and pieces of worthless matter, the "chaff" that floats ashore, "left by the tide." In "Out of the Cradle," the sounds of the sea awaken the creative impulse. But in the next poem, "As I Ebb'd with the Ocean of Life," that impulse, called "the electric self," ruminates indecisively over the surf's debris, seeking "likenesses" or "types" in the flotsam of the sea's surges. The triumphant poet of "Out of the Cradle" is replaced by a vacillating speaker who perceives his "arrogant poems" as "blab," like the "wash'd-up drift" of weeds and wood that piles up in meaningless patterns on the windrows of the seashore. The customary bravado of the solitary singer is tempered in the next poem, "Tears," by a "shapeless lump" who acknowledges in the darkness of night the depth of his despair, a feeling that reflects the concluding view in "As I Ebb'd with the Ocean of Life" of the "random," "capricious" drift of "a limp blossom or two" out of the chaotic welter of experience. But the pile of drift contains the seeds of resurgence, and in the next poem, "To the Man-of-War-Bird," the speaker soars into the "air cerulean," sailing in spirit above even the high-flying frigate bird. From that vantage point, the cosmic vision returns, the bits and pieces of sea-drift no longer disparate and dead. The next six poems flow. Shipwreck threatens; clouds and storm make their appearance. But the prevailing spirit is one of release, ascent, and union. In the final poem of "Sea-Drift," "After the Sea-Ship," the mystery of ebb and flow is evident in the withdrawal and then return of waves in the wake of the ever-voyaging ship.

In most respects, the ocean waves serve the same symbolic function as Whitman's master symbol, the field of grass. They are dynamic, multitudinous, and one—a field of crests waving gently as spikes of grass, each simple, single and unique but a part of the whole. And the wave, like the grass, is associated with creative power. In "Had I the Choice," the poet affirms that understanding the secret of "the undulation of one wave" would give his poetry greater power than emulating the techniques of Homer, Shakespeare, and Tennyson would—an attitude that reflects Emerson's view that the main attribute of the imagination is "to flow." The sea is as intimately linked to the poetic faculty in other poems, as in "Out of the Cradle Endlessly Rocking," where "the outsetting bard," standing on the verge of land and sea, absorbs their secrets and translates them into the substance of his songs. The poet as "pilot" guides "ships" that stray from their course in "What Ship Puzzled at Sea." "In Cabin'd Ships at Sea," one of the "Inscriptions" that introduce *Leaves of Grass*, invokes the same destiny for the poet's volume of poems as that of "ocean's poem," to open up "the boundless vistas and the horizon far and dim" to all who read it.

The emblematic association of the sea—its sounds and rhythms, its mystery, magnitude, and beauty—with the process of poetic creation is not limited to Whitman's works. In "Terminus," Emerson portrays his declining years in the familiar terms of coming into port at the end of an ocean voyage. Although he decries the "legacy of ebbing veins" inherited from his sires, the poem is less concerned with the physical debilities of old age than with the loss of creative energy. The preacher of unlimited possibilities in his earlier years now accepts "the God of bounds" as "Fancy departs." "Amid the Muses, left . . . deaf and dumb," he portrays himself somewhat ironically as capable of finishing up a few minor projects but starting no new ones. In Bryant's "Day-Dream," the sea nymphs' monologue is a complaint on the loss of their place in an unimaginative, utilitarian world. But the speaker's evocation of the sea nymphs is a powerful imaginative act inspired by "the eternal flow" of the sea, and the nymphs' fear of oblivion is temporarily allayed by the poem that revives them.

Longfellow also uses the sea to explore the nature of poetry and of the creative process. In "Elegiac Verse," he attributes the origin of metrical patterns in poetry to an attempt by some ancient poet to replicate "the motion and sound of the sea" in his verses. The "mighty undulations" of Milton's poetry are compared to the cadences of the sea in "Milton." In "Possibilities," the great poet is portrayed as an admiral commanding "stately argosies of song" on a voyage to uncharted lands. The voyage may not always be a smooth one. In "Becalmed," Longfellow portrays the uninspired poet as a motionless ship

at sea "waiting the auspicious gales" that will "fill the canvas of the mind."
"The Sound of the Sea" affirms that the source of those inspirations is mysterious and divine, that they arise as spontaneously as "the first wave of the
rising tide." But Longfellow suggests in "Seaweed" that the act of creating,
even when aided by inspiration, still involves the inner turbulence of struggle
to give material form to the vision. The older poet will view his achievement
in terms of that struggle. In "The Broken Oar," a poet seeking "some final
word" with which to close the volume of his creative efforts selects the inscription on a broken oar that comes ashore as he ponders that problem. The
inscription reads, "Oft was I weary, when I toiled at thee."

The quintessential image of the sea's creative power is found in Holmes's
"Chambered Nautilus." Sometimes disparaged for its explicit didacticism,
the poem remains both a remarkable description of a natural phenomenon
and an apt analogy to human aspiration. The inhabitant of the shell has the
persistence of Thoreau's artist of Kouroo, the ascendant spirit of Whitman's
singer, the "finely organized" nature of Emerson's poet. In the evolution
of the shell, Holmes captures the essence of human creativity. The shell is
merely the thing made and, like the poem or the picture, the symbol of the
spirit in the maker. The butterfly created by Owen Warland in Hawthorne's
"Artist of the Beautiful" performs the same function in that story. The purpose of making the shell is to rise above it, to leave it behind, to be freed
of it. Released from the bondage of matter that restricts like the shell, the
builder in Holmes's poem is not unlike Whitman's "man elate above death,"
the journeyer no longer pushing off on "The Ship Starting," but as in "Joy,
Shipmate, Joy," welcoming the release from anchorage and the unlimited
voyage that comes in death.

As a "ship of pearl, which . . . sails the unshadowed main," the chambered nautilus is, from the outset, journeying toward its ultimate destiny.
Longfellow, Whitman, and Thoreau provide equally suggestive images of the
human journey in "Ultima Thule," "Facing West from California's Shores,"
and "Life." In the first poem, the "unending endless quest" for "the land of
dreams" continues, despite the detours caused by life's currents and storms.
"Facing West" reflects the view elaborated in detail in "Passage to India" of
the gradual westward movement of civilization from its origins in the East.
The journeyer, described as "a child, very old," looks "toward the house of
maternity" realizing that "the circle [is] almost circled." But the journey is not
nearly completed, for it is, as previously noted, to "more than India." Facing
"west" but looking "east" toward his origins, the voyager—an Everyman of
the past, the present, and the future—still has the greatest journey to take—
the journey through death to the source of life. In "Life," Thoreau explores a

number of experiences that seem metaphorically appropriate to the course of life. The striking image in the first part of the poem, a horseman who wishes to be "welded" to an "unresting steed" in an endless journey, seems unrelated at first to the more conventional image of the ship at sea in the second half. But the centaur invoked here is perhaps the figure of Chiron, the teacher of the Greek voyager Jason. The poem's voyage-quest for a spiritual golden fleece, symbolized by the images of the cloud and the eagle, takes place in a vessel "with an unsanded bow" plowing "the shoreless seas of time." As in "Ultima Thule," the prize is not as valuable as the journey.

The metaphor of the sea journey as the journey of life is the broadest application of sea poetry to human experience. In poems such as those above, the mythic and mystical dimensions of the journey invoke the motifs of heroic exploration and quest. Even in poems like these in which the destination is not reached or may be referred to in only the vaguest of terms—a place of origins, an ultimate place, a timeless place—the goal of striving for that destination seems appropriate to the sea in its magnitude and expansiveness. The voyaging mind, as Jones Very suggests in "The Sight of the Ocean," seems to find "a space for its flight" in the vast prospect of the ocean. But while the metaphor is the same, each voyaging mind is unique. Longfellow's is the voice of experience looking back on the early conditions of the voyage. Whitman's retrospect is tied to the whole pattern of human history. Thoreau, relentlessly bent on the journey ahead, never looks back.

Other voyages produce their own distinctive modulations, some as comprehensive in scope, some considerably more restricted. Among the more prominent patterns to be noted is that the sea journey is seldom dissociated entirely from the land. More often than not, the land is the antithesis of the sea, especially in poems in which the sea is associated with conflict and adversity, danger and uncertainty. The land takes on the values of home and hearth, becomes a refuge of peace, love, and stability. In "Put Off Thy Bark from Shore," Frederick Goddard Tuckerman portrays the voyager as a "landsman on the deep," mentally exhausted by the search he seems compelled to continue but always looking for "some shore of rest." James Gates Percival focuses on the early stages of life in "Voyage of Life." Childhood is characterized as a period of alternating calms and breezes, but the passions of youth bring on stormy seas that only subside after the voyager resolves to sail for "the peaceful breast" of the land—emblem of the voyager's love-mate. In Holmes's "La Maison D'Or," life's journey is one of "toil and struggle" on "the restless sea," but the "restful mountains" on the seashore offer a glimpse of "eternal peace." In Sargent's "Rockall," the speaker longs for the same steadfastness in faith as the small isle that faces with implacable strength the fury of the sea. James Russell Lowell's wonderfully evocative "The Sirens" is the song of sea nymphs who entice the mariner to leave "life's gloomy sea" for their "green isle" where he will "rest forevermore."

In this small sampling of poems, the land is a haven from the disorders of excessive thought, sexual passion, tragic strife, religious doubt, and loneliness. But in only one of these poems does the voyager come ashore, for most recognize that the song of Lowell's sea nymphs is a call to an illusory sanctuary. Most of the "voyage of life" poems of the period reflect Auden's view of the romantic perspective on the sea: shore life, while attractive, is trivial; the sea—wild, lonely, but still vital—is the real situation; the voyage, the true condition of man. The call of the sea is the call to the challenges and the hazards of an unpredictable and ever-changing voyage. Those who fail to heed the call fail at life itself.

Another common metaphorical voyage in American sea poetry of the nineteenth century is the voyage taken by the "ship of state." Although the vessel has different names, including the *Union, Columbia,* the *'76,* it has certain recurring attributes. It is always well made, its keel having been constructed with special care. It is an outgoing vessel, on a voyage recently begun, to a destination still unknown but with a special cargo intended for a special purpose. It has experienced stormy times, has survived them, and will survive future perils at sea, in part because of the integrity of its construction, in part because it is protected by providential favor. It sails in waters that seem boundless and free. Its captain is wise and brave, its crew, heterogeneous in origin but united in purpose. The perils it has overcome in Simms's "Song" are the perils of Old World tyranny. Numerous later versions describe the rough passage through the storm of the Civil War. One of them, Lowell's "On Board the '76," is a tribute to "the Singer" of the ship (William Cullen Bryant) who "remanned" the crew in the early "dark hours" of the war. Another one also commemorates the achievement of an individual—during the Civil War—and is the best-known poem on the ship of state: Whitman's "O Captain! My Captain!" portrays the agony of a bittersweet victory. With the end of the Civil War, the ship of state has come into port; but because of Lincoln's assassination, it anchors with its captain dead on deck. Other significant features of American history form a part of the journey. Longfellow's "Building of the Ship" reflects the great influx of immigrants to America in the 1840s in the image of the ship's hull containing materials from "every climate, every soil." Parenthetically, it might be noted that the journey west was often accomplished in a prairie "schooner."

Like other "ship of state" poems, Whittier's "Mantle of St. John de Matha" deals with the Civil War; but it associates the events of that time with a much

earlier period of history. Drawing from the legend of John of Matha (1160–
1213), who was canonized for ransoming Christians being held captive in the
Moslem world, Whittier portrays the fate of America in the religious terms
of spiritual trial, divine intervention, and deliverance. John of Matha's red,
white, and blue mantle, said to have been bestowed on him by an angel, is the
connecting link. Seven hundred years of European history are bridged as the
God of the medieval world takes the helm of the rudderless American ship.

To the extent that Whittier's poem anchors American origins in European history and traditions, it goes against the then-prevailing current of
deemphasizing British and European influence on American culture. Chas-
tizing his contemporaries for groping "among the dead bones of the past" in
Nature, Emerson encouraged the American scholar of 1837 to see that "our
day of dependence, our long apprenticeship to the learning of other lands,
draws to a close." John O'Sullivan stated the case more bluntly two years later
in the influential "Great Nation of Futurity." On America's relation to the
nations of Europe, he insisted that "we have, in reality, but little connection
with the past history of any of them, and still less with all antiquity, its glories,
or its crimes." Despite the prevalence of that view, advocated enthusiasti-
cally throughout the 1840s by nationalistic literati such as the Young America
disciples, a significant body of nautical poetry using historical materials concerns itself with the issue of American origins in Old World sources.

Sometimes the occasion for the poem is an unearthed artifact of the past.
Whittier's "Norsemen" is inspired by a fragment of a statue found near the
Merrimac River in Massachusetts. The "mystic relic" casts a spell on the
observer, and in that state he has a vision of the primeval world that the Scandinavian mariner came to centuries earlier. Predictably, Whittier abandons
the historical past in speculation on "an immortal origin" for America. But
before he does, he presents a vivid portrait of the Viking as a bold voyager.
Using another set of artifacts, Longfellow takes the portrait one step further
in "The Skeleton in Armor," where he sees evidence of Scandinavian origins
for American culture in the Newport tower popularly known today as the Old
Stone Mill and in the armor-encased skeleton of a warrior discovered on a
hill outside Fall River in 1836. Whittier's dream-vision evokes a generalized
portrait of the Viking. But as the warrior's "skald," Longfellow renders more
vividly the traits of the warrior, who speaks his own story. What emerges is a
portrait not essentially of Nordic sensibility but of American character. Bold
and brave in youth even to the point of recklessness, the warrior abandons
the life of an adventurer for the love of a young woman. But because she
is "a Prince's child," he is rejected by her family. Always rebellious toward
authority, the warrior rejects the family presumption of social superiority

and elopes with the young woman. They escape the oppression of tyrannical royalty by sailing to America where they lay the foundation for a more egalitarian society with a family of their own. The Viking's tower is a memorial to his search for freedom and a symbol of his success in bringing it to America.

Lowell's "Voyage to Vinland" (1850) is the most detailed treatment of the origin of an American character in the qualities of Scandinavian explorers of the continent. Part 1, "Biorn's Beckoners," portrays the eventual discoverer of America as born for a special destiny. As a youth, he is restless, unwilling to settle for the ordinary course of life and work for "base gain of raiment, food, and roof." A dreamer, he seeks "the larger life" led by the heroes of old and feels the call to "wring / Some secret purpose from the unwilling gods." In part 2, "Thorwald's Lay," that purpose is clarified. Biorn's Scandinavian dream becomes an American dream. Inspired by Thorwald the skald, he resolves to set sail on the "unpathwayed seas" and thereby establish a basis for the "first rune in the Saga of the West." In part 3, as Biorn's vessel approaches "Vinland," the prophetess Gudrida chants a vision of its future glories. Masses of men, she sings, will come to Vinland from different nations and social classes. "Empty-handed" but willing to work, they will subdue the land and prosper. The vision has most of the elements of Crevecoeur's romantic description of eighteenth-century America in "What Is an American?," including allusions to the settler as "the New Man." It incorporates Emersonian self-reliance and emphasis on the present rather than the past. It predicts the defeat of the gods of the Old World by inhabitants with the same fire and vision as Biorn. "Wilderness tamers" like him, they will create a great civilization. Then the poets will come and celebrate with "birth-carols" the feats of the "man child" of the New World.

While the Liefs and Biorns of Scandinavian legend play a prominent role in sea poetry dealing with American origins, Columbus is not ignored. Lowell provides one of the more interesting portraits in "Columbus." Standing alone on deck at night, feeling closer to the stars than to the "herd of earthen souls" who have shipped with him, Columbus ponders the question of human greatness. Like Biorn, he is voyager and visionary, and he has the same deep conviction of a special destiny. From the time he was a boy, he listened to "the legends of the sea" and, among other tales, "heard Biorn's keel / Crunch the gray pebbles of the Vinland shore." In mid-sea, he remains resolute in the face of his crew's skepticism, impelled by the "great design," confident of success, for "to have greatly dreamed precludes low ends." His firm belief in the existence of a new world is bolstered by the realization that Europe has become "effete." For a moment he wonders, of the "untried world" he journeys toward, "Shall the same tragedy be played anew?"

Whitman appears to provide an answer to the question in "Passage to India" when he addresses Columbus and says: "(Ah Genoese thy dream! thy dream! / Centuries after thou art laid in thy grave, / The shore thou foundest verifies thy dream)" (sec. 3). Born "for purpose vast," America, like her founder who is characterized as "History's type of courage, action, faith," is "Gigantic, visionary." But while America, with its transatlantic cables and transcontinental railroads, is portrayed as central in 1870 to the particular place to which human history has journeyed, it is not the final destination or nearly the best. "Passage to India" takes up the theme of American origins in an earlier voyage, but it transcends that theme, for the spatial voyages of the medieval West were preceded by the spiritual ones of the ancient East. The journeys to and by America are connected to a global journey with mystical origins and a mystical destiny, one "where mariner has not yet dared to go." In this respect, "Passage to India" is a "journey of life" poem. But "life," for Whitman, is that condition that comes after "Time and Space and Death," in which we can now confidently exclaim, "O farther, farther, farther sail!"

JOSEPH FLIBBERT

Chapter Seven

HERMAN MELVILLE

Herman Melville never left the sea. In a life of authorship that began with imaginative transformations of his own nautical experience and ended with an imagined story of sacrificial death in the British navy, the sea, in idea when not in setting, was seldom absent from his writing. He dove deeply into a nautical world of power and mystery, discovering, inventing, and revealing a sea of meanings stretching far beyond the immediate into psychology, culture, metaphysics, and literary expression itself. As he did so, he found narrative voices and expressive forms for a sea-based consciousness.

In his perceptive study of American writers, D. H. Lawrence calls Melville the greatest poet of the sea. Melville himself said that much matter-of-fact writing about the practicalities of nautical life had greatly damaged "the poetry of salt water"; in his own best work, though, he poetically evokes the mystery and majesty of the oceanic world. Deeply aware, in ways that his practical-minded predecessors and contemporaries in nautical narrative seemed not to be, of the sea's paradoxical qualities, Melville created a realm that is marvelous, alien, and portentous—a challenge to comprehension.

Yet Melville's experience had fitted him for a practical knowledge of the subject. Coming from a family whose men, on both sides, frequently took to the sea for their livelihood, he had sailed as a "boy" aboard a merchant vessel on the busy New York–Liverpool route and, within a few years, had gone whaling to the Pacific and been a sailor on a navy frigate. These activities gave him material for his writings—strange seas, strange creatures, exotic societies, ships' officers and crews, friends and enemies. During the early years of authorship he supplemented his firsthand knowledge with what he read in narratives of exploration, scientific reports, the accounts of seamen, and chronicles of marine disasters.

He first made use of his experiences in *Typee* (1846), which reshaped the conventional faraway adventure story by adding to its traditional mix of travelogue and autobiography the symbolically suggestive tale of an unsuccessful return to paradise. The narrative begins at sea with a longing for the shore after six months on a whaler without sight of land, but one-fifth of the way into the book, Tommo, its narrator, jumps ship; the narrative then remains ashore for all but a few more pages. No sea story this. In those early chapters of this first book, however, Melville accomplished a lot pertinent to his later nautical writing: descriptions of life under the soothing trade wind, tall tales of the sea, complaints about the conditions of the sailor's life, sly puns and innuendoes, lightly veiled political criticism, tonal shifts in narrative voice, and particulars of both personal and communal experience at sea—all of which he would develop in books to come. Then, too, though his preface insists on his "anxious desire to speak the unvarnished truth," a claim common to nineteenth-century nautical narratives, scholarship has shown that much of that supposed "truth" actually comes from imagination and from various printed sources.

Tommo is a perceptive young man, but rather at a loss to articulate fully his perceptions. Melville manages, though, as readers have noted, to imply much more than can be stated by Tommo, who has left his ship (it is aptly named the *Dolly*) like a child running away from a harsh home where little is allowed and where deprivation and the threat of punishment reign. Running away to the Typees, Tommo does not by that act become a liberated adult but, suffering from a mysterious wound which suggests psychological and sexual impotence, is made once again a virtually volitionless child—this time under the sway of the solicitous but perpetually watchful and capricious natives. Yet in contrast to his hosts, who often seem childish, he is a restive adult, hemmed in by a paradise ultimately no more congenial to him than was the ship. His second running-away—back to sea, of course—is more desperate than was the first.

Melville cautions the reader that *Omoo: A Narrative of Adventures in the South Seas* (1847) is not intended as a sequel to *Typee*, but the book's own captivity and escape action (humorous though it may be) does hint connections between the two. Rescued from the Typees by the whaling ship *Julia*, the narrator (named Paul or Typee) signs on as a seaman. Though Melville said that his story gives "plain, matter-of-fact details" of nautical life and its hardships, a political subtext of corruption, incompetence, victimization— and their mutinous consequences—seems to shape the sea story that occupies the first twenty-nine chapters. The latter part of the book, in which the narrator becomes an "omoo" or wanderer in the islands along with his friend Doctor Long Ghost (Peter), before he ships aboard another whaler, is similarly deceptive. Like *Typee, Omoo* is not the ingenuous young man's narrative it has often been taken for.

From the beginning of his career, then, Melville was subverting the kind of book he was ostensibly composing, "a book of unvarnished facts," like John Byron's *Narrative* or the numerous similar accounts of seamen's travels. Suspecting that Melville was inventing at least parts of his first two books (though they were far from seeing the extent of the subversion), some critics asserted that the author was disguising fictional romances as factual narratives. Melville said he then resolved "to show . . . that a *real* romance of mine is no Typee or Omoo, & is made of different stuff altogether" (*Letters* 70). As a result, *Mardi: And a Voyage Thither* (1849) is indeed different, and far more complex than either of its predecessors. Its opening chapters present an archetypal sea adventure in which the narrator, who will become known as Taji, and his friend, Jarl, desert their whaling ship to set out on the open sea in one of its boats, sailing west without compass or quadrant. In such an often retold small-boat story, like that of the *Essex* survivors after their ship was stove by a whale or that of Captain Bligh and his loyal crew members set adrift in an open boat by mutineers, the sea dominates in its presence and power. The men encounter it in all its moods, from doldrums to storms, and they fight its sharks. Encountering a drifting ship, they board it to find a man and woman, the survivors of a massacre. The ship has been wandering helplessly on the open sea; but now, with four aboard, it can sail a purposeful course westward. The plans are wrecked when a storm destroys the vessel and the woman, leaving the survivors marooned once more in their whaleboat.

This portion of the book, except for its hyperbolic, even giddy tone, resembles a traditional sea story, but then comes a turn into colorful romance, satire, allegory, and philosophical thought. In chapter 169, titled "Sailing On," Melville in effect announces a subject and theme of much of his subsequent work. Readers have noted the reverberations in the narrator's suggestive assertion that he is willing to sail beyond the utmost physical hazards into another world entirely: "the world of mind." Most obviously in *Moby-Dick*, but elsewhere too, that new, *meta*physical world of voyaging, however we read it, will be most frequently identified with or conceptualized in terms of the sea.

Although the difficulties of *Mardi* cost Melville more than a year of labor, the book was badly received. He recognized that some writers "have a certain something unmanageable" in them that goes beyond pleasing readers and must be indulged, as in *Mardi;* but for practical reasons he turned aside and produced another ostensibly "plain, straightforward, amusing narrative"

(*Letters* 86). *Redburn: His First Voyage* (1849), actually a tale of initiation into the seagoing life, is threaded with hints of other meanings—autobiographical, sexual, cultural. From the vantage point of long experience as a sailor, Redburn gives an account of his youthful, initiatory voyage on a merchant ship bound for Liverpool. He sees his young self as an outcast, "a sort of Ishmael in the ship" (ch. 12), harshly treated by the captain, his superiors, and his shipmates. His great hopes for his future, and his idealistic view of the grandeur, beauty, and sublimity of the ocean make him one of the worldwide company of young men who set sail with a sense of wonder, later tempered by experience—such as an encounter with a drifting, wrecked schooner, with long-dead bodies lashed to the rail. Unprepared for the sailor's day-to-day existence, he finds the work hard and unrewarding, though he comes to respect the skill of the true sailor who "not only knows how to reef a topsail, but is an artist in the rigging" (ch. 26)—a phrase that also suggests the sailor as author.

Redburn is the only book in which Melville treats, in detail, life in the forecastle. This subject of many other sailor authors is memorable here not only in its picture of Wellingborough Redburn's painful experiences as a naif among tough, mostly unsympathetic seamen, but in the double vision with which the first-person narrator is able to evoke sympathy for his harassed young self while suggesting his own former priggishness and foolish sense of class superiority. Some noteworthy individual portraits also distinguish this treatment of the sailor life. Developed at some length, and particularly memorable, is the sickly, cursing, jeering, misanthropic Jackson, deferred to by everyone. Despite Joseph Conrad's avowed dislike of Melville's work, Jackson seems a likely source of James Wait in *The Nigger of the "Narcissus."*

Melville wrote that he considered *Redburn* a job done for money instead of the kind of book he most wanted to write. He offered the same harsh judgment of *White-Jacket; or, The World in a Man-of-War* (1850), based on his one cruise in the navy, and written in the then-popular mode of reform literature. He produced *Redburn* and *White-Jacket* in a few months in 1849 and felt that "no reputation that is gratifying to me, can possibly be achieved by either of these books" (*Letters* 91). A year after *White-Jacket*'s publication, Melville was to complain in a letter to his then-closest friend, Nathaniel Hawthorne, about the tension between the kind of book he was moved to write and the kind that readers demanded. He could not entirely write the "other" way, and as a result, "the product is a final hash, and all my books are botches" (*Letters* 128).

But most readers find his account of a young man facing the inhumane discipline of naval life and appalled by both the inequities and iniquities of that life no botch. Unifying much of the book, seriously and tongue in cheek, are its two main metaphors: the ship-as-world and the world-as-ship—in which context naval reform becomes a theme with macrocosmic meaning. Melville was actually no pioneer in that subject. He was familiar with and made use of such accounts as William McNally's *Evils and Abuses in the Naval and Merchant Services Exposed,* Samuel Leech's *Thirty Years from Home,* John Sleeper's *Tales of the Ocean,* Nathaniel Ames's *Mariner's Sketches,* and Mercier and Gallop's *Life in a Man-of-War.*

Here, as in those works, the ship's company is a satirical human microcosm. Though White-Jacket finds a few of his shipmates to be good and admirable men, more come under withering attack. The skipper of the *Neversink,* Captain Claret, is continually half-drunk; in a tempest his orders have to be countermanded by an inferior officer—the sort of punning, revelatory situation in which the narrator revels. Bland, the master-at-arms, a charming and even dainty person, is also an "an organic and irreclaimable scoundrel." Worst of all is the pompous, heartless Cadwallader Cuticle, the surgeon of the fleet, himself an assemblage of prosthetic devices, who callously performs an unnecessary amputation that kills the patient.

In contrast, White-Jacket feels lucky to belong to the crew of the maintop, led by the much-admired Jack Chase—a real-life shipmate and true sailorman, whose impact on Melville is suggested by his dedicating to this "Captain of the Maintop," forty years later, the story of the doomed "handsome sailor," Billy Budd.

The book's blatant but telling satire accompanies lengthy exposition as White-Jacket becomes a mouthpiece for Melville's democratic views. The chapters on flogging and oppression, in which White-Jacket is driven to the verge of murder-suicide by the threat of a flogging for a dereliction of which he is innocent, indict practices that contravene civil immunities and deny the very humanity of their victims. During Melville's voyage on the *United States,* 163 floggings were logged; that bitter fact no doubt lies behind the statement that for the sailor, "our Revolution was in vain; to him our Declaration of Independence is a lie" (ch. 35). Such criticism also occurs in substantial dramatizations of salient points. Chapters 85–87, on "the massacre of the beards," admiringly delineate the character of Ushant ("you shan't"), the venerable captain of the forecastle, who defies the captain's demeaning order to shave in spite of the flogging he receives. As admirable as Ushant himself appears, however, an attentive reader may well suspect that the writer does not fully concur with young White-Jacket's admiring depiction of the other sailors willing to mutiny rather than shave.

A parallel distinction between author and narrator ends the book. The last chapter begins, "As a man-of-war that sails through the sea, so this world that sails through the air." It goes on to moralize on all that has been implicit in the trope throughout the book—and more, since the world-ship is now construed in cosmic, metaphysical terms: "We mortals are all on board a fast-sailing, never sinking world frigate, of which God was the shipwright." The chapter is a sermonic plea for harmony, cooperation, acquiescent belief in a "Lord High Admiral" God who will eventually right human wrongs, even if after long ages still to come. Besides, our real home is in heaven, not here. This comforting faith may indeed be young White-Jacket's idealistic response to tribulation and injustice; for Herman Melville himself, though, it is surely ironic. *White-Jacket* is the fullest, among the early works, of Melville's social, economic, and political critiques, all of which reflect to some degree not only the reform impulse of the time but also his own powerful critical intelligence—often expressed obscurely or subversively so as not to alienate the readers on whom his career aspirations rested.

Late in the book, in a now-famous passage, White-Jacket tells of his near-drowning, as, falling from a yard, he sinks into the sea, weighed down by his self-fashioned white jacket, an ambiguous symbol of his identity (ch. 95). Cutting himself free of the encumbering jacket he rises to life again. With the symbolic death and rebirth of the sailor, Melville's maturing concept of himself as an author and of his character/narrator stood on the verge of producing the Ishmael of *Moby-Dick.*

But first he made a journey to England in 1849–50 to secure advantageous terms for the publication of *White-Jacket.* His encounters with publishers, writers, and artists, his sightseeing in London and, briefly, on the Continent, and his frequent visits to art exhibits and galleries offered a wealth of new thoughts and literary materials. He read eagerly and bought books to carry home. The sea voyage, itself, taken this time as a passenger, not only produced some important human interactions but also must have refreshed his experience of the sea in its actual and mythic dimensions; his journal of that voyage is revealing. When he returned, he was prepared to compose the book he would describe, cautiously, to a publisher, as "a romance of adventure, founded upon certain wild legends in the Southern Sperm Whale Fisheries, and illustrated by the author's own personal experience" (*Letters* 109). In the New York *Literary World* for November 22, 1851, Evert Duyckinck called it "a remarkable sea-dish—an intellectual chowder of romance, philosophy, natural history, fine writing, good feeling, bad sayings."

Moby-Dick; or, The Whale (1851) is, of course, much more than either description admits. For Ishmael, the friendless wanderer and narrator who flees the solid ground of the "terraqueous" world, a whaling voyage is, among other things, an emotional cure, made necessary by his experience of the confinement and the repressive, self-denying values of the land, the "civilized hypocrisies and bland deceits" (ch. 10) of prejudice and conformity. And he goes to sea because it opens "the great flood-gates of the wonder-world" for him (ch. 1). The encounter with this "wonder-world" compels the reader, along with Ishmael, "to grope down into the bottom of the sea," into "the unspeakable foundations, ribs, and very pelvis of the world" (ch. 32). While *Redburn* and *White-Jacket* emphasize the life of the ship, *Moby-Dick* dives deeply into the sea itself—for in several senses "that unsounded sea . . . is Life."

Ishmael, who develops and matures in the narration by his older self, apprehends the sea in all its moods and movements. Once, as he stands masthead watch, its mildness bewitches him and he "takes the mystic ocean at his feet for the visible image of that deep, blue, bottomless soul, pervading mankind and nature" (ch. 35). This transcendental serenity that he seems to need so desperately makes itself felt elsewhere in mild and gentle scenes: a whaling crew, breaking through a thrashing circle of leviathans, comes on a scene of nursing mothers, friendly baby whales, and "leviathanic amours" in an oceanic Eden (ch. 87); when Moby-Dick first appears to the crew of the *Pequod*, in chapter 133, he swims before them with "gentle joyousness" and a "mighty mildness of repose."

But Ishmael, more thoughtful and complex than Tommo, Redburn, or White-Jacket, understands that the serenity of the sea and its marine life is partial and deceptive. Its *otherness* is dangerous, and he advises, "heed it well, ye pantheists." Even a perfectly calm sea, while supporting the body of Pip the cabin boy, drowns in its loneliness, his soul. In this compelling treatment, Melville moves far beyond the areas charted by earlier sea fiction, deriving some of his power from the conscious application of the tradition of the sublime. He had read Edmund Burke on the sublime and beautiful, knew the works of Shakespeare and the earlier Romantic writers, and was well acquainted with the devices of landscape art practiced by Salvator Rosa, Claude Lorrain, J. M. W. Turner, and the Americans Thomas Cole, Benjamin West, and Washington Allston. At times he portrayed the ocean in the tradition of the "horrific sublime," emphasizing its power, immensity, darkness, and solitude.

Consequently, the sea seems to show its truest nature in its storms: at the first lowering for whales, Ishmael's boat is swamped in a sudden squall and its crew is nearly lost; in the Sea of Japan, where calm should reign, the ship encounters "the direst of all storms, the Typhoon." As if to confirm Captain Ahab's assertion that "all visible objects, man, are but as pasteboard masks,"

Ishmael is aware that beneath the gentle surface of the sea lies another world of power, "universal cannibalism" and "inscrutable malice." The vision of the gigantic whale gently swimming is only an incomplete impression: hidden are "the full terrors of his submerged trunk" and the "wrenched hideousness of his jaw."

This disjunction between appearance and reality, the sense of a hidden, inner life, also derives from a perception of life as biology. The many chapters on the physical whale—what has been called the "cetological center" of *Moby-Dick*—attest to that perspective, earlier evidenced several times in *Mardi*, as well as by the underwater brush with life itself in the famous fall from the yardarm scene in *White-Jacket*. Though Melville quotes from Darwin in the "Extracts," his position in *Moby-Dick* seems closest to nineteenth-century "natural theology," in which the book of the world, read scientifically, reveals the work of God just as does the book of the word (the Bible). But the deity implied by the evidence of the biology, zoology, and cetology is no comfort to the traditional believer: " 'Queequeg no care what god made him shark,' " says the islander after almost having his hand taken off by one, " 'but de god wat made shark must be one dam Ingin' " (ch. 66).

On this ocean and in pursuit of its most formidable creature, the sperm whale, sails the *Pequod,* a real whaler, a seagoing microcosm, and a mythic symbol, encompassing all the human races as well as various cultures, beliefs, trades, strengths, and weaknesses. The ship is admirably adapted to its trade by virtue of its fittings, tools, supplies, and its efficient disposition of cramped spaces for its diverse activities. In the epic tradition, the book describes at length the tools of the crew, with chapters about the hempen line, the harpoon, the crotch. Much attention is given to the crew's work, the cutting in, the removal of the whale's blanket of blubber to slice and try it. Here, the book follows the course of popular whaling accounts, both fiction and nonfiction, in particular J. Ross Browne's *Etchings of a Whaling Cruise.* Melville had complained that plain recitals of particulars could impair the charm of writings about the sea, but he worked his blubber artfully. All the mundane actions of whaling and the anatomizations of whales are larger than themselves, pregnant with figurative meanings. And at least some of the exposition is linked to the developing narrative. It is necessary that the reader know about the size and strength of leviathan and that it be proved that a ship can be stove by a whale. The line, so carefully described as a potent danger, will be the noose that pulls Ahab to his death. The huge cask of the whale's head will almost become Tashtego's tomb. The tail, celebrated for "subtle elasticity" and "exceeding grace," will cause destruction among the *Pequod*'s whaleboats.

The weakness of humanity is evident as it tries to master the power of the ocean by its confident striving. Ahab may declare himself "lord of the level loadstone" and believe in his power over the sea, but he cannot triumph over it. Encouraged by knowledge and ego, "baby man may brag of his science and skill" but is no match for the elements, since "for ever and for ever, to the crack of doom, the sea will insult and murder him, and pulverize the stateliest, stiffest frigate he can make" (ch. 58). The destruction of the *Pequod,* its captain and crew, is the stern answer of the natural world to the puny humans who float on the surface of the grand oceans.

The world of the whale hunter and the epic pursuit of Moby Dick on the seas of life is also the interlocked stories of the quests of Ahab and Ishmael. Ahab, searching for the infinite, is a seagoing Faust. He cannot be swerved from his monomaniacal purpose of bringing the infinite to an accounting by attacking what he considers its "inscrutable malice," represented for him by the white whale.

Ishmael believes Ahab to be "a mighty pageant creature, formed for noble tragedies" (ch. 16), and the *Pequod* a fit stage for tragedy. As a sailor, he sees everything in terms of nautical metaphors; and as the informing sensibility of the book, through whom, usually, the reader is admitted to its knotty intricacies, Ishmael is himself an intricate character. He loves the mysterious and exotic but is also fascinated by the ordinary processes that occupy so much of the whaler's time—reading in each one a metaphor of human life. He wants to understand, if possible, the creatures he kills. Reflecting Melville's reading of naturalist writers, he tries to classify whales, to portray them, to explain them; and he renders this attempt in language sometimes simply expository but more often intense, rhetorical, and poetic, as the chase of the whale becomes the pursuit of understanding itself.

Though drawn, like the other seamen, into Ahab's "quenchless feud," Ishmael nevertheless learns lessons that will lead to his salvation. In "The Lee Shore" he sees the magnificence of his shipmate Bulkington, newly landed from a four-year voyage, who sets to sea again. Ishmael compares him to a "storm-tossed ship," exemplifying the "mortally intolerable truth; that all deep, earnest thinking is but the intrepid effort of the soul to keep the open independence of her sea; while the wildest winds of heaven and earth conspire to cast her on the treacherous, slavish shore." Like Bulkington, Ishmael too has found the land "scorching to his feet." Bulkington is a glorious ideal—but the truth he exemplifies is "intolerable." Under the influence of his friend, the "soothing savage" Queequeg, Ishmael learns to accept the land world rather than attack it (as he wanted to do before going to sea). And on a tranquil day, in chapter 94, as he and his shipmates squeeze lumps out of spermacetti, the erstwhile loner is transported by "affectionate, friendly, loving feeling[s]" which engender in him a vision of "angelic" yet accessible joy:

> I have perceived that in all cases man must eventually lower, or at least shift, his conceit of attainable felicity; not placing it anywhere in the intellect or the fancy; but in the wife, the heart, the bed, the table, the saddle, the fire-side, the country.

Although he is drawn to the intolerable truths of landlessness and the pursuit of the infinite, Ishmael also comes to see that "in the soul of man there lies one insular Tahiti, full of peace and joy" (ch. 58). This island is not the land, whose values he rejects. Rather it suggests a place of psychological refuge lying somewhere in Ishmael's inner sea. But solitary islands in the sea, no matter how peaceful, cannot in themselves heal the wounded soul. Ishmael has absorbed the wisdom found in Hawthorne's tales of head and heart. Like Ethan Brand or Doctor Rappaccini, Ahab damns himself by putting human considerations aside to pursue a perilous ideal, thus losing, as Hawthorne asserts, his "hold of the magnetic chain of humanity." Ishmael recovers his humanity by joining hands and turning toward the attainable.

Critics who were inclined to be sympathetic recognized at once the great merits of Melville's newest book, though most took the occasion to point out what they conceived to be its evident faults. A half-century and more later, they had not come to terms with it. W. Clark Russell, the English author of sea adventures and a great admirer of his American counterpart with whom he exchanged letters, thought it his "finest performance" but obscure in parts, with a "transcendental mysticism" poorly fitted to the sailors who expressed it.

While he was creating *Moby-Dick*, Melville wrote to Evert Duyckinck, "I love all men who *dive* . . . the whole corps of thought-divers, that have been diving & coming up again with bloodshot eyes since the world began" (*Letters* 79). Judging by his next book, Melville's own thought-diving had left him with the bends. Ironically promising Sophia Hawthorne a "rural bowl of milk" to replace his usual "bowl of salt water," he quickly composed the deeply autobiographical *Pierre; or, The Ambiguities* (1852), a tale as emotionally violent as *Moby-Dick*, and exploring love, lust, adultery, illegitimacy, incest, suicide, and murder. Even in this earthbound story, though, repeated sea tropes evidence its author's analogical sea-consciousness. Among several sorts of application, two stand out. One emphasizes Pierre's response to the erotic beauty of the "castaway" Isabel: "every downward undulating wave and billow of Isabel's tossed tresses gleamed . . . like a tract of phosphorescent midnight sea"; and Pierre's "whole soul was swayed and tossed by super-natural tides" (bk. 8, 2). The other, involving Pierre in a Bulkington-like plunge into stormy seas of thought, characterizes his disastrous encounter with painful truth: "Truth rolls a black billow through thy soul" (bk. 3, 6); "the deeper and the deeper that he dived, Pierre saw the everlasting elusive-ness of Truth" (bk. 25, 3). As a result of his agonizing struggle in the sea of truth, on the very last page Pierre is "the deluge wreck—all stranded here."

Few readers forget the tonally contrasting, strangely beautiful scene in book 26, in which, aboard a ferry, Isabel has to be physically restrained from leaping overboard to go "away, out there! where the two blues meet, and are nothing." But storm, shipwreck, and disaster are inescapable in this story. Sadly and appropriately enough, this all-too-disturbing book was a commer-cial failure.

To the firm of Harper, his publishers, he next proposed a volume, "300 pages, say—partly of nautical adventure, and partly—or, rather, chiefly, of Tortoise Hunting Adventure," and received an advance for it. The work ma-terialized as something less than Melville had envisioned: not a complete book at all but a series of magazine sketches, "The Encantadas, or Enchanted Isles" (*Putnam's Monthly Magazine*, 1854). This revision of his aims consti-tutes, perhaps, a certain failure of Melville's constructive powers. He had always been able to build, on the skeleton of his own adventures, the flesh and muscle of a continuous fictional narrative. The tortoise story, set in the Gala-pagos Islands, offered materials as promising as his stay with the Typees, his participation in a mutiny, his early voyage to England, his unfortunate enlist-ment in the navy, and his experiences aboard whaling ships. He had a first-person narrator to tell stories of deserters, mutineers, and castaways, and to brood over the hellish landscape of the islands. The Dog-King of Charles's Isle and the hermit Oberlus of Hood's Isle provided stories that could have been realized at greater length than the sketchy narratives granted them.

The treatment of the subject that he chose, however, provides a grim poetry and dark vision of the sea and some of its islands. Melville empha-sizes the powerful influence of the visual arts on his imagination by taking the pseudonym Salvator R. Tarnmoor for magazine publication of the sketches. Salvator Rosa, the seventeenth-century Neapolitan painter of land and sea-scapes, was a popular artist during the eighteenth and nineteenth centuries, his canvases described approvingly by American travelers in Europe. Mel-ville had already alluded to his sea-pieces in drawing the demonic Jackson in *Redburn;* now it was clear that he intended to create landscapes and seascapes in the style of Salvator, ironically reminding the reader that tarns and moors would not be his subject matter.

Ocean and island meet harshly, dangerous to ships caught by treacherous winds. The humans are solitaries, runaways, and castaways, usually the dregs of civilized society. There could be no greater contrast than between the Edenic glades of the Marquesas Islands and the wasteland of the Galapagos chain. Melville's "Pisgah view from the rocks" in the fourth sketch reveals a grim, charred scene, the sort of view that emerges as a theme in his later writings, accompanying even more ambiguity.

The novella "Benito Cereno" (1855) is an example. Drawing on the narrative of Captain Amasa Delano of Massachusetts (1817), as well as from his own memories of the highly publicized *Amistad* slave mutiny (1839), Melville brilliantly constructs a complex, disturbing, and subversive morality tale. Character becomes theme in his creation of a fictional Delano, naive American, "a person of a singularly undistrustful good nature, not liable, except on extraordinary and repeated incentives, and hardly then, to indulge in personal alarms, any way involving the imputation of malign evil in man" (*Piazza Tales* 47). Captain Ahab had been permitted to make his voyage a vengeful expedition partly through his mate's "incompetence of mere unaided virtue or right-mindedness." A similar but inexcusable incompetence, supported by Melville's sardonic double negative, "undistrustful," makes Delano unwilling to see what is before him on the decrepit Spanish slave ship, to the aid of whose apparently neurasthenic captain he has come: the evidence of enslavement, violent mutiny, and murderous deception.

The ships, their boats, and the sea are the setting for this ugly drama, but the sea is dominant. In the early morning, its leaden grayness is that of "Shadows present, foreshadowing deeper shadows to come." Yet the setting is not the open sea but the harbor of a small uninhabited island, its waters calm and shadowy. Likewise, the prevailing tone of the story is quiet, mostly understated, though charged with menace, and the violent action that comes as a release is briefly disposed of.

Intent on the riddle of action and personality, Melville allows Delano's behavior and thoughts to expose his racism and self-satisfaction, and he shapes his narration to evoke the reader's prejudicial preconceptions. Documents are presented, as they had been by the real Delano, ostensibly "as a key to fit into the lock of the complications"; they fail to do so, instead illustrating the gap between documentation (supposedly objective but actually self-interested) and something closer to truth—a problem he would return to in *Billy Budd, Sailor.*

In next serving up the "Revolutionary narrative of the beggar" in *Israel Potter, His Fifty Years of Exile* (1855), Melville dealt more whimsically with sober or invented facts, drawing on a short "Life and Remarkable Adventures" of his hero. The book contains chapters of comedy, adventure, melodrama, and thoughtful criticism, but the comedy predominates. The satirically addressed dedication to "His Highness the Bunker-Hill Monument" makes Melville's intention clear. He asserts that Potter deserves the tribute of biography; but Potter's own narrative, a pathetic story, undercuts the claim. His service at Bunker Hill is less than heroic. Captured and sent to England, he works for sympathizers with the colonists' cause by delivering secret papers to a sententious and stingy Benjamin Franklin. Deserving of biography, as the antiheroic Potter is not, Franklin receives ridicule. A second American hero, John Paul Jones gets parallel treatment in Melville's upside-down history. Widely admired for his bravery and resolution, he is a fair target for creative parody.

Analogously (though readers have admired the book's sea scenes), the battle between the *Bonhomme Richard* and the *Serapis* is treated in language disturbingly off-key for any serious recreation of a famous historic event. In the early stages of the struggle, the ships maneuver "like partners in a cotillion." Dropping hand grenades on the *Serapis* from their yardarms, the American sailors are "busy swallows about barn-eaves and ridge-poles," and the gun crews, stripped to the waist, are like fauns and satyrs. The British gunners are mechanical in their discipline: "The Parcae were not more methodical; Atropos not more fatal; the automaton chess-player not more irresponsible" (ch. 19 continued). The *Richard* wins the gory battle but, gutted by fire, sinks, leaving the narrator to ask, as Melville had done from his first book, "in view of this battle one may well ask—What separates the enlightened man from the savage?"

Melville thus upsets conventional reactions to the Revolutionary War, its heroes, and its most prominent battles in his "Fourth of July story." The date is one for celebration, but when Israel, home after fifty years of exile, debarks in Boston on the Fourth, he is nearly run down by patriots celebrating the heroes of Bunker Hill. The world he had known a half-century earlier is hardly recognizable. Like Rip Van Winkle before him, he is a stranger in a strange land.

The mocking spirit of *Israel Potter* flourishes in *The Confidence-Man: His Masquerade* (1857), a witty but obscure allegory on swindling as universal fact. In "Bartleby, the Scrivener," Melville had alluded to Monroe Edwards, an infamous confidence trickster of the 1840s, whose life, trial, imprisonment, and ingenious escape attempts had been widely publicized. Now, recalling such stories, he created another microcosmic shipboard world whose passengers exist to cheat or be cheated. The *Fidèle*, a Mississippi river steamer bound from St. Louis to New Orleans, is the stage for a series of pointed conversations about charitable schemes, followed by confidence games. Moving from the open sea of the earlier books and the enclosed harbor of "Benito Cereno," Melville brings the world of water close to shores, where his actors can enter and depart. The scenery unrolls like the gigantic panoramic paintings of John Banvard and other artists, picturesque and transient while the protean con man shifts shapes and diddles the boat's passengers.

Reviewers found the book anecdotal, or too long, or absurd, and its poor reception, in conjunction with Melville's undermined health, drove him from the writing of prose fiction. Unsuccessful lecture tours attempting to capitalize on the journey he took to Europe and the Holy Land in 1856–57 concluded one effort to change his profession. In 1863 he moved the family to New York, where, after the Civil War, he was able to secure employment in the custom house.

Meanwhile, as Thomas Hardy was later to do for similar reasons, he turned from prose to poetry. A visitor in 1859 described him as a disappointed man, "soured by criticism and disgusted with the civilized world and with our Christendom in general and in particular" (*The Melville Log* 2:605). But poems were accumulating, enough to assemble a volume for which he could find no publisher. The conflict between the states, though, gave him the subject for *Battle-Pieces and Aspects of the War*, published in 1866. Compared to other writers on the war, his position was anomalous. Unlike John W. De Forest and Thomas Wentworth Higginson, he did not participate in its battles; unlike Whitman, he did not go to the battlefield or attend to the wounded. Nor, apparently, did he feel moved to write till the fall of Richmond in April 1865, but then he composed quickly and fluently a poetic war diary that allowed him to articulate his feelings about the catastrophic event.

In a prefatory note, his metaphor for his accomplishment emphasizes the instinctive: "I seem, in most of these verses, to have but placed a harp in a window, and noted the contrasted airs which wayward winds have played upon the strings." But careful choices of themes are implied by his reading of newspaper accounts and *The Rebellion Record,* and by what he includes and omits. He is obviously moved, for example, by the destruction of the whaling ships scuttled to block Charleston Harbor. The old sailor who laments the loss of "The Stone Fleet," like Melville himself, finds the act "a pirate deed" that fails, since "Currents will have their way; / Nature is nobody's ally." With a Shakespearean use of nature's mirroring of distemper in human affairs, "Misgivings" presages the coming war as "ocean-clouds over inland hills / Sweep storming"—concluding in one of Melville's finest lines that "the hemlock shakes in the rafter, the oak in the driving keel." Melville's patriotic sense of the war's rightness is nearly invalidated by the gravity of death, for there is "no knowledge in the grave." Since triumph is inseparable from horror, his prose "Supplement" says, "Let us pray that the terrible historic tragedy of our time may not have been enacted without instructing our whole beloved country through terror and pity."

As an observer of this calamitous drama, the poet often finds correlatives for his feelings in poems of sea battles. The "old order" of struggle between wooden ships has been displaced by battles of the ironclads. In "The Temeraire," Melville evokes "some one craft to stand for the poetic ideal" of a naval past, drawing on his knowledge of J. M. W. Turner's great painting, "The Fighting *Temeraire*," and creating an old Englishman to comment on the "deadlier lore" of the present. The lethal conditions of present-day battle are anatomized in "A Utilitarian View of the Monitor's Fight," a deliberately ugly, antiheroic poem, where, without passion, the warriors are operatives, and all goes "by crank, / Pivot, and screw, / And calculations of caloric"; as a result, "War's made / Less grand than Peace." The historic battle of the *Monitor* and the *Merrimac* in 1862 haunts Melville's vision of the war.

These poems signal a change in Melville's perception of the oceanic world. The experiences of Tommo, White-Jacket, and Ishmael were supported by his own participation in an old and settled order of sailing ships, even though that order was changing when he took to the sea. The introduction of steam engines and ironclads swept away this familiar world, and when Melville returns to writing of the sea some twenty years later, his mood is somber and elegiac.

The poems of *John Marr and Other Sailors* (1888) recall the bygone age of sail. John Marr, disabled in a sea fight, moves from sea to seaports and then inland in 1838. He marries and then suffers the loss of his wife and infant child. His "void at heart" persists in the landlocked world, where his neighbors care nothing for the sea or his experience of it. His former shipmates become "subjects of meditation," and, in his memory, affectionately take on "a dim semblance of mute life." He expresses his desire and need in the untitled prefatory poem addressed to them: "To see you at the halyards main— / To hear your chorus once again!" In thus creating a sailor-poet as persona and narrator, Melville once more returned to the device he had often employed. Like Redburn and Ishmael, among others, John Marr recounts past experience from the vantage point of an uncertain present, depending on "memory's mint"; and he can be linked with Melville by the poetic use of the symbolic "harp in the window." In *Battle-Pieces* it implied his instinctive responses to war; in "The Aeolian Harp," it signals Marr's ambiguous response to the troublous sea. Depicting shipwrecks, Melville describes Marr and himself:

> Dismasted and adrift,
> Long time a thing forsaken;
> Overwashed by every wave
> Like the slumbering kraken.

Hurt, wounded, like Marr, he searches his past for its healing powers.

The shipmates of Marr's poems are aspects of his own nature and thought. In "Bridegroom Dick," the old Melvillean tar reminisces regretfully on halcyon days "Ere the Old Order foundered" and sailors "learned strange ways" under the impact of new sea-going technology. The aged petty officer of "Tom Deadlight," dying on board his British *Dreadnought* in 1810, and wandering in his illness, is a representative of the Old Order, as is the heroic Jack Roy, "captain o' the *Splendid's* main-top." Ned Bunn, in "To Ned," recalls "Authentic Edens in a Pagan sea." As these are familiar figures to Marr, so they are to Melville, for Tom Deadlight and Jack Roy elicit the naval past of Jack Chase and other friends; Ned Bunn, one of the "Typee-truants" of earlier times, is Melville's weary, regretful celebration of Tommo's Toby. In these poems, thus enriched by substantial personalities from his past, Melville assesses his personal loss in the new age of steam.

This collection engages Melville's deepest emotions as well as his philosophical concerns and his aesthetic ideals. The sea, after all, is his element; even if it is a "destructive element," he, like Conrad's Lord Jim, must be immersed in it, and his best poems come from these considerations. Among the outstanding ones are some from later in the book in the sections entitled "Sea Pieces" and "Minor Sea Pieces," in which threat and destruction set the tone. "The Berg," subtitled "A Dream," in its collision of warship and iceberg, depicts both nature's implacability and human irrationality. "The Maldive Shark" suggests human revulsion against natural, biological functioning, as in the picture of this "pale ravener of horrible meat."

Fully as important as the sea's undoubted harshness, though, is its life-enhancing power, a persistent Melvillean theme. White-Jacket, nearly drowned by his fall into the sea, finds rebirth there as well. Ishmael, buoyed up by the caulked coffin, survives the general destruction of the *Pequod* and its crew. In his distress, John Marr, a castaway in the heart of the vast American continent, like the castaways and exiles on the inhospitable coasts of the Encantadas, reflects on this paradoxical aspect of the sea in "The Tuft of Kelp":

> All dripping in tangles green,
> Cast up by a lonely sea
> If purer for that O Weed,
> Bitterer, too, are ye?

Having relived, in memory, the embittering and radiant experiences of the nautical life, John Marr can assert his cure in the closing "Pebbles" at the end of his imaginative journey: "Healed of my hurt, I laud the inhuman Sea— / . . . / For healed I am . . . / Distilled in wholesome dew named rosmarine." Rosmarine, the rosemary of the seashore, is for the remembrance that softens the pitiless breath of ocean and brings relief to the suffering sailor/poet.

Of course, John Marr is not Melville and expresses imperfectly a part of Melville's revelatory disclosure of all the sea means to him. But Marr, like Melville, is a poet. Since *Moby-Dick*, Melville had not created a persona capable of an understanding that would match his own profundity, preferring to speak as the omniscient author and creator. John Marr is nearly as mythopoetic as his creator and speaks with philosophic maturity reflecting his author's lifelong encounter with the sea.

Melville's last fiction, the unfinished *Billy Budd, Sailor* (1924), written in part while he composed the poems of *John Marr and Other Sailors*, began as a poem with a headnote. Like "Tom Deadlight," it used the prose note to describe the situation of the sailor, who then spoke his mind in verse. The sailor, Billy, awaiting execution for his part in a mutiny, reminisced about his past and his shipmates. But Melville's meanings developed in the course of extended revision. One of them is mutiny. Well acquainted with the execution of three men aboard the U.S. brig *Somers* in 1842, Melville compares that incident to the fictional situation aboard the British man-of-war *Bellipotent* in 1797. His cousin, Guert Gansevoort, had reported the talk of mutiny on the *Somers* and then presided over the hearing that convicted the men. Melville himself had been involved in an uprising aboard ship and had treated the incident in *Omoo;* in *White-Jacket* he had attacked the punitive American Articles of War and depicted a near-mutiny over the "massacre of the beards." In *Moby-Dick*, Starbuck contemplates a mutinous act while Ahab carefully plans his actions so as to prevent the crew's rebellion against his usurpation of the voyage's legitimate purposes, and *Moby-Dick*'s *Town-Ho* story is centered in mutinous action. The overthrow of authority, then, had been on Melville's mind at least from the beginning of his writing career. Only once, though, does mutiny succeed (temporarily)—on the *San Dominick* in "Benito Cereno."

Like Hawthorne in "The Custom-House," Melville's narrator works backward from an imaginary document and looks behind the facade of history. In the summer of 1797, after mutinies at Spithead and the Nore, precautions are taken aboard the ships of the British navy. Rear Admiral Nelson's power of personality is called on to have a settling influence on a recently rebellious crew. During impending battles, officers sometimes stand with drawn swords behind their own gun crews.

In this context, Melville draws his three major characters into fateful confrontation. Billy Budd, the Handsome Sailor, impressed from the vessel *Rights of Man*, known as a peacemaker and beloved by most of his shipmates, is physically powerful, noble of birth, an unsophisticated "upright barbarian" with a simple flaw—a stutter. Claggart, the master-at-arms, is as devious as anyone who plies that trade aboard a ship; but worse, he cannot be well comprehended since his is "an evil nature . . . innate" (ch. 11). Captain Vere, whose determining role grew in late revisions, and through whom Melville examines the exercise of authority by what would seem to be an extraordinarily competent and sensitive officer, is an excellent sailor; he is also intelligent, intellectual, and well-read; a man of settled convictions, he believes in law and order above all.

Claggart, unable to control the evil in himself, reports Budd as a possible mutineer. Vere responds impatiently and acts rashly by letting the accuser face the accused with only himself as witness. Budd, shocked by the false accusation, in agony at his inability to speak and defend himself before his captain, strikes out with his fist.

Vere recognizes that the death of Claggart is a "divine judgment," but knows that under the Articles of War, merely striking a superior is a capital crime. Whatever his feelings or the humane reactions of his court, private conscience must, to him, yield to the public code. He convinces the reluctant court to condemn. His own conscience seems clear; wounded and dying some weeks after Billy's execution, he pronounces the boy's name but not, apparently, with remorse.

Billy Budd, first and foremost a sailor, victimized by his own virtues and defects, allusively linked to Jesus, dies sacrificially. He is venerated by his shipmates, who regard a chip of the spar from which he was hanged as they would "a piece of the Cross." He impresses them so deeply that another foretopman recreates his image in a ballad that will ensure his immortality.

Melville's narrator is one who seeks truths to stand against the falsities of life and reports the event in his ambiguously subtitled "inside narrative" as a corrective to the brief, twisted account in a naval journal, in which (as under military law) victim and victimizer are made to change places. He labels as rumors tales of Claggart's supposed criminal past and refrains from offering an invented story to account for the feelings of the master-at-arms toward Billy. For him, truth "uncompromisingly told will always have its ragged edges" (ch. 28). Omniscient at times, he can tell the surgeon's suspicions about Vere's mental state; but he will not create the scene of Vere's last interview with the doomed Billy. An integral part of the story, perhaps even the voice of Melville himself, he is an inconsistent bearer of its message—though such judgments are compromised by the unavoidable fact of an unfinished manuscript.

Beyond the forms of naval life, the world of ocean has its own significant spell. After Budd's burial, gun crews stand at their posts, officers give reports, music and religious rites tone the ordered life aboard ship. But beyond its bounds still lie the mysteries of the sea, the air, the sky, truth itself.

In his poem "In a Bye-Canal," Melville wrote, "Fronted I have, part taken the span / Of portents in nature and peril in man" (*Timoleon*). The portents of nature give his sea fiction and poetry the shape of his special, profound vision. He was a seer, and this quality permitted him to create a body of work unmatched by his contemporaries. His experience of the mariner's life was the bedrock of his stories and poems, but they go far beyond a record of experience. In "Hawthorne and His Mosses," Melville, speaking in praise of his friend, also characterized himself: "there is no man in whom humor and love are developed in that high form called genius; no such man can exist without also possessing, as the indispensable complement of these, a great, deep intellect, which drops down into the universe like a plummet" (*Piazza Tales* 242). The natural world of ocean and island, shark and whale, was in several ways his universe, in which struggle its solitaries and occasional noble brothers, its whalemen, ship's captains and officers, lowly sailors and boys, renegades, castaways, and thieves—and in which floats "the ungraspable phantom of life." When Ishmael finds himself frustrated to explain how the whale seems "the gliding great demon of the seas of life," he remarks that to comprehend it "would be to dive deeper than Ishmael can go" (ch. 41). Yes; in Melville's universe full comprehension is neither accessible nor, finally, desirable. But his genius was for diving deeper, always deeper.

HASKELL SPRINGER AND
DOUGLAS ROBILLARD

A Portfolio

AMERICAN SEASCAPE ART

The visual tradition of American seascape from the seventeenth century to the present is immensely rich and varied, and it is widely available in numerous publications (for two useful surveys, see John Wilmerding, *American Marine Painting,* and Roger B. Stein, *Seascape and the American Imagination*). The following portfolio is not a minisurvey of masterpieces; it is intended, rather, to suggest some of the ways in which the visual tradition intersects with the literary works discussed in this volume, and the pervasiveness of the language of the sea not only in oil paintings and sculpture for a limited elite audience but also in graphic and vernacular forms that reached a larger public.

The selection here emphasizes the relevance of visual images to the literature of the sea in the United States in a variety of ways: through similarity of subject matter or theme, including the direct illustration of a literary text or some biographical link to an author; affinities of vision or approach to a particular maritime motif; related aesthetic strategy in giving shape to a vision of the sea, the coast, the voyage. It is important to recognize that the visual artifact is never a passive "reflection" of the literature or the literature of the art or both of some static notion of "American culture." The differences in media are always critical: words in poetry or prose forms, oil paintings, watercolors, engravings, scrimshaw. Each has its possibilities and limitations, its conventions of usage and its history of forms, and these are always significant in assessing how and for whom sonnets, sea chanteys, novels, ship portraits, figureheads, and newspaper cartoons have functioned within American culture. The value of the following portfolio lies as much in recognizing the differences between what literature and visual art can and do accomplish with the sea as subject, as it does in discovering the affinities.

ROGER B. STEIN

1. Thomas Smith (dates unknown), *Self Portrait.* 1670–90. Oil on canvas, 24½ × 23¾ inches. Worcester Art Museum, Worcester, Massachusetts.

Although the identity of the artist, the date, and the specific subject matter of the naval battle scene in the window view have never been fully determined because written documentation on this early work is limited, its centrality to late seventeenth-century modes of understanding is clear. It is an emblematic portrait, in the graphic tradition of picture with accompanying interpretive text that came down from the Renaissance and was familiar to colonial Americans in both its secular and religious typological forms. In this painting, the poem beneath the skull shapes the meaning of the whole:

> Why why should I the world be minding
> therein a World of Evils Finding.
>> Then Farwell World: Farwell thy Jarres
>> thy Joies thy Toies thy Wiles thy Warrs
> Truth Sounds Retreat: I am not sorye,
>> The Eternall Drawes to him my heart
>> By Faith (which can thy Force Subvert)
> To Crowne me (after Grace) with Glory.

The sea battle in the window view is probably linked to a particular political event in the subject-sitter's life, in the maritime conflicts between the Dutch and the Algerians (whose flags are visible on the ship and fortress). But in larger terms, this naval conflict stands for strife in the world; and the poet-painter's hope is defined typologically through the allusions to Psalm 8:5; Hebrews 2:7, 9; and 1 Peter 5:4. They ask us to move from the living self portrayed through death (the skull, a *memento mori*) to achieve "grace" and transcendence of the sea that stands for the worldly life. At the same time, both the strength of the aging subject's visage and the images of the world around him suggest the vividness of the struggle, his attempt—like that of so many writers—to embody and give metaphoric life to the idea of the worldly voyage of the self. In this sense, the Thomas Smith *Self-Portrait* shares with both the prose literary tradition from William Bradford to Cotton Mather, and poetry from Edward Taylor to Richard Steere and beyond, the idea that ultimate spiritual deliverance comes from a deep engagement with the maritime world.

Figure 1

2. John Singleton Copley (1738–1815), *Watson and the Shark.* 1778. Oil on canvas, 71 ¾ × 90 ½ inches. Ferdinand Lammot Belin Fund, © 1993 National Gallery of Art, Washington, D.C.

One of John Singleton Copley's first major presentations to a London audience after a twenty-year career as a skillful portraitist in the colonies, this large-scale exhibition piece crystallizes the issues of its historical moment. Aesthetically, it employs the monumental pictorial language of the Grand Manner of European history painting to depict a heroic struggle in order to define a 1749 incident in the young life of Brook Watson when he was attacked by a shark in Havana Harbor, located with topographical precision along the horizon line. The pyramidal structure of the rescuing boat and the placement and gestural language of the figures borrow from Renaissance and classical pictorial and sculptural prototypes. The foreground visionary contrast between Man and Nature, clear and obscure, are in the new language of the Burkeian Sublime. In this sense, it shares the emerging exploration of different and sometimes conflicting aesthetic languages of the sea with such writers as Philip Freneau, in both "The British Prison Ship" and his West Indian sea lyrics, or the epic pretensions of Joel Barlow's *Columbiad.*

But the attempt to universalize this melodramatic incident from ordinary life has its counterthrust in the contemporaneity of the dress (no classical togas or other facets of "general nature," despite Sir Joshua Reynolds's principles) and in the critical presence of the black man whose rope forms an almost umbilical connection with Watson, a motif that Melville would later exploit so richly in the Ishmael-Queequeg relation. In 1778, however, Watson was a Tory, a wealthy merchant who had been involved in the colonial slave trade. Liberals in Parliament were arguing for abolition of the slave trade, with the Tories pointing out the contradiction between Americans' political rhetoric of freedom from enslavement to British tyranny and their support of black servitude. Critics today are divided as to the ultimate meaning of the black man in this picture. Some have inferred conclusions from Copley's Tory family connections, others emphasize the black's benedictory and saving gesture and his consequently equivocal position (Copley not wishing to jeopardize his search for a broad British audience and patronage). In this political sense too, the legacy of the pictorial image is shared with the American literary tradition as it attempted to work out the contradictions of race in our maritime experience.

Figure 2

3. Thomas Birch (1779–1851), *United States and Macedonian, 25 October 1812.* 1813. Line engraving, hand-colored. 19⅞ × 25¾ inches. Benjamin Tanner, engraver. Published Philadelphia, 1814, after the painting by Birch (1813; Historical Society of Pennsylvania, Philadelphia). Beverly R. Robinson Collection, U.S. Naval Academy Museum, #51.7.481.

Thomas Birch was the leading painter of the many naval engagements of the War of 1812, from the Battle of Lake Erie to the notable "duels" between the *Constitution* and the *Guerriere,* the *Wasp* and the *Frolic,* and the one portrayed in this image. He made numerous replicas in oil, and the images were quickly translated into beautifully crafted engravings by Tanner and others and rough woodcut versions to illustrate articles in magazines and books for a larger public. The reproduction of these images continued after the war, and even British manufacturers produced "transfer" images on porcelain plates and jugs and inexpensive cotton scarves for the American market.

The precise detailing of the Tanner image facilitated the understanding of the narrative by an audience knowledgeable—and hence demanding—in maritime particulars. James Fenimore Cooper, in his two-volume *History of the Navy of the United States of America* (1839), described the moment with the same dramatic rendering of details he brought to his maritime fictions:

> At length the stranger's mizzen-mast came down over his lee quarter, having been shot away about ten feet above the deck. He then fell off, and let his foresail drop, apparently with a wish to close. As the ships got near together the shot of the American vessel did fearful execution, the forecourse being soon in ribands, the fore and main-topmasts over the side, the main-yard cut away in the slings, and the foremast tottering. (vol. 2, ch. 5)

Yet beyond the narrative intensity, images such as this served as nationalistic emblems of American prowess at sea. They functioned as the seventeenth-century Dutch paintings that served Birch as pictorial models had for that bourgeois maritime power. There is no doubt, even without the engraving's detailed caption, as to which flag is triumphant, and the viewer's vicarious participation is ensured by the point of view: we are centered right at the waterline, wedged between the two ships in the grandeur of nature's space. For the audience as for the participants, sea and sky are not passive backdrop but active agents in the unfolding drama, a lesson that was to be explored with a range of skills by the literary romances of the coming years.

United States and Macedonian

Figure 3

4. Edward Augustus Brackett (1818–1908), *Shipwrecked Mother and Child.*
1850–51. Marble. 29½ × 75¼ × 34¼ inches. Worcester Art Museum, Worcester, Massachusetts. Gift of Edward A. Brackett.

The motif of drowned woman appears in literature over the centuries and became a romantic archetype with Bernadin de St. Pierre's immensely popular and frequently illustrated French novel *Paul et Virginie* (1789). Longfellow drew on this motif in "The Wreck of the Hesperus," and Daniel Huntington was among the illustrators of that poem in collected editions of Longfellow from the 1840s on. In 1873, Winslow Homer borrowed from Huntington's image for his *Harper's Weekly* illustration of a recent maritime disaster, *Cast Up by the Sea; The Wreck of the Atlantic.* Harriet Beecher Stowe used the same motif in *The Pearl of Orr's Island* (1862), and Lafcadio Hearn explored its fictional potentialities in *Chita* (1889).

Exhibited at the Boston Athenaeum first in 1852 and yearly thereafter until 1866, the sculpture was praised initially for its realism. It was later placed in Mount Auburn Cemetery, and Edward Augustus Brackett finally gave it to the Worcester Art Museum in 1906. When it was first exhibited, *Shipwrecked Mother and Child* seemed in some senses an implicit if idealized memorial to Margaret Fuller, who had been lost at sea with her child and husband off Long Island on July 19, 1850. The tragedy of Fuller's death, at the height of her power as writer, becomes more complex in the face of her passionate intensity: intellectual, spiritual, and political. In 1850, as liberal supporter of the Italian revolution and as feminist, she was returning to the United States with her new husband and child. The image is powerfully voluptuous, undulating like the sea, and a far cry from the sweet little "sleeping children" images that adorned American cemeteries.

The classical white marble allowed Brackett to display the nude female form in its link to the sea visually in a way that was in general unacceptable for American writers of the nineteenth century, although here the presence of the child limits and controls the woman's erotic power by signifying her role as "mother" to its audience. Hawthorne, sea-fearing son of a ship captain who never returned, dealt with these relationships circumspectly. He seems to have preferred idealizing the woman as carved figurehead in "Drowne's Wooden Image" (1844). Though the sea holds forth the possibility of escape and eros for Hester Prynne, her sensuality is revealed in the forest rather than at the seaside in *The Scarlet Letter,* and one year later Clifford Pyncheon links the sea with his own (male) suicide in *The House of the Seven Gables* (1851). William Cullen Bryant played rather tamely with women, the sea, and freedom in a late fantasy, "Sella." Emily Dickinson was to explore more boldly a range of other possibilities. But the full implications of linking the erotic female nude with the sea and death, and in tension with her role as mother, would await Kate Chopin's symbolic portrayal of this material in the closing scene of *The Awakening* (1899).

Figure 4

5. Anonymous. *Meditation by the Sea*. c. 1850–60. Oil on canvas, 13½ × 19½ inches. Gift of Maxim Karolik. Courtesy, Museum of Fine Arts, Boston. © 1993.

This strange little work, by an unidentified artist, seems unconnected to any other known works. Its rough date is based on the clothing of the little figure in the foreground. The work appeared on the art market only in the 1940s. It was exhibited in the 1943 Museum of Modern Art's early "Romantic Painting in America" exhibition, then at the Tate Gallery in London in its "American Painting" show in 1946, by which time it had been purchased by the Karoliks and donated with a superb, large group of other American paintings (including the Lane image that follows) to the Boston Museum of Fine Arts. In the years since it has become a well-known icon, included in countless exhibitions and books on American art, "folk art," and seascape, an interesting case of the power of an image—in this instance a vernacular image by an artist of limited academic or technical skill and training—to elicit responses by the compelling quality of its vision (the title itself is a modern addition).

Like many vernacular images, this work combines sharp and precise local seeing (the tightly manicured waves and weird cloud forms, the stance and dress of the foreground figure) with a strong sense of abstract design. Its "timeless" quality is a function of its visual clarity and lack of atmosphere as well as the absence of "facture," the evidence of brush strokes. Such qualities have made the work appealing to a modernist generation trained to look for abstract harmonies of line, form, and color. But like most vernacular works, it also draws conceptually on the ideas and expressive systems of its time and place. The exaggerated vanishing point perspective intersects with the sharp line of the horizon, two conventional languages of distance. Within this bare structure the artist balances right and left signs of the natural and the human: the hints of a struggling tree form on the cliff, the line of rocks at the beach's end, and the ships on the horizon (a convention of seascape art), like the balancing cloud forms, parenthesize pictorially the central horizon line that is the essence of sea space.

Although the two tiny figures approaching us along the beach below the cliff edge offer an alternative possibility of shared human activity, the central figure is not doing but being, clasping himself as he contemplates the waves and we contemplate him and the openness of the sea. "Yes," as Ishmael asserts in the opening chapter of *Moby-Dick*—and just before he goes on to describe the romantic landscape of the visual artist—"as every one knows, meditation and water are wedded for ever."

This work reminds us that the alternative to active sea voyaging by artist and writer is the land-based contemplative act, the meditation that transforms maritime perceptions into acts of the mind, that wrestles with the meaning for a perceiving consciousness of the openness, the undifferentiated vastness of sublime space. From William Bradford's peroration at Plymouth, inviting the reader to stand with him "half amased" at having "passed the vast ocean, and a sea of troubles," to Emily Dickinson's subjunctive "As if the Sea should part" (#695) or "Exultation is the going / Of an inland soul to sea" (#76); from Robert Frost's "Once by the Pacific" to the Key West poems of Wallace Stevens—one dominant mode of American maritime literature and art has been this shore-based confrontation with the meaning of the sea.

6. Fitz Hugh Lane (1804–1865), *New York Harbor*. 1850. Oil on canvas, 36 ×
60 inches. Gift of Maxim Karolik for the Karolik Collection of American
Paintings 1815–1865. Courtesy, Museum of Fine Arts, Boston. © 1993.

For the visual as for the literary artist, the harbor view is one of the two poles
of his or her art, to be contrasted to the open sea. Where the latter defines
the site of struggle of humans in ships against the illimitable, the undifferentiated, the sublime, the harbor represents the place from which the voyagers
set out and to which they return, the complex human community of men
and women, the ties that bind (for better and for worse)—which generate the
voyage and receive its rewards. It is "all that's kind to our mortalities," as
Melville put it (*Moby-Dick*, ch. 23), his New Bedford and Nantucket, with
all their provincialities and comicalities; or the childhood memory world of
Longfellow's Portland in "My Lost Youth"; Emma Lazarus's place of refuge
for the "tempest toss'd" immigrants, or the "pillowed bay" of Hart Crane's
"Harbor Dawn" (*The Bridge*).

The aesthetic language of these two imaginative centers is equally distinct:
the sea is the boundless, without measure, and art seeks to bring it under a
degree of control through the action of the ship and its hierarchies of form
and order; the harbor is by contrast the bounded and gridded social space,
centered, within which the artist orders and locates ships and their cargo.
Seascape space is centrifugal; harbor space is centripetal; and literary and
visual artists work with or against those conventions.

Harbor views exist in American art from early maps and charts, graphic
and painted views of ports from Charleston, South Carolina, to New York
and Boston in the early and mid-eighteenth century, to twentieth-century
images of industrialized ports, or recent, meticulously researched recreations
by John Stobart of American ports during "The Great Age of Sail" for nostalgic consumers of historic revivalism. The finest of all such harbor painters
was Fitz Hugh Lane. If Thoreau could boast "I have travelled a good deal
in Concord," Lane could say the same of his hometown, Gloucester, and its
complex coastline, its inner and outer harbors, from which he constructed a
visual universe of great variety and complexity. Yet despite his limited mo-
bility (he was crippled in childhood, probably by polio), he also traveled to
and pictorially recorded the Maine coast, Boston, and in this 1850 image,
New York harbor.

Touted in recent years as a "luminist," Lane was in his own time not
a nationally prominent artist, though he was acclaimed in Boston and his
hometown, respected for his meticulous realism, his precise nautical detail
(his vessels are easily identifiable by knowledgeable viewers). But what has

attracted the modernist "luminist" reading of Lane is not only the almost airless, bell-jar clarity of his atmosphere but also the astonishing abstract geometrical patterning of his pictorial space. His 1850 view of New York Harbor not only juxtaposes clearly identifiable sail and steam vessels of varying kinds (the propeller tug next to the merchant ship in the foreground, the paddlewheeler with the brig in the middle ground), but locates them in delicately balanced pyramidal structures. These organize and clarify the harbor space in both two and three dimensions, through the contrast of heavily gridded horizontals and verticals against which play the varied draped sail shapes and the openness above and between which leads us into deep space of both the town beyond and the sky above. The formal play with nautical details is Lane's pictorial language for human order. Yet especially in 1850, the very quietness of Lane's image, the harmonious linking of sail and steam, seems like the calm at the center of the storm of maritime capitalism and sectional strife.

Figure 6

7. Felix Octavius Carr Darley (1822–1888), *The Death of Scipio*. 1859. Steel engraving by Alfred Jones, 3 ¾ × 3 ½ inches. Title page vignette illustration for James Fenimore Cooper, *The Red Rover* (1828), in the collected edition (Philadelphia, 1859).

One of the ways in which visual artists made the literature of the sea available to a wide public was through illustration of particular literary texts in regular book format, gift book annuals or the gift book editions of poetry and prose (usually issued for the Christmas market), and in the popular illustrated periodicals beginning in the 1850s. These images were usually commissioned by a publisher once the text was completed, sometimes immediately, sometimes many years after, to capture a new or continuing market for the work. The illustrator was to some extent constrained by the text to be illustrated, the audience to be reached, and the editorial control exercised over his or her choices. These constraints varied considerably; freedom to construct images depended on the reputation of the artist and on the particular publisher.

Furthermore, the artist did not usually control the final image. Most often it was translated from the more or less precisely rendered drawings put into engraved or lithographic form by craftsmen within the publisher's or printer's establishment. Once issued, illustrations were frequently reused and reissued—with or without permission—in subsequent and collected editions and many other contexts. Depending on the reputation of the artist, his or her identity as illustrator may have been acknowledged in image, caption, or title page, but frequently it was omitted, especially as illustration came to be seen as a "lower" form of art. However, illustration is never automatic; it always involves a *choice* of which aspects of its subject to treat and how to treat them. As such it is by definition interpretive.

All of this is clear in the work of F. O. C. Darley, perhaps the best-known literary illustrator of the middle decades of the nineteenth century—of Irving and Hawthorne, as well as Cooper. For the 1859–61 Townsend edition of Cooper's collected works, Darley did two illustrations for *The Red Rover*: a frontispiece of the Rover as Byronic pirate in the midst of battle, and the title page image here from the penultimate chapter. As this image shows, Darley was not skilled in nautical details; his primary language was figural. The death scene gathers the central characters at the simplified mainmast: the cannon and the chaplain (war and peace), Scipio in the arms of Dick Fid, the kneeling Wilder whose life he has saved, and, standing above, both the rebel pirates and the Rover. In a book that dramatized the appeal of the sea as the site of romantic rebellion, Darley measures the costs. The image restores social order through its formal figural relationships, by pitting Christian piety and a traditional deposition (from the cross) motif that links black and white and the representative of the legitimate order, Wilder, against the background of the disgruntled rebels. For an 1859 audience, the Christian martyrdom of the black man recalled as well *Uncle Tom's Cabin* (1852), and thus reinterpreted Cooper's drama for an audience on the brink of a quite different civil war.

THE RED ROVER.

"The limb stiffened and fell, though the eyes still continued
their affectionate and glaring gaze on that countenance he had
so long loved, and which, in the midst of all his long endured wrongs,
had never refused to meet his look of love in kindness."

Chap. XXXI Page 522

Figure 7

8. James Hamilton (1819–1878), *The Pack off Sylvia Headland.* 1856. Steel engraving by A. W. Graham, after a sketch by Elisha Kent Kane. 3½ × 7 inches. Illustration for Kane, *Arctic Explorations in the Years 1853, '54, '55* (Philadelphia, 1856), vol. 1, facing p. 180.

The visual exploration of the polar regions by American artists generally followed the literary record, although some images appeared in the published account of the Wilkes South Sea exploring expedition that served as a source for Poe's *Narrative of Arthur Gordon Pym* (1838). The artistic language of the sublime by artists, defined primarily in terms of the American wilderness in earlier years, was extended in the 1850s and thereafter to the poles. In this respect too, James Fenimore Cooper was a pioneer in the mode in his penultimate romance, *The Sea Lions* (1849), in which the skeptical ship captain Roswell Gardiner is converted to trinitarian Christianity aesthetically by his Antarctic vision of "the immensity of the moral space which separates man from his Deity" (ch. 17).

In the 1850s, Frederic Church traveled to the northern regions in search of subjects, after his success with both *Niagara* (1857) and *The Heart of the Andes* (1859), and the trip was recorded by his ministerial companion Louis Legrand Noble in *After Icebergs with a Painter* (1861). The finest of the early representations of Arctic navigation came in the books by Elisha Kent Kane, which record the two Grinnell expeditions in search of Sir John Franklin in 1850–51 and 1853–55. Kane made numerous sketches, which were then turned over to artists, engravers, and lithographers who prepared them for the amply illustrated editions of these popular accounts.

The Irish-American James Hamilton of Philadelphia was the key visual artist in recreating the images for the second two-volume work. Among his earliest graphic works are tiny woodcut images for John Frost's *Pictorial History of the American Navy* (1845). Later he would do large seascapes of *The Capture of the Serapis by John Paul Jones* (1854), *Old Ironsides* and *The Ancient Mariner* (both 1863), and smaller versions of the *Monitor-Merrimac* battle in 1874. Hamilton's style was heavily influenced by J. M. W. Turner's sublime visions, with their immense dramatic vortices above turbulent seas. *The Pack off Sylvia Headlands* is one of many such pictorial reorganizations of Kane's vision, dwarfing both men and ship beneath the sweeping forms of cliff and cloud. The triangle of horizon, ship mast, and darkened shaft of impending storm, rather than imposing human order on the space, as was the pattern in the Grand Manner tradition (look at Copley's *Watson and the Shark*) becomes an ironic statement of the triumph of nature over men in ships.

In this manner the artist does create a sublime pictorial order in such a scene. The Arctic attracted a number of maritime painters such as William Bradford, Albert Bierstadt, and Thomas Hill in the later nineteenth century. Among the literary constructions of sublime Arctic space are the contrasting visions of Sarah Orne Jewett and Jack London. In the Captain Littlepage episode of Jewett's *Country of the Pointed Firs* (1896), the old mariner's recollection of his Arctic journey becomes visionary, surreal, "the waiting place," a sign of a lost past of metaphysical male adventuring in a harbor community defined now by its women. Jack London, by contrast, in *The Call of the Wild* (1903), *White Fang* (1906), and other Arctic tales, played out schematically the Darwinian biological imperatives that had informed *The Sea-Wolf,* in pure animal dramas against the sublime space—a vision anticipated by Winslow Homer's winter coastal images such as *Fox Hunt* (1893) or *Below Zero* of the following year.

Figure 8

9. William Holmes (dates unknown) and John W. Barber (1798–1885), *The Fast-Anchored Ship*. Wood engraving. 2⅞ × 2½ inches. From *Religious Emblems* [1846]; reprinted in *The Bible Looking Glass* (Philadelphia, 1875), pp. 109–110; Special Collections, Alderman Library, University of Virginia.

The facsimile pages illustrated here reproduce one image/text combination in the most popular of American emblem books. These pages were first issued in 1846, but frequently reprinted in the United States and in England, and rebound in combination with other such collections, as well as with reprints of Bunyan and of Thomas Grey's "Elegy in a Country Churchyard," in a variety of bulky editions under various titles throughout the century.

Where Thomas Cole's famous large-scale series of four paintings of *The Voyage of Life* (1839–40; second version 1842) were the most famous emblematic images of their kind and were turned into relatively expensive large engravings, these popular images reached a much wider religious audience, with their specific textual citations and allegorical certainties. The emblematic and typological tradition of understanding the sea had come to New England in the seventeenth century with such books as John Flavel's *Navigation Spiritualized* and was used by literary and visual artists as diverse as Edward Taylor and William Bradford, Cotton Mather and Thomas Smith in his *Self-Portrait*, Thomas Cole and Fenimore Cooper in *The Sea-Lions*. Melville's Father Mapple continues the tradition, as does Longfellow in "The Building of the Ship" and other poems.

Recently George Monteiro and Barton St. Armand have demonstrated the importance of William Holmes and John W. Barber's version to Emily Dickinson, and of *The Fast-Anchored Ship* in particular to her "Wild Nights." Dickinson both uses and transforms this orthodox Christian tradition, playing wittily and ironically with its certainties, as would Stephen Crane after her; but the persistence of these emblem books in the popular tradition through the nineteenth century, perhaps especially for the education of children, attests to the continuity not only of a specifically Christian tradition of maritime understanding but also of an emblematic habit of mind, of a way of knowing through the intersection of the visual and the verbal.

HEBREWS,
Chap. vi:
verse 19.

MARK,
Chap. xi:
verse 22.

1 PETER,
Chap. i:
verse 5.

JAMES,
Chap. v:
verse 15.

ROMANS,
Chap. xii:
verse 12.

LAMENTANS,
Chap. iii:
verse 26.

ACTS,
Chap. xx:
verse 14.

PSALM
cxlv:
verse 19.

THE FAST-ANCHORED SHIP.

Both sure and steadfast. Heb. vi: 19.

Lo! where the war-ship with her tattered sail,
Tho' late escaped the fury of the gale,
An anchor safe within the bay she rides;
Nor heeds the danger of the swelling tides:
Though high aloft the furious storm still roars,
Below, she's sheltered by the winding shores.
The church of Christ a war-ship is below,

She spreads her sails to meet her haughty foe;
Satan assails her with his furious blasts,
Her sails are riven, broken are her mast,
A night of darkness finds her in some bay,
She drops her anchors, and awaits the day;
Faith, Hope, and *Prayer,* her anchors prove,
With *Resignation* to the powers above.

This engraving represents a ship riding by four anchors. To escape the rage of the storm at sea, she has sought shelter in the bay. Her sails are torn, and cordage damaged; she needs to undergo repairs. The gale still howls fearfully overhead; but protected by the land, she rides comparatively in smooth water.

The Church of God may be com-

pared to a ship, and to a ship of war, built by the great Architect who made heaven and earth—first launched when Adam fell overboard—chartered by divine love to take him in, with all his believing posterity, and convey them to the port of glory.

Jehovah is her rightful owner; Immanuel is her captain; the Holy Spirit is her pilot; the Holy Bible is both

chart and compass; self-examination is her log-book; her pole-star is the star of Bethlehem. Under her great Captain, the ministers of religion take rank as officers; besides whom, there are a number of petty officers. Her crew consists of all those who "follow the Captain." Passengers, she carries none—all on board are "working hands."

This world is the tempestuous sea over which she makes her voyages. It is a dangerous sea; rocks, shoals, and quicksands hide their deceitful heads beneath its dark blue waves; mountainous billows roll, furious storms descend, and treacherous whirlpools entice only to destroy.

The voyage is from time to eternity. The good ship never puts back; well stocked, she carries bread of life, and waters of salvation, in abundance; no "southerly wind" ever afflicts her. The Church is a ship of war; she carries a commission authorizing her to "sink, burn, and destroy" whatever belongs to Beelzebub, the great enemy of mankind, and to ship hands in every quarter; therefore Beelzebub, being a "prince of the power of the air," comes out against her, armed with the four winds of heaven, and attacks her as he did the house of Job's eldest son.

Bravely does she behave amid the storm.

were it not that there is treachery on board; some "Achan" compels her to "about ship." She runs into the bay of Promise, and casts first of all the anchor of *Hope.* Though "perplexed," she is "not in despair." Hope is as an anchor to the soul in the day of adversity. Hope, however, is not sufficient; another anchor divides the parting wave, even that of *Faith.* Faith takes hold of the promises made to the Church in her times of trial, especially this one: "Call upon me in the day of trouble, and I will deliver thee." *Prayer,* consequently, "is let go" next. Ah! now she "takes hold on God," now the vessel rightens; now she is steady. Nevertheless, she is not yet delivered. What more can she do? There is yet one more anchor on board: *Resignation,* last of all, is received by the yielding wave. The good ship has done her duty; now she may lie still, and wait for the salvation of God. Soon it comes; heavenly breezes fill her flowing sails; she is again under weigh for the *port of glory.*

"Where all the ship's company meet
　Who sailed with their Savior beneath,
With shouting, each other they greet,
　And triumph o'er trouble and death.
The voyage of life's at an end,
　The mortal affliction is past,
The age that in heaven they spend,
　Forever and ever shall last."

Figure 9

10. Albert Bierstadt (1830–1902), *The Departure of Hiawatha.* c. 1868. Oil on paper, 6⅞ × 8½ inches. National Park Service, Longfellow National Historic Site, Cambridge, Massachusetts.

The visual art of the Great Lakes, relatively unknown in our time before J. Gray Sweeney's important Great Lakes Marine Painting of the Nineteenth Century exhibition in 1983, follows the patterns of American marine painting generally. It includes such types as naval battle scenes of the War of 1812 like Thomas Birch's various versions of the Battle of Lake Erie, ship portraits of the various sail and steam vessels that plied the lakes, or harbor views of Detroit by William James Bennett as early as 1836 or of Duluth by Gilbert Munger in 1871 and Albert Bierstadt about 1886. We also find images of marine disaster and a variety of coastal views, including those by John Douglas Woodward, Alfred Waud, and William Hart of Lakes Michigan and Superior in the popular Appleton's picture-text production, *Picturesque America* (1872, 1874).

A special group of images interpret Henry Wadsworth Longfellow's famous *Song of Hiawatha* (1855), that poetic combination of history, anthropology, and myth in a Great Lakes setting. They range from tiny graphic images in the many illustrated editions of the poem to the oil paintings it inspired: large canvases by Thomas Moran and small studies, such as that by Thomas Eakins. Among the small works is one with a special biographical link: the image that Bierstadt created and presented to Longfellow at a dinner he gave for the poet in London on July 9, 1868. Attended by some eighty guests, including Admiral David Farragut and Parke Godwin of the *New York Post,* and British notables such as poet Robert Browning, painter Sir Edwin Landseer, and Liberal politician William Gladstone, the much-publicized event was intended by Bierstadt to boost his visibility (and his patronage, he hoped) through association with the immensely popular poet.

The image he painted for the occasion is based on canto 22, "Hiawatha's Departure":

> Turned and waved his hand at parting;
> On the clear and luminous water
> Launched his birch canoe for sailing. . . .
> And the evening sun descending
> Set the clouds on fire with redness, . . .
> Westward, westward Hiawatha
> Sailed into the fiery sunset,
> Sailed into the purple vapors,
> Sailed into the dusk of evening.

Bierstadt locates us with the tiny-scale Native American community on the shore, while Hiawatha is in the direct line of the setting sun. The "inland sea" almost loses its horizon as all planes converge on the burning light, which is at once "West" as direction, sunset, and the end of a world, constructed by both Longfellow and his illustrator as an obliteration of self in the transcendent "luminous" glow. In 1868, however, and in 1872 when the passage was again quoted by Constance Fenimore Woolson in her text for the Hart image in *Picturesque America*, railroads, white settlement of the West, and the U.S. Army were pushing Native Americans toward reservations. An image such as this thus participated in the convenient myth of the conflict-free disappearance of the avatar of the "vanishing race."

Figure 10

11. Antonio Jacobsen (1850–1921), *Augusta Victoria.* 1889. Oil on canvas, 22 × 36 inches. Courtesy of the Mariner's Museum, Newport News, Virginia.

Ship portraiture was a distinct genre in visual art, practiced by port painters in cities around the globe: in Marseilles, most notably by the Roux family of artists; in Canton, by Chinese painters for a worldwide trade; in Amsterdam, Bremen, London, or Greenock, Scotland. In the United States, it was practiced in Salem at the beginning of the nineteenth century by Michel Felice Corné, and in Boston by Robert Salmon and Fitz Hugh Lane. In Philadelphia, Thomas Birch was among the earliest, the brothers James and John Bard specialized in Hudson River steamboats, and of course there were numerous ship painters in New York City and Hoboken, New Jersey, across the river, where Antonio Jacobsen lived and worked for over forty years.

Ship portraits were often commissioned by the owners or captains of vessels, sometimes in quick watercolor sketches done while a ship was visiting in port, sometimes in oil in several replicas to be hung in captains' homes or the shipping firm's offices, or in lithographic form in multiple copies by Currier and Ives, Sarony and Major, and other firms. Images fashioned by such artists as James Butterworth cashed in on the fame of record-breaking clipper ships in the 1850s or winners of the America's Cup or other yachting races from the 1850s into our own time. The ship portrait expressed the pride and power of ownership or command. It could be a form of commercial advertising, a way of drumming up trade for mercantile or passenger vessels. Knowledgeable audiences judged the works on their meticulous accuracy to proportions, to hull and rigging, and to the flags that denoted country and company, nationality and ownership. The demand for such visual *vraisemblance* was the same that engaged the nautical descriptive skills of a Fenimore Cooper or a Melville. But in these images, human figures are frequently minuscule, when they are at all visible on deck or in the shrouds. The standard format is a full broadside image, parallel to the picture plane and occupying most of the horizontal space, occasionally with fore and aft views at the left and right edges, or a tiny ship to mark the horizon.

Given these conventional demands, ship portraiture tended to become static and repetitious, and it was the outstanding artist who could hew to the demands of a patron for accuracy and also create a dynamic image, a sense of volumes moving in atmospheric maritime space. Antonio Jacobsen was such an artist. Painting twenty-six hundred vessels between 1873 and 1919, many in multiple versions, he completed between forty and sixty paintings per year in most years. His importance lies in part in the way he makes the transition in his work from sailing vessels to steamers, applying the same conventions to the newer types of oceangoing ships for the merchant fleets of the many nations whose vessels came through the port of New York, where Jacobsen took precise sketches in his notebooks for translation into large oil canvases. The *Augusta Victoria* was one such, a 7,661-ton screw steamer whose portrait Jacobsen took in her first year for the Hamburg-American line. She carries auxiliary sails, but the belching smoke from her three stacks makes clear the source of power that impels the simple, rigid, horizontal form of her hull through the waves.

The writer had more difficulty making the transition from sail to steam, as Edward Sloan and others have pointed out. Travelers resented the noise and smoke, while Melville lamented an associated ugliness in "A Utilitarian View of the Monitor's Fight," and, with Felix Riesenberg, Joseph Conrad, and others, decried the loss of romance, as they tried to reconcile their aesthetic aims with the new technology of the sea. The full horror of steam for the men who worked these machines—by contrast to passengers lolling on deck or port painters constructing water-level distant visions—would be realized by Eugene O'Neill in his grim expressionist fantasy "*The Hairy Ape*" (1922).

Figure 11

12. Albert Pinkham Ryder (1847–1917), *Jonah.* c. 1885. Oil on canvas, mounted on fiberboard. 27 ¼ × 34 ⅜ inches. National Museum of American Art, Smithsonian Institution, Washington, D.C. Gift of John Gellatly.

One of the greatest and most unusual of American seascape paintings, Albert Pinkham Ryder's *Jonah* occupies a special place in the history of American art and culture. Thematically the Old Testament story was early understood as a biblical type of death and resurrection, and figures as such in colonial sea narratives and in sermons, including both the seventeenth-century British John Ryther's *Seaman's Preacher; Consisting of Nine Short and Plain Discourses on Jonah's Voyage. Addressed to Mariners,* which went through several American editions, and those of the nineteenth-century Father Taylor, the Sailor Preacher in Boston's Seaman's Bethel, the apparent model for Melville's Father Mapple in *Moby-Dick.* As these uses indicate, its appeal lay in the direct testing of individual faith in the face of the maritime sublime beyond the reach of the organized church.

Ryder's painting, begun in 1885, exhibited in 1890 but reworked (as was Ryder's frequent habit) before 1895, was as renowned in its own time as in ours. It recasts the romantic heritage of J. M. W. Turner's *Slave Ship* (c. 1840) and Eugène Delacroix's *Christ on Lake Gennesaret* (1853), both available and well known in America in the late nineteenth century, for a later age. His technique is the opposite of the meticulous linear clarity of Copley's *Watson and the Shark,* with which it has been frequently compared. The encrusted forms dissolved in light that glows from within through glazes built up over time, the awestruck Jonah, rubbery boat, and curving fish form through which we move toward the extraordinary anthropomorphic Deity at the upper edge are visionary, archetypal, rather than literal renderings.

Though linked to romantic predecessors and an inspiration to twentieth-century abstractionists (Ryder's paintings had a special place in the 1913 New York Armory show of modernist art), the painting is of its own moment: a rejection of both realist techniques and imperialist expansionism, a fin-de-siècle anti-Darwinian search for a spiritual and aesthetic center that can be found in many of the vague poetic seascapes of the period. Ishmael's description of the "boggy, soggy, squitchy picture" of "chaos bewitched" in "The Spouter-Inn" chapter of *Moby-Dick* seems to prevision Ryder's work both in technique and theme; Emily Dickinson's visionary poems parallel in their own way his imaginary voyages. But the popularity of the eccentric Ryder's paintings to late-nineteenth and turn-of-the-century patrons and audiences with wholly different agendas argues for a more complex understanding of the seascape tradition in American art and literature during these fin-de-siècle years, one in which an inward-turning antimodernism intersects with the outward-turning responses that we have defined as a shift from realism to naturalism.

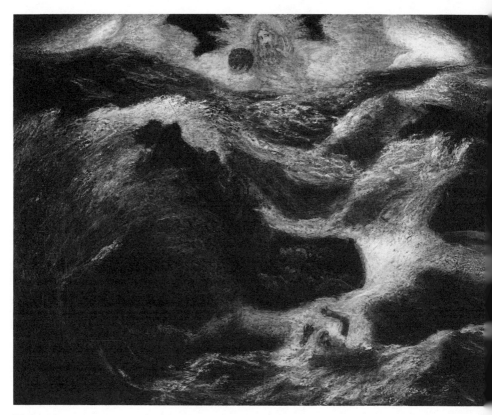

Figure 12

13. Winslow Homer (1836–1910), *Northeaster.* 1895. Oil on canvas, 34⅜ ×
50¼ inches. The Metropolitan Museum of Art, New York. Gift of George A.
Hearn, 1910 (10.64.5).

The sea, the coast, and the voyage were central themes throughout Winslow
Homer's artistic career, and his work intersects with parallel developments
in literary seascape. His early graphic work for *Harper's Weekly* and other
popular periodicals (1850s–1870s) recorded vacation life at coastal resorts
such as Newport, Rhode Island, and Long Branch, New Jersey, interpret-
ing the accompanying verbal texts. His frequently exhibited oil paintings of
the 1860s–1870s explored further these and other places, including Long
Island and northern New England locations that Howells, Thaxter, and other
fiction and travel writers were using as settings, in a pictorial style similar
to that of Eugene Boudin and the French Impressionists. In the 1870s, he
began an extraordinary series of light-filled watercolors of Gloucester and of
Florida and the West Indies that extended into the first decade of the twentieth century, eagerly bought by individuals and museums that relished both
their skillful technique and their optimistic, leisured version of a seemingly
untroubled world.

But especially in the 1880s, after a crucial trip to the tempestuous North
Sea coastal fishing communities of Britain, the sense of tension and conflict among men, women, and the forces of nature became a dominant strain
in his work. His late oils, mostly done at his home at Prout's Neck on the
Maine coast, are testaments to this darker vision, and *Northeaster* is characteristic. He reduced the elements to a minimum, eliminated the narrative,
and frequently all traces of human presence, boldly simplified and frequently
decentered the structure, and then employed the rich resources of paint to
capture the texture and feel of rock, sea, and sky, while insisting through the
facture that what we see is not some transparent "reflection" of "reality" but a
constructed pictorial representation. The godless universe of struggling Darwinian "forces" for which naturalist novelists such as Jack London struggled
to find adequate literary forms through abstract assertions and Übermen-
schen's narrative gestures Homer translated into symbolic, understated, and
frequently ironic pictorial constructions that were quite literally "elemental." A close literary parallel, in both theme and technique, is to be found in
Stephen Crane's poetry and "The Open Boat."

Northeaster became a touchstone for the next generation of artists: both the
Ashcan realists and more traditional illustrators of sea narratives like N. C.
Wyeth and those who were exploring the implications of modernist abstrac-

tion. In the process, our most subtle painter of the complex relations between human beings and the sea was reconstructed aesthetically to justify a "masculine" and literal notion of realism that was proclaimed to be the essence of a "truly American" art.

figure 13

14. Frederic MacMonnies (1863–1937), *The Barge of State; or, The Triumph of Columbia.* Colossal scale; figures over 12 feet tall. Staff (plaster and straw). The Lagoon, World's Columbian Exposition, Chicago. 1893. Illustration from *The Dream City: A Portfolio of Photographic Views of the World's Columbian Exposition* (St. Louis, 1893).

The World's Columbian Exposition took place in the heartland of America, but symbolically it proclaimed the U.S.'s imperial destiny as an echo of earlier empires, a destiny defined by its sea power. Its site on Lake Michigan was sculpted into lagoons plied by gondolas (one recalls Ruskin, and after him Howells's and James's earlier laments for the death of beautiful Venice as a seapower). The lagoons were framed by classical architecture, with the monumental sculptures of Liberty by Daniel Chester French at one end and Frederick MacMonnies's colossal nautical confection at the other.

At the prow of this pseudo-Spanish barge stands a winged Fame, at the helm a similarly winged Father Time. The propulsion is provided by eight beautiful maidens in various states of dishabille, allegorizing on one side the Arts (Music, Architecture, Sculpture, Painting), and balanced on the other side by Industries (Agriculture, Science, Industry, Commerce). A triumphant Columbia bearing a torch is perched precariously on her throne above, in defiance of gravity and all principles of maritime engineering. Their temporary material imitating white marble, which glowed especially when lit at night, MacMonnies's nautical women recall midcentury idealizations such as Brackett's *Shipwrecked Mother and Child* or the many female figureheads at the prows of commercial and naval vessels. Yet clearly here women have been turned into specific representations of American national purpose, to idealize and legitimize collectively in abstract terms the new global aspirations of American (male) expansionism.

Frederick Jackson Turner announced at Chicago in 1893 the closing of the American land frontier, but Alfred T. Mahan was articulating at just this point a new version of maritime nationalism. Thus from the Columbian discovery the United States had moved to a future that was to be acted out within the decade in the Panama Canal, Cuba, the Philippines, and Teddy Roosevelt's sending the American fleet on a worldwide tour to flex our naval muscles: a major change in scale and type from the maritime nationalism of the first half of the century. Political or social Darwinist premises such as Mahan's would be developed in some naturalist writings by, for example, Morgan Robertson, Felix Riesenberg, and Eugene O'Neill.

Figure 14

15. John Marin (1870–1953), *Maine Islands*. 1922. Watercolor on paper. 16⅞ × 19¾ inches. The Phillips Collection, Washington, D.C.

John Marin was one of a number of American avant-garde artists whose work was supported and marketed by the photographer and artistic entrepreneur Alfred Stieglitz in the 1910s and after in his New York City galleries. The Stieglitz circle combined experimentation with the formalist aesthetic techniques of European postimpressionism and cubism and a strong commitment to promoting American art and artists. Marin's art exemplifies these twin tendencies in both his celebrations of New York City (including a series of images of both the Woolworth Building and the Brooklyn Bridge) and in his work on the Maine coast north of Portland, where he spent nearly forty summers.

Marin painted in oil, but his best-known work is in watercolor, which links him to another Maine artist and the greatest watercolorist of the previous generation, Winslow Homer. Marin's coastal settings, frequently involving views of islands in "the country of the pointed firs," has led critics to relate his vision to that of Sarah Orne Jewett, although she was a native of the region rather than one of the many artistic and literary summer visitors to the area from the mid-nineteenth century up to our own day.

Like Jewett and other literary artists, Marin focused on place as a conscious aesthetic construction. The frame is not a transparent window into the world "out there" but an active agent, which divides, orders, blurs and clarifies, links and separates through planes and washes that call attention to the act of seeing and making. But where Jewett's controlling consciousness is the sophisticated urban narrator seeking to make human connections with the coastal community of men and women in a once prosperous maritime economy, Marin focuses our attention not on historical time but on the momentary effects of atmosphere, of light and color, on a mostly unpeopled world. The drama of a Marin watercolor lies in our movement in space from shore to sea and sky, the crystalline beauty of the image eternally fixed by the artist's pen and brush.

figure 15

16. Marsden Hartley (1877–1943), *Eight Bells' Folly: Memorial to Hart Crane.* 1933. Oil on canvas, 30⅝ × 39⅜ inches. Collection of Frederick R. Weisman Art Museum, University of Minnesota, Minneapolis. Gift of Ione and Hudson Walker.

Although his earliest and last years were spent in his home state, Maine, Marsden Hartley, like Marin, was of the first generation of modernist artists and writers who traveled between New York and France (and in Hartley's case Germany as well). He spent time in the studios and salons of Alfred Stieglitz, Gertrude Stein, and Mabel Dodge, and knew Matisse, Picasso, and Kandinsky, and the work of T. S. Eliot, William Carlos Williams, and Hart Crane, among others; he observed them working out theories of abstraction, of imagism and primitivism, as well as new approaches to the question of what constituted specifically American art and literature. All of these shaped Hartley's artistic career as a painter and as a frequently published poet and essayist, as did his early reading of Emerson, Thoreau, and Whitman—also important to Marin—and writers of the mystical tradition.

His attachment to the sea and the New England coast was deep and strong, fed both by the tragic losses of male friends and companions and by the powerful influence on his pictorial vision of the work by both Winslow Homer and Albert Pinkham Ryder (whom he met in Ryder's last years). His last paintings of the Maine coast are very close in feeling to Homer's representational version of the power of the sea in *Northeaster. Eight Bells' Folly* explores the more abstract symbolism and mystical strain of Ryder.

At the beginning of his Guggenheim Fellowship year in Mexico City and the end of Crane's, Hartley befriended this deeply troubled fellow artist and homosexual, whose suicide one month later from his homewardbound ship he was helpless to prevent. His painting of the following year, with its mystic stars and numbers—8's (signifying also eight bells, the noon time when Crane jumped) and 33 on the ship's sails (Crane's age at death)—also draws on Ryder's motifs of moonlit ships framed by abstracted cloud forms. The terror of the shark from the deep seems like an elision of Copley's *Watson,* Ryder's *Jonah,* and perhaps Melville's "Maldive Shark" (we recall Crane's own threnody, "At Melville's Tomb"). Although this biographical memorial is not a literal illustration of or a gloss on a particular poem of Hart Crane's, the two artists shared modernist searches for an adequate symbolic language and for artistic structures that could hold in tension both their mystical, transcendental yearnings and their tragic versions of the sea as place. Both artists struggled between epic and lyric impulses, between the search for the overarching Whitmanic meaning of the word out of the sea and the recognition of Williams's "no ideas but in things."

gure 16

17. Jackson Pollock (1912–1956), *Full Fathom Five.* 1947. Oil on canvas with nails, tacks, buttons, key, coins, cigarettes, matches, etc. 50⅞ × 30⅛ inches. The Museum of Modern Art, New York. © 1994. Gift of Peggy Guggenheim.

With the Jackson Pollock image, we arrive at full-scale abstraction only hinted at in the work of Marin and explored symbolically in Hartley's *Eight Bells' Folly.* The representational and referential are almost totally denied, except in the title allusion to the song from Shakespeare's *Tempest;* and yet the poem offers a fundamental clue to Pollock's vision. As a young painter in the 1930s, he had done some small expressionist seascapes: simple, bold, and thick with paint. "The only American master who interests me is Ryder," he declared in 1944, a statement that links him not only to Ryder's technique of building up layers of paint to form a complex and glowing web but also to Ryder's visionary approach, his search for a dream-induced, deeper reality than can be found in the quotidian representational world.

Pollock's *Full Fathom Five* asks how the seascape vision can be reinterpreted in the modernist era, given the fundamental ruptures between image and referent effected by formalist abstraction. His "abstract expressionist" response was to probe inward, to seek the psychological meaning of his own experience—keyed to the *Tempest* son's search for the father—and to translate that back outward into an image through active engagement with the pictorial means: the richness of paint in which is imbedded objects of daily life. Pollock travels beyond the sublime vortices of J. M. W. Turner (and James Hamilton after him), beyond the modernist questioning of T. S. Eliot's "Dry Salvages" or Wallace Stevens's "maker's rage to order words of the sea" in "The Idea of Order at Key West." His art here moves back to the primeval chaos of the first day before the Creator drew that horizon line separating the land from the waters and sky (in the Judaeo-Christian version of the creation myth), where—with Whitman—death and rebirth in the sea are one.

Figure 17

Chapter Eight

REALISM AND BEYOND

American literature of the sea during the last third of the nineteenth century looks remarkably different from that of the golden age of sail of Cooper, Dana, and Melville, of Longfellow and Whitman. If the metaphysical quests for sublime truth in the sea and the celebration of maritime competence and national destiny are not wholly missing from the sea literature of 1860 and beyond, surely the terms were transformed, and a new set of strategies emerged for shaping writers' visions of the sea, the coast, and the voyage.

Of course the emergence of steamships and commercial and industrial expansion had preceded the Civil War. Atlantic crossings under continuous steam power were available as early as 1838 and were a part of life in the 1840s and 1850s. But the onset of the war transformed shipping and shipbuilding, rapidly refocusing technological and organizational skills and investments to the production of naval and commercial steam vessels. The naval war itself was an affair of blockades and blockade runners on the high seas and up the Mississippi. The Southerners were not, as Alfred Mahan was later to remark, a seafaring people; they lacked a navy, and their population was too sparse to defend their extensive coastline.

The initial impact of the war comes to us, metaphorically filtered, as a storm at sea that beat young men about in the surf in *The Education of Henry Adams* (1907, 1918). Yet the war as young Adams experienced it in London was an intricate series of diplomatic maneuvers on the part of his father, Charles Francis Adams, to keep the British government from recognizing the Confederacy and, in the "Battle of the Rams" chapter, from supplying the Confederacy with metalclad gunboats, steam vessels, and bold raiders. But if the duel between the notorious *Alabama* and the Union *Kearsage* in 1864 off the coast of Cherbourg seems an echo of the grand naval duels of the War of 1812 between the *Constitution* and the *Guerriére* or the *United States* and the *Macedonian*, apotheosized in words and pictures in the earlier period, it was the exception rather than the rule in both the literary and visual history of the war. Melville's *Battle-Pieces* (1866) did celebrate some of the naval encounters, and his "Temeraire" is a long lament for the passing of the older grandeur of naval sailing ships; but Melville also recognized the transformation that the Civil War had effected in his "Utilitarian View of the Monitor's Fight," surely one of the ugliest poems in the English language. In its parodic rhetoric, clanking rhythms, and forced rhymes it ironically dramatized the loss of the older, heroic vision.

Of all the writers of the sea emerging at just this moment, the key figure and by far the greatest was Emily Dickinson. Her poetry stands at a critical juncture in the history of American literature of the sea. Behind her lay the grand romantic quests for the sublime, the male voyages into space in search of ultimate goals and commercial conquests, shaped by voices that declared "Call me Ishmael" or "My name is Arthur Gordon Pym" or even the more nostalgic reminiscences of Longfellow's "A boy's will is the wind's will," all of them songs and celebrations of selfhood. Dickinson was sounding a distinctly different note. Her life was circumscribed by inland Amherst, Massachusetts, by her failure to experience conversion at Mt. Holyoke Female Seminary during a period of religious revival, and by her situation as a woman; and yet out of skepticism, irony, and the woman's voice, this "inland soul" fashioned some of the most profound sea poetry in our entire literature.

Dickinson's art makes clear that a powerful literature of the sea does not depend on direct experience on the high seas, but on consciousness; and the sea offered her a rich and varied language: the lee shore, the storm at sea, the spar, the leaky boat, and shipwreck—different nautical situations give shape to different challenges. Her sea language is not, however, merely a series of facile tropes to decorate an argument; it elucidates her awareness of her culture's terms for defining a range of spiritual and secular concerns. Thus she writes to her friend Abiah Root from Mt. Holyoke in March 1846: "I feel that I am sailing upon the brink of an awful precipice, from which I cannot escape & over which I fear my tiny boat will soon glide if I do not receive help from above"; and again in late 1850: "The shore is safer, Abiah, but I love to buffet the sea—I can count the bitter wrecks here in these pleasant waters, and hear the murmuring winds, but oh, I love the danger! . . . Christ Jesus will love you more. I'm afraid he don't love me *any*!" Dickinson had available the conventions of both high art and popular literature of the sea: the language of the sublime, the typological and emblematic devices for linking ordinary experience to ultimate meanings (and frequently in specifically

Christian terms), and a compressed poetic form based initially on common hymn rhythms as well as the models offered by metaphysical poets of the seventeenth century.

Like Thoreau, she made of her local space an extended nautical metaphor for world traveling; like Poe, she dramatized the terrors of the abyss and the maelstrom and the white unknowableness of an alien universe; and if with Emerson she explored at times the correspondences that consciousness finds between self and soul, between the natural world and the supernatural, it was the struggling, skeptical side of Emerson that moved her, in an ironic kinship with Melville.

And yet, ultimately, her sea poetry looks different from all of these. With her, we cross a boundary into the subjunctive, the conditions contrary to fact. Her confident assertion that "I never saw a Moor—/ I never saw the Sea—/ Yet know I how the Heather looks / And what a Billow be" (#1052) insists—even grammatically—that sea knowledge is not finally dependent on a past or present indicative experience. In a later poem she asserts: "If my Bark sink / 'Tis to another sea—/ Mortality's Ground Floor / Is Immortality—" (#1234). The opening two lines are borrowed from optimistic Ellery Channing's declarative "A Poet's Hope" (1843), already turned problematic when Emerson used them as the poetic conclusion of his essay on "Montaigne; or, The Skeptic." The final two lines are hers. They attempt to spatialize and localize, in "houses founded on the sea," as Emerson put it in his "Montaigne," to map existentially the journey toward that ultimate destination which she can name but not necessarily reach: "Exultation is the going / Of an inland soul to sea, / Past the houses—past the headlands—/ Into deep eternity—" (#76).

Dickinson's sea poetry defines the transformation that would occur in the last third of the nineteenth century. Spiritually and metaphysically, it is a poetry of doubt and skepticism, not of affirmation. For Whitman, crossing Brooklyn Ferry and the laying of the Atlantic cable become means for experiencing union with one's fellow passengers, with the city and the nation, even with an imperialist dream of commercial destiny; and his injunction to "farther farther sail!" leads to spiritual fulfillment in a merely rhetorical question: "O daring joy, but safe! are they not all the seas of God?" ("Passage to India"). For Dickinson, by contrast, the answer is distinctly problematic; the romantic mythic voyage is experienced most frequently as loss, not attainment. Even the individual victories are known not as lived historical experience but as facts of consciousness, with which, through a series of ironic strategies, an extraordinary capacity to balance opposites, to create dissonances between form and content that embody the failed affirmation, the poet exercised her own sort of control over doubt and skepticism.

Finally one must recognize that the strategies of consciousness in Dickinson's poems are frequently gendered ones, and the persona of the speaker that of an innocent girl adrift (#30), or a proper young lady who "started Early—Took my Dog— / And visited the Sea," which is figured as a potentially threatening male (#520). In her most erotically charged poem, "Wild Nights—Wild Nights!" (#249), the speaker is herself the bold lover, passionately engaged, and the sea identified with female sexuality. "Wild Nights" has also been read as a religious allegory, and the speaker's desire to "moor— Tonight— / In Thee" as an act of belief, in the language of the popular emblem tradition still very much alive in Dickinson's youth. But here as elsewhere Dickinson at once uses and transforms the tradition. She challenges the easy dualism of the orthodox believer who navigates the world but always with "wean'd affections," recognizing that the spiritual port will relieve him or her of the burden of the perilous voyage. She also stands against the alternative symbolic identification of the sea with the power of the divine and of male seafaring. That is, she challenges the transformation of the erotic, life-generating energies of the sea into a metaphysical force whose "right worship is Defiance" by an Ahab or other male power-seekers, or its redefinition in terms of a Darwinian natural process that late-nineteenth-century male culture then turned into imperialism. Dickinson's female voice strikes a decidedly different note, whose implications will be felt in the work of other women writers in the coming decades.

No other poet emerging at the opening of the Civil War made the sea such a central subject and language in some twenty or more years of writing. Of course Whitman's and Whittier's sea verse extended to 1892, the year of their deaths, and Melville's later years involved not only the writing of *Billy Budd, Sailor* but also his verse-pilgrimage *Clarel* and his prose-poetry collection *John Marr and Other Sailors* (1888). Other writers created individual sea poems of power and intensity in the years after 1860: Sidney Lanier's "The Marshes of Glynn" (1878), a southern coastal meditation, reminiscent of Whitman, George Santayana's briefer New England version in "Cape Cod" (1894), or the bleaker "Song of the Wave" (1897) of George Cabot Lodge. The sea shaped the popular verse of J. T. Trowbridge's "At Sea" (1869), with its Christian affirmation appealing to the magazine audiences for whom he wrote, including *Our Young Folks*, which he edited. One thinks also of Eugene Fields's "Wynken, Blynken and Nod" (1892), written for an even younger set, or Richard Hovey's popular "Sea Gypsy," who wishes to cast off on a

topsail schooner "for the Islands of Desire." In political terms, Emma Lazarus's "The New Colossus" (1883) celebrated the Statue of Liberty, which
welcomed "the wretched refuse of your teeming shore / . . . the homeless, tempest-tost" immigrants to New York Harbor. At the other extreme
is Joaquin Miller's 1892 "Columbus," whose injunction to "Sail on! Sail on!
and on!" seems, retrospectively, to be directed as much toward American
maritime imperialism of these years as to the Chicago Columbian Exposition
that was its occasion, as it was the occasion for many other poems.

The greatest American literary achievements of the period, however, were
in prose rather than poetry. Young William Dean Howells had been rewarded for his campaign biography of Lincoln with the consulship at Venice,
and the book that resulted, published in 1866 as *Venetian Life*, suggests the
national transformation perceived by Henry James, who, in his 1879 book
on Hawthorne, said that the Civil War had "introduced into the national
consciousness a certain sense . . . of the world being a more complicated
place than it had hitherto seemed, the future more treacherous, success more
difficult." Howells sorted his way through the romantic literary and visual
heritage of Venice, the great "Queen of the Seas," including Byron's *Marino
Faliero* (1821), Fenimore Cooper's *Bravo* (1831), and Ruskin's *Stones of Venice*
(1851–53). Defining himself as "newly from a land where everything, morally
and materially, was in good repair," he "rioted sentimentally on the picturesque ruin, the pleasant discomfort and hopelessness of everything about
[him] here" (ch. 2). This self-indulgence could not be sustained, however,
for Venice was a city under siege, the captive of the Austro-Hungarian Empire, and the evidence of that political oppression was everywhere to be seen.
Venetian Life thus became Howells's initial recognition that the terms of life
and art had changed in the 1860s, that one must reject as a delusion romantic,
sentimental values and the art that sustained them.

The novel, the realist fiction and the realist credo that justified it, thus became the task of Howells, his friends Mark Twain and Henry James, and a
host of others in the years that followed. Unlike the romance, which had been
a primary vehicle for American sea fiction, the novel turned primarily to the
urban world of men's work and moral choice, of the complexities of class and
caste, and the domestic relationships of marriage and family lived within the
houses that gave them social form.

This is the world that Howells brought so vividly to life in *The Rise of
Silas Lapham* (1885); and because Howells was such a precise imaginative
historian of the present, his novel illuminates the situation of American sea
literature during the period. The Brahmin Cory family live on Beacon Hill,
in a home built by the Cory grandfather's success as a merchant in the East

India trade. The nouveau riche Laphams are building on the water side of Beacon Street in the new Back Bay, and the second floor girls' bedroom has a lookout window from which a few small boats and a larger schooner with sails furled are visible—Howells's image of the change in Boston's commercial function from sea trade to the marketing of Lapham's paint, and from active seafaring to passive views for marriageable young women.

The seaside is the place to which one escapes in the summer. In *The Rise of Silas Lapham*, the Corys' declining fortunes have forced them to sell their home at Nahant on the near North Shore above Boston to outsiders, and the Cory women go off to Bar Harbor, Maine, on Mt. Desert Island. The Lapham family, by contrast, stay at a summer hotel and then rent a cottage at Nantasket on the South Shore, to which Silas daily commutes from his office by the frequent little steamers that service such summer visitors. Howells deftly captures the new versions of post–Civil War seascape. The seaside is the place for summer vacationing, where wealthy women and children set up temporary residence and businessmen visit, and these seaside spaces become social symbols.

Patrician Boston was in fact moving to the fashionable North Shore in the post–Civil War years, facilitated by the railroads that made daily commuting possible. New Yorkers and Philadelphians traveled to Long Island and to the New Jersey shore, the locus of some of Rebecca Harding Davis's sea fictions. Places that in the 1830s–1860s had been the scenes of sublime seascape views by romantic artists became in the postwar decades the residences of literary and visual artists and the fictional and poetic settings of some of the finest American sea literature.

But the business of the novel in these decades lay elsewhere, for the most part. Tom Cory may woo the Lapham girls a bit in walks by the sea, but in the central issues of social and economic rise and fall, with their concomitant moral and emotional implications, the male self is tested not by lone confrontations with the sublime sea nor by the hierarchy and brotherhood of the ship, but in the competitive marketplace of business and its uptown counterpart, the domestic marriage marketplace. It is no wonder then that the sea plays a subsidiary role in the drama of realist fiction, or that its older role becomes sometimes comic. At the same time that Howells was working on *Lapham*, he also did the versified text for an opera bouffe in the manner of Gilbert and Sullivan, called *A Sea Change; or, Love's Stowaway*, which takes place on a transatlantic steamship. Howells knows the conventions of the sea romance and can burlesque them as he can the voyage into nothingness, but here it is pure froth. The sea facilitates the marriage game of wealthy Bostonians on vacation.

Henry James was already a frequent transatlantic passenger as a child. As a young man, he spent time at Newport, Rhode Island, a major commercial sailing port in the eighteenth century, but by the 1850s a middle-class spa. His third published story, "A Landscape Painter" (1866), is set on the New England coast with the artist taking views—conventionally described—of the "pretty little coast." James was not notably a writer of sea fiction, though his very first published story, "A Tragedy of Error" (1864), is set in a French seaport town and the protagonist Mme. Bernier's despair at her husband's imminent return—that he will learn of her "misconduct" with another man—leads to an exchange between the lovers that is phrased in terms of shipwrecks, floating spars, and battling with the waves and drowning. The ironic twist of the plot signaled in the title derives from the mistaken assassination of her lover, rather than her husband, in the water between ship and shore. In 1888, James published his rather dark version of the steamship romance story, "The Patagonia," which moves from the insufferable heat of Boston, a "desert" (and a young man who has just returned from Mt. Desert Island), to an ocean journey in which his careless seduction of a young engaged woman leads to her drowning-suicide, as much through the threat of exposure by the vicious gossip of the passengers, including the intrusive narrator, as through any high ideals of love.

But as James advanced from his early, gothic, plot-defined pieces to the challenges of the social scene, the locus of his tales moves primarily to urban settings, sometimes set off by pastoral scenes, and the sea functions as a rich metaphoric language rather than as spatial setting. There are, of course, exceptions to this pattern. The last third of *What Maisie Knew* (1897) takes place in Folkstone and Boulogne on the English Channel, but the "freedom" that the parents and stepparents achieve is *not* a function of the voyages proposed and carried out; it is freedom to pursue their selfish desires, meaning ultimately, to abandon their child. The final note and the bitterly ironic freedom that consciousness achieves is Maisie's "Oh I know," as she stands on deck, "in mid-channel, surrounded by the quiet sea," but her knowledge is of the shore world. As with Huck Finn aboard his Mississippi raft, the voyage offers only a momentary respite from the recognition that "human beings *can* be awful cruel to one another." For "The Aspern Papers" (1888), James transposed his factual sources from Florence to Venice as the incarnation of "the romantic"; but the city's evocative powers are deeply and ironically undercut by literary avarice and emotional sterility. By contrast, in *The Wings of the Dove* (1902), Venice is both a key setting and the place in which a fierce sea storm rages in symbolic evocation of human turmoil. The language of the sea offers James in this novel a central metaphorical structure (one of many) for his understanding of his heroine.

Predominantly, though, James drew on the sea again and again to define a state of mind or consciousness. The issue was clear to him as early as 1881 in *Portrait of a Lady:* that Isabel Archer, the young woman "affronting her destiny," must learn that that destiny exists not in sublime space, in flying over life as observer, or even in sailing her metaphorical ship, but within society. For all its bitter ironies, life is figured as social enclosure. The language of the sea comes in her final scene at Gardencourt, when Caspar Goodwood tries to persuade her that "the world's all before us—and the world's very big." She tries to resist, saying, "the world's very small," but, as James goes on,

> it was not what she meant. The world, in truth, had never seemed so large; it seemed to open out, all round her, to take the form of a mighty sea, where she floated in fathomless waters. She had wanted help, and here was help; it had come in a rushing torrent. (ch. 55)

But this language of the sea is a trap, a momentary sentimental relapse into romantic consciousness—unbounded, sublime, erotic, subjective, and "metaphysical." James's droll narrative commentary immediately intrudes to collapse the fantasy, which, in its subjectivity, contradicts all that she and we, her readers, have learned about the nature of "realist" life; it must be finally and firmly rejected.

Mark Twain offers an even more extreme version of the metaphysical seascape as late-nineteenth-century trap. His early sea voyages in *Innocents Abroad* (1869) or the Sandwich Islands sections of *Roughing It* (1872) are the means of conveying us from one social environment to another. They are not usually intrinsically meaningful, nor do they offer to the tale's controlling consciousness clues to self-definition. Twain's description of a storm in chapter 7 of *Innocents Abroad* is conventional, full of verbal clichés about "tempestuous and relentless seas," the "blackness of darkness," ships climbing toward heaven, and the "shriek of the winds," held in check only by his extraordinary sense of rhythmical control, his ability to shape traditional material into something emotionally convincing by the power it has on our aural sense. Fine as the passage is, it does not shape the succeeding book in any significant way.

Twain's most productive territory was not the sea but the river; his great consciousness is Huckleberry Finn, who, for all that he finds a kind of primitive joy on the river, has as his main task—one must account it his great failure—to find some accommodation with the world of the shore. This is not the place to explore those strategies of consciousness, but only to point out that when they failed Twain, as they so frequently did in his later years, crippled with personal tragedy and financial disaster, his despair expressed itself in a series of fragmentary seascape fantasies recently collected as *Which Was the Dream?* "The Enchanted Sea Wilderness" suggests in its very title its Coleridgean origins: the dream, the sea-as-land, the old Sargasso Sea/ Atlantis myth revivified. "An Adventure in Remote Seas" is a pirate story of mutiny, and "The Great Dark" an extraordinary, Poesque fantasy of the self become microscopic and entering a world that resembles *The Narrative of Arthur Gordon Pym* more than anything else. Twain's final statement and his ultimate seascape is the *Mysterious Stranger* manuscripts, which end where the romantic consciousness had always feared we might end—in solipsism: "alone in shoreless space, to wander its limitless solitudes without friend or comrade forever" as a pure but "homeless Thought, wandering forlorn among the empty eternities" (ch. 34 in the "No. 44" version of the ms.).

The language and form of metaphysical seascape, which Dickinson had explored skeptically and ironically, existed for these late-nineteenth-century male writers of fiction primarily as a sign of despair, a reversion to romantic modes of consciousness that signaled the failure to define and resolve the social dilemma that realist fiction had staked out as its appropriate territory. And, indeed, one can find a similar pattern of imagery clustering around Lily Bart in Edith Wharton's *The House of Mirth* (1905), frequently used as an emblem of her social—and finally her emotional—despair in the world of New York society. But the pessimism of Mark Twain, and, to a degree, that of his friend Howells, was politically based as well. Both were anti-imperialists by the 1890s, strongly critical of American capitalism—Howells was for a time a utopian socialist—and opponents of the expansionist drive that had taken U.S. naval power to Cuba and the Philippines and led to the Spanish-American War.

The theorist of this drive was Admiral Alfred T. Mahan, whose *Influence of Sea Power upon History, 1660–1783* offered the framework for imperialist policy. The book was published in 1890, and three years later, at the Chicago World's Columbian Exposition, Frederick Jackson Turner announced that there was no longer a definable line of frontier settlement in the West and that thus a major phase of American history was over. But the "White City" fairgrounds at Chicago were constructed on the Lake Michigan shore around a grand series of pools, plied by Venetian gondolas and other craft, with a huge sculpture of "The Barge of State" (or "The Triumph of Columbia") by Frederick MacMonnies in which women are used allegorically to define American maritime values (see figure 14). The nautical symbolism underlined what Mahan had been saying: that the sea is the great frontier and that a nation's power comes from its control of the seas.

The significance of Mahan for American literature of the sea in the late nineteenth century lay in his recasting of the sea voyage historically as an instrument of national policy, and his central emphasis on sea power as the expression of American national character, the rechanneling of competitiveness and the love of money into the language of male sea-venturing. A strong dollop of the social Darwinian sense of struggle helped Mahan to transform the old metaphysical quest for deity or the inner and dangerously solipsistic search for individual selfhood into a national instrument for capitalist control of world markets, rather than a brotherhood of men squeezing hands in the sperm bucket.

Stephen Crane sought, as reporter, to follow these imperial adventures. When the ship *Commodore* on which he was traveling to Cuba with a cargo of arms sank, he fashioned "The Open Boat" as an ironic commentary on the older idea of the brotherhood of the sea. The potential rescuers at the lifeguard station, so highly touted in the press and in pictures, ignore the four men in a boat; the strongest, Billie the Oiler, does *not* survive but dies exhausted in the shallows, not in courageous action at sea. The story comes to us through the eyes and consciousness of the landsman-correspondent, the writer; as a result, the repeated poetic fragment about Africa that he remembers is understood as exotic claptrap, and the old sublime romantic pseudo-poetic refrain that rants about the "mad gods" is seen to be an inadequate rhetoric, unauthentic to the flat gray colors and quiet cadences of a diminished maritime reality. In "The Open Boat" as in his gnomic poems, Stephen Crane is the imaginative son of Emily Dickinson: the skeptical ironist, who shapes art out of the dissonance between form and content, between failed metaphysical ends and a compressed and antiromantic rhetoric. It would take a naturalist like Jack London to translate Mahan's *realpolitik* vision of male power at sea into the fantasy fiction of *The Sea-Wolf.*

As our earlier discussion of Emily Dickinson suggests, however, there was another whole line of development, a countercurrent—which focused on the implications of the sea for women in the late nineteenth century. Some women traveled the high seas, not only as elegant passengers on ocean liners but also as the wives of ship captains, and recorded their experiences in their journals for the eyes of their husbands and families, though rarely for the public press.

The published literature of the sea by women, by contrast, focuses on the coast and the maritime community, not the voyage; and the model for this genre was Harriet Beecher Stowe's *The Pearl of Orr's Island,* published in

1862 at the high point of Dickinson's own creativity. In the opening chapter, we witness, with an old man and a young wife, tempest and shipwreck. The sea kills the returning husband; the wife dies in childbirth immediately thereafter, and her daughter Mara finds a counterpart in the baby named Moses, thrown up by the sea in a subsequent shipwreck.

These elemental strokes of the plot become the romance structure, fed by explicit echoes and borrowings from Shakespeare's *Tempest*, that is superimposed in the first half of the novel on a richly textured rendering of the daily life of a maritime community, realistically portrayed and seen in large part from several female points of view: the inhabitants, their customs, beliefs, and patterns of speech—all that came to be associated with the idea of regional or "local color" fiction—based on Stowe's years in and around Brunswick, Maine. At one early point, the young children Mara and Moses drift in a small canoe out to sea, a recurrent motif in sea literature of the period—one thinks of Emily Dickinson's early "Adrift! A little boat adrift! / And night is coming down!" (#30), or of Lucy Larcom's reminiscences of Beverly, Massachusetts, in her *New England Girlhood* (1889), or of the scene of the vacationing children saved by the fisherman in Elizabeth Stuart Phelps's *Madonna of the Tubs* (1886). As Moses grows, he turns from book learning to boatbuilding and then ships out on a sailing vessel.

The realistic rendering of the ordinary texture of daily life in New England in the first half of the book would excite and inspire writers of this and the next generation. May Mather would mimic pieces of the plot in her brief "The Story of the Sea" for *Our Young Folks* in July 1868; and Sarah Orne Jewett acknowledged Stowe as an inspiration for her own fictionalizing of Maine coast material. But in the second half of *The Pearl*, Stowe focuses on the romance triangle between headstrong Moses, the ethereal, sensitive, and sickly Christian martyr Mara, and the healthy, bold, and temporarily coquettish Sally. Stowe's schematizing and spiritualizing of the novel's form has important gender implications. Male and female roles are carefully channeled: Moses's adventurousness is encouraged, and his destiny is to go before the mast, to reach out into the larger maritime world. Women are to be the repositories of community values, of both practical skills and transcendent spiritual concerns. For Stowe, the interlocking of traditional gender ideology with Christian belief ensures the stability of the maritime community.

Elizabeth Stoddard's exactly contemporary *The Morgesons* (1862) renders its imaginative New England universe even more exclusively from the female point of view of her protagonist Cassandra, who is alternately melodramatic, sexually passionate, bitter, and mordantly ironic, as she moves from her seacoast world inland and back to the coast, and experiences her limitations as a woman in a man's world. The sea is again and again, in this novel as in her later *Temple House* (1867), identified with passionate life, by contrast to the corrosively restrictive social patterns of New England culture. Cassandra's father, Locke Morgeson, has been a whaling captain: Argus Gates in *Temple House* is a retired sea captain. In such a context, Stoddard's novels ask what outlets there are for the passionate, life-demanding nature of a woman beyond her metaphoric identification with the sea. In *The Morgesons*, again the literary form seems split between an ideal structure of love and marriage and the possibilities open to a bolder female consciousness in exploring its emotional and spatial universe. Clearly, though, Stoddard refuses the rigid spiritual closure that Stowe demands of her coastal community, or that Larcom imposes on her autobiographical *New England Girlhood*, when the death of her father, another former sea captain, and ensuing poverty force the Larcom women inland to the Lowell mills for work.

How to define the lives of women in the working maritime communities of New England that faced economic decline in the later years of the nineteenth century was one challenge to these women writers, but the concomitant rise of the northeastern coast as the haven for vacationers offered other possibilities only hinted at by Howells in *Silas Lapham*. The life and writings of Celia Laighton Thaxter illuminate both. Her childhood was spent on the Isles of Shoals, rock outcroppings about nine miles off Portsmouth, New Hampshire, where her father served as lighthousekeeper. By 1848, he had opened Appledore House as a summer hotel, and the resort attracted a wide variety of the literary, artistic, and intellectual lights of New England, including, in the early years, Hawthorne and Richard Henry Dana, Jr. James Russell Lowell and Whittier, among others, would publish poems about Appledore.

Most important, Laighton's daughter Celia, whom Hawthorne in his notebooks dubbed "Miranda," married and moved off the islands to live in the Boston area and become a writer of poetry. Her first published poem, suggesting her response to this change, was printed by Lowell in the *Atlantic Monthly* in 1861 as "Landlocked." She continued to publish poetry for the next thirty years, much of it centered on the sea as subject. But encouraged by James T. Fields to turn also to prose, she wrote essays that began to appear in the *Atlantic Monthly* in 1869 and were gathered together in book form in 1873 as *Among the Isles of Shoals*. It is the *Walden* of the coastal world.

The opening sentence alludes to Melville's "enchanted islands," but the book that follows is a far cry from his "Encantadas," as it is also from Miranda's *Tempest*. With a fairly sustained intensity of expression, Thaxter evokes the experience of the place: its rocks and vegetation, its sounds and silences—the crashing of the waves, the cries of the birds, the dead calm before a storm—and the profusion of fish in its waters. We move through the seasons, experiencing the inwardness of the winter months, and the "unspeakable bliss" of the coming of spring. The point of view is that of the woman and the childhood self she evokes in reminiscence. The Isles of Shoals are known through legends and tales, and they are understood historically, as fishing stations in the colonial and revolutionary past and in their drunken degradation in the early nineteenth century. Thaxter has an especially sharp eye on the lives of the working families, differentiated into the fishermen who go to sea and the women who often stand on the headlands, waiting out the storms in which the ships are sometimes wrecked and sometimes preserved by a whimsical nature. These women are transformed by the grim demands of their work lives into exhausted shadows of their younger selves.

Thaxter aspired to be drawn "to the heart of Nature," but her nature worship was hard won, unsentimental, and not at the expense of her adult awareness of the sea's indifference and its capacity to bend, to break, to annihilate. Out of these contradictions she created her artistic voice, and *Among the Isles of Shoals* is its finest expression.

Equally important, she made her summer home and garden on Appledore the center of an aesthetic community of writers, artists, musicians, and composers. Two years after her death, *The Poems of Celia Thaxter* were collected and edited by another friend and neighbor from nearby South Berwick, Maine: Sarah Orne Jewett. Jewett came to Appledore sometimes with her close companion, Annie Fields, the publisher's widow and a poet and classical scholar in her own right, as well as the center of a literary salon in Boston during the winter season. All of these writers and artists contributed to the literature of the sea in coastal New England, and Thaxter's Appledore summer cottage was in many ways a catalyst to their creativity. By contrast to the harsh "brotherhood of the sea" learned by men aboard ship, Thaxter's Appledore modeled an alternative: the seacoast community shaped by women.

One sees it notably in Elizabeth Stuart Phelps's *The Madonna of the Tubs* (1886), a brief tale of Helen Ritter, a wealthy and aesthetic young urban woman who is vacationing at a coastal resort that in some of its particulars seems modeled on Appledore. She befriends a poor fisherman's wife, a laundress to the summer visitors. When she later learns that this "madonna of the tubs" has lost her fisherman husband at sea, she comes at Christmas bearing gifts and a loving sympathy that reaches across the class barriers to heal what the sea has wounded. Ultimately, though, Phelps cannot sustain this kind of female bond as a full alternative; the missing husband returns, and Helen Ritter is quickly paired off with a young man of her own class.

Sarah Orne Jewett had already mined a similar vein in her early stories, gathered into book form as *Deephaven* in 1877. The frame is that of two young, wealthy Boston women spending their summer on the coast, living in the beautiful old mansion of a dead aunt but visiting among the locals. Old Captain Sands sounds the keynote, that "everything's changed from what it was when I used to follow the sea" ("Cunner-Fishing"). Deephaven is a dozen miles from the nearest railroad, an isolated backwater that never fully recovered from the Embargo of 1809. Shipbuilding is a thing of the past, and there are more women than men, many of whom were drowned at sea. The strength of *Deephaven* lies in its evocation of this fading world in the voices of its inhabitants; its weakness lies in its point of view. The two rich outsiders are sympathetic to their subjects but forever aliens to their world, and the book teeters always on the edge of that distancing, picturesque "quaintness" that was to sabotage so much local-color fiction of the period.

Jewett struggled with this problem in her later fiction, like the novel *A Country Doctor* (1884), the last third of which is set in the coastal "Dunport," or *A Marsh Island* (1885), set in the vividly evoked landscape of tidal streams and rushes behind the barrier dunes of Plum Island in the northeastern corner of Massachusetts. Her greatest achievement lay in the form she had probed earlier in *Deephaven*, in that quiet masterpiece of 1896, *The Country of the Pointed Firs*. Dunnet Landing on the Maine coast is a more fully imagined world than Deephaven, and the book is brilliantly held together by the point of view of the unnamed narrator, a writer from the city who has returned for the summer and lives with Mrs. Todd, the local herb woman. Mrs. Todd becomes her friend and her entree into the community. The narrative consciousness both shares the lives of the townfolk and stands outside, shaping and relating, mostly without condescension, her vision of this coastal world. For despite the emphasis of some modern criticism on the pastoral elements in the book, it is the presence of the sea that organizes the book spatially, and the expansion and constriction of human lives is defined by sea experience.

The early story of Captain Littlepage involves his journey to the Arctic and being locked into that icy world. This Poesque visionary excursion into the sublime, in which the boundaries of self, of inner and outer, are blurred and its narrator-protagonist loses control, are framed at the outset by his comments that the Dunnet Landing of the present is constricted:

In the old days, a good part o' the best men here knew a hundred ports and something of the way folks lived in them. They saw the world for themselves, and like's not their wives and children saw it with them. . . . they got some sense o' proportion. (ch. 5)

Littlepage's insight informs the book. We are again and again glimpsing indications of the characters' contacts with that larger world of the past. Thus the narrator's imaging of Mrs. Todd standing isolated against the sky and the sea as a caryatid or, recalling her lost love, as a Theban Antigone on the plain is not merely the aesthetic decoration of an outsider; it suggests the larger imaginative and spatial universe that these women have inhabited.

Although Dunnet Landing is a decaying world from which the young and the able-bodied men have fled—to the city, to the West—it is still a place whose very language is drenched in sea knowledge: "You may speak of a visit's setting in as well as a tide's" (ch. 12). The tale that unfolds during one visit is of "poor Joanna" who, jilted by a lover, chooses to commit the unpardonable sin of isolating herself on Shell Heap Island, "a dreadful small place to make a world of," but an apt emblem of her life (ch. 13).

There are no "insular Tahitis" in Jewett's intricately woven world of land and sea. The bonds of sisterhood are forged by older women for whom marriage is not the resolution that links them to land as spiritual beacons while men venture forth to sea, in the easier polarities of the romantic voyagers and of traditional gender ideology. *The Country of the Pointed Firs* redefines the terms of American literature of the sea at the end of the nineteenth century, when the grand voyages of the Captain Littlepages are things of the past, half-crazed memories or fantasies that the women ballast with a firm "sense o' proportion," of seeing things "as they be," and of human commitment in a series of quiet gestures. They are female equivalents of Ishmael's "Squeeze of the Hand," but the sisterhood achieved is less explicitly homoerotic than that of Ishmael or Whitman's narrator or the eroticized power fantasies of Jack London in the years immediately following. In her own personal life, Jewett was a politically conservative daughter of old New England, lapsing into nostalgia in her historical romance *The Tory Lover* (1901); but through the female network that she formed most closely with Annie Fields and with the Appledore world of Celia Thaxter, in her best fiction Jewett offered a radical critique of American sea fiction of the previous century by insisting on the central role played by women in the maritime world.

In Jewett's fictional world, the passions are carefully ordered. Her women have known great joys and great losses from their encounters with both men and the sea, and the scars are visible. Their task is to live through these experiences with dignity. For the fully passionate engagement between women and the sea, one must turn finally from New England to the South. In American literature of the sea, "south" had usually stood for some passionate encounter: one thinks of Marvell's "Bermudas" or Royall Tyler's *Algerine Captive* or Philip Freneau's Carribean poems, Poe's *Pym*, the North/South polarities of *Moby-Dick* and "Benito Cereno," and countless other works. As these examples suggest, the passionate, the heart-over-head, is frequently linked to the erotic and to some racial encounter with blacks, Hispanics, or Melanesians, whether as the violence northerners fear or the forbidden sexuality they desire in some posited "primitive" world.

Lafcadio Hearn explored this vein in his Louisiana and Carribean writings of the 1880s, before abandoning the Western world altogether in 1890 to spend his remaining years in Japan. His finest sea fiction is *Chita*, published in *Harper's Monthly* in 1888 and in book form the following year. The first part, "The Legend of L'Ile Dernière," is a powerful descriptive evocation of the great Gulf storm that wiped out that Louisiana island resort on August 10, 1856. Drawing on rich newspaper resources, Hearn recreates the experience of the sea overwhelming this sandy stretch while the gorgeously bedecked visitors dance in the hall and the captain of a vessel offshore tries to stay afloat and save those washed out by the tidal wave, with a coda on the vicious "Spoilers" who come after to scavenge the wreckage and the bodies.

The second and third parts tell the story of Chita, saved from the sea on which she had been floating with her dead mother and nurtured by a Spanish fisherman and his wife on a nearby island. The city, New Orleans, is the corrupt, selfish, and disease-ridden foil for the beautiful, luxuriantly tropical island and the surrounding sea, which is, for Hearn and for Chita as she grows, alternately "the witchery of the Infinite," the source of fables and of beauty, and the place of menace and death. The "huge blind Sea" that sings a "mystic and eternal hymn" or preaches "the Elohim-Word of the Sea" puts her mind in touch with "the Soul of the World"; but the menace she feels, before she learns to swim, is that "the sea seemed to her the one Power which God could not make to obey Him as He pleased." Although Chita seems to have imbibed from her birth a sense of racial purity—she wants to pray to a *white* Virgin—Hearn's tale pulls her toward the life-giving values of the dark-skinned folk with whom she lives, who are in respectful but close touch with the power and energy of the sea. Hearn celebrated the vitality of the blacks and mulattoes in his Caribbean tales as well, including perhaps his finest story, " 'Ti Canotié" (1890), about two poor Creole boys adrift between the islands, but it is in his fragmentary fiction *Chita* that Hearn's richly encrusted, aurally evocative prose forges a tentative link between the sensuousness of a southern sea and the growth of a young girl's consciousness. It remained for Kate Chopin to develop the potentialities of this genre, but in some startlingly different directions.

The Awakening (1899) also takes place in the "exotic" racially mixed world of Louisiana, and within this world Chopin formulates a stunning challenge to the older gender ideology of women and the sea in American literature. It begins as another Gulf Coast version of the vacation resort story, with a parrot screeching "*Allez vous-en.*" And indeed Leonce Pontellier, the New Orleans businessman out at Grande Isle to join his wife Edna and children for the weekend, does immediately go away, leaving Edna with handsome, young Robert Lebrun.

The Awakening is not, however, a traditional seduction novel, nor is it even an American *Madame Bovary*, despite Chopin's critical allegiance to Flaubert, Maupassant, and later French writers. It is rather, from one point of view, a political novel in which the female protagonist rebels against the restrictions of her given roles as upper-class wife and mother, and tries to imagine and then find the forms through which a woman might control her own life and choices in a patriarchal society. It is also a regional fiction whose two settings are the city and the sea, although Chopin's ironic and symbolic treatment undercuts easy polarizing in spatial terms of social forms and sensuous, passionate values.

If *The Country of the Pointed Firs* and related works, which Chopin knew and admired, move us beyond or away from the absolute or sublime spiritual and metaphysical goals of earlier American literature of the sea, *The Awakening* rejects altogether a Christian context—whether Edna's father's Presbyterianism or her lover's Catholicism—in a fin-de-siècle reimagining of the erotic as the central generative significance of the sea.

The sea gives shape to Edna's vision and symbolic form to the book that Chopin creates about her. As the refrain that frames the novel has it,

> The voice of the sea is seductive; never ceasing, whispering, clamoring, murmuring, inviting the soul to wander for a spell in abysses of solitude; to lose itself in mazes of inward contemplation.
>
> The voice of the sea speaks to the soul. The touch of the sea is sensuous, enfolding the body in its soft, close embrace. (chs. 6 and [with some variation] 39)

Such a passage obviously contains strong echoes of Melville and especially of Whitman's "Out of the Cradle Endlessly Rocking," and behind both is the Emersonian doctrine of correspondences. The sea also finally offers to Edna, as it had to Whitman, "the low and delicious word death," as her final act of walking naked into the water and swimming out to sea makes clear. In this sense, *The Awakening* seems to affirm the absolute values of selfhood, a selfhood apart from the community and opposed to the city, and so a kind of Dickinsonian bathing in the infinite. It seems to be the embodiment of Dickinson's "Captivity is Consciousness—/ So's Liberty" (#384).

Chopin thus recaptures the earlier transcendentalist vision deliberately and unequivocally for the female consciousness. It is at once a personal and a social act, and therein lies its ambiguity for readers. As a social act, Edna seems to confront the trap of gender through suicide. Social rebellion ends in an act of self-annihilation in what Philip Freneau called "Dread Neptune's wild unsocial sea." Edna Pontellier does not, however, submit to the male sea-god Neptune any more than she does to what Melville and Dickinson saw as the problematic Judaeo-Christian Jehovah. She becomes in the final image, as Sandra Gilbert has pointed out, the goddess Aphrodite, Venus, born on an island set apart from the city where men make history and use their power to limit women's freedom. The sea in *The Awakening* exists as a liberating symbolic speech that projects beyond its origins in local-color vacationing, beyond the male-defined understanding of class and race and gender. By appropriating the sea to her white, upper-class female self, Edna rejects the identification of the sea's erotic generative power only with the dominated "primitives" and especially their females (such as the book's Mariequita) who had always been accorded these illicit powers.

Finally, in Edna's encounter with the sea, Chopin declares the cleansing beauty of female sexuality in a way that was an affront to the dominant ideology of her culture. Listen, by contrast, to George Engelmann, president of the American Gynecological Society, in *The American Girl of Today: Modern Education and Functional Health* (1900):

> Many a young life is battered and forever crippled in the breakers of puberty; if it crosses these unharmed and is not dashed to pieces on the rock of childbirth, it may still ground on the ever-recurring shallows of menstruation, and lastly, upon the final bar of the menopause ere protection is found in the unruffled waters of the harbor beyond the reach of sexual storms. (9–10)

The Awakening rejects the harbor as the woman's place, as it had been the spiritual "home" for conventional sea writers over the centuries. Chopin rejects Dickinsonian "mooring" in either God or a man. Her Aphrodite heads finally to sea alone and without even a ship on which, with Whitman, to "farther farther sail" on a passage to India; alone and "without a friend or comrade near," as Mark Twain's mysterious stranger wanders into endless space. She moves as goddess-woman, not as the disembodied "Pure Thought" that is Twain's solipsistic despair of the world of male power. If one must still judge her act as a socially despairing suicidal gesture, aesthetically—indeed, poetically—in her use of the sea Chopin may also have been reaching toward what Wallace Stevens would later call "The Idea of Order at Key West."

ROGER B. STEIN

Chapter Nine

FICTION BY SEAMEN
AFTER MELVILLE

American sea fiction did not end with *Moby-Dick*, nor did young men cease to enter literary careers in the traditional way—following Cooper, and especially Dana and Melville by gaining their experience as working seamen. Even after the heart-breaking reception of his great book, Melville continued to work with the sea, publishing "The Encantadas" and "Benito Cereno." "Benito Cereno" is certainly one of the greatest American sea stories, but in its treatment of the sea and the sea captain we can see how great a change had come over Melville and, by extension, perhaps, the American view of the sea. In 1850–51 he had found in the simple democratic seaman and his voyaging mind an inspiring basis for his incomparable celebration of the sea and—because in *Moby-Dick* these are one—life. But in his story of 1855, there is no simple seaman with a voyaging mind, only the defiant Babo, who, though the story's single truth-seer, remains silent. Moreover, Captain Delano is so congenially blind to the darkness of life that he lacks the mythic voyager's essential motive, to discover light; and, most tellingly, Melville signals his new darker mood at the beginning of the story by remarking that the "sea . . . seemed fixed." In Melville's imagination over the next three decades the sea would be dominated by imagery of sea warfare during the Civil War and by imagery of the Dead Sea (in *Clarel*).

By 1855, the golden age of American sea fiction had come to an end, along with that "extraordinarily concentrated moment of expression" that F. O. Matthiessen described in *American Renaissance*. The "Christian belief in equality and brotherhood" (656) that figured so centrally in *American Renaissance*, particularly in Matthiessen's treatment of Melville's work, was soon overshadowed by the approaching war. The Civil War took its toll on the spirit of maritime America, and with it, a number of other developments in American history would lead us into a long period of "nautical darkness," as Richard Matthews Hallet described it in *The Lady Aft* (1915). The opening of the Panama Railway in 1855, the collapse of the whaling industry, the rapid development of steamships and their eventual eclipse of sail, the expansion of the Western frontier, the opening of the Transcontinental Railway in 1869, and the opening of the Panama Canal in 1914: these and other forces changed American maritime experience for all time.

But far from subduing the American imagination's fascination with the sea life, these changes played a significant role in bringing about a resurgence of American sea fiction that began in the late 1880s and extended into the 1930s. Seeing the absolute disappearance of the wind seaman's life; being drawn to that life not only for its considerable actual rewards, but for its ideal possibilities as immortalized in the American imagination by Dana and Melville; and wishing to resist the forces of industrial America as they were gathered in the steamships—a number of Americans born between the 1860s and 1890s went to sea as working seamen before the mast and then entered the literary world in the traditional way that Dana and Melville had established. Jack London (1876–1916) is by far the most famous of these men, but before him there were Morgan Robertson (1861–1915) and Thornton Jenkins Hains (1866–?). And, in part stimulated by the literary successes of these men, others followed: Arthur Mason (1876–?), Felix Riesenberg (1879–1939), Bill (Bertram Marten) Adams (1879–?), Lincoln Colcord (1883–1947), Richard Matthews Hallet (1887–1967), and Archie Binns (1899–1971).

Of these men, all but Binns were born during Melville's lifetime, but although they self-consciously followed Dana and Melville in their writings of the sea, the work they produced (beginning in the late 1880s) envisions a sea that is fundamentally different from that in which Moby Dick had swum. This sea change, evident in Melville's own last work (*John Marr and Other Sailors* and *Billy Budd, Sailor*), arose not so much from the proliferation of steamships and the general decline of the American merchant marine, but from the scientific revolution that began with the *Origin of Species* in 1859. The vision of the sea that Melville expressed in "The Monkey-rope," that "that unsounded ocean you gasp in, is Life," and in his cetological meditations on life (based as they were on pre-Darwinian biological thought; i.e., "natural theology," as it is called by historians of science)—this vision of life was displaced by the new understanding of how life had originated in the sea and evolved through natural selection. The "nautical darkness" in late nineteenth- and early twentieth-century American writing resulted even more from this loss of traditional values than from the gradual disappearance of the great sailing ships. (That the sailing life did not end abruptly in the

1850s, as is often thought to be the case, is clear from the fate of the *Glory of the Seas:* launched in 1869 but soon defeated in her competition with steam shipping, she endured a long, inglorious decline before being burned for her copper in 1922.)

It is therefore possible to account for the resurgence of American sea fiction that began in the late 1880s in a number of ways: first, by noting that the tradition *existed*—as originated by Cooper and then transformed and democratically revitalized by Dana and Melville. Also, the steamships' threat to traditional values provoked a defensive and often nostalgic revival of interest in the sailing life; and for similar reasons, sea stories—especially those of a simpler past—found a large receptive audience during these years, one similar to the one to which the regionalists had appealed. But most important, the spiritual darkness that was produced largely by Darwinian thought called forth a response such as that suggested in another context by Theodore Roethke in his poem "In a Dark Time": that "In a dark time, the eye begins to see." In American sea fiction of this period, the impulse to see finds its natural expression in the idea of the voyage, whose ancient, mythic appeal is its promise of renewal, discovery, light. And in the sea stories of these years it seems particularly clear that, as in the history of literature and myth in general, the voyage idea is essentially a strategy for the survival of the troubled imagination—a motive no less powerful than the physical necessities that must have led to the actual voyages of antiquity. Moreover, as every literal or literary voyager's embarking is an implicit act of faith, the long series of voyages that constitutes the tradition of American sea fiction has contributed a sense of affirmation and hope to American writing in general that is not so constantly present in other fields—e.g., in our literature of the frontier, the South, or the regionalists' New England; our literature of American industry and business, and of the urban experience in general; certainly not our literature of war.

In *John Marr* and *Billy Budd*, Melville responded to the late nineteenth-century darkness in essentially the same way that he had in concluding *Clarel* (1876), by denying that "Darwin's year, / Shall . . . exclude the hope—foreclose the fear." In "John Marr," the story of an old seaman approaching death on a Western prairie, far from the sea and the shipmates of his youth, Melville associates the old sailor's loss with that of the "exterminated" Indians and consoles him with "a reminder of ocean"—a sense that the rolling prairie is like the sea, and that only "the ocean that washes Asia" can limit the western advance of "civilized life" in America. But John Marr's chief reward is that through "meditation" he can muster his old shipmates and claim their "fellowship" again. And in *Billy Budd*, a story "in the time before steamships,"

BERT BENDER

he wrote his elegy for the simple "Handsome Sailor," and sent him "fathoms down" into the watery world. From these depths, both a Jonah (as imaged in Father Mapple's sermon: "ten thousand fathoms down, . . . the weeds . . . wrapped about his head") and a Lycidas, he will perhaps "Emerge," like Clarel, "from the last whelming sea, / And prove that death but routs life into victory."

Despite his severely tested faith, Melville apparently ended his writing career by clinging to the advice he had offered in chapter 97 of *Mardi* ("Faith and Knowledge"): "let us hold fast to all we have; and stop all leaks in our faith. . . . The higher the intelligence, the more faith." But in the American sea fiction from the 1890s to the 1930s there are repeated crises of faith (which are, typically, coincidental with crises of navigation) in which the simple sailor is bereft of his traditional faith and left to struggle for both his physical and spiritual survival in new seas of meaning. In one such crisis, for example, having seen his own ship (a great steamer) run down a helpless sailing ship, and having been abused by the ship's officers and rejected by a lover, Morgan Robertson's hero in *Futility* (1898) sees his fate as "part of the great evolutionary principle, which develops the race life at the expense of the individual." Then, feeling "unfitted to survive," he denies the existence of "a merciful God—a kind, loving, just and merciful God." He gives way to "a fit of incongruous laughter" and, staring "into the fog," concludes: " 'The survival of the fittest,' . . . 'cause and effect.' It explains the Universe—and me" (ch. 6). And in *Lightship* (1934), Archie Binns (the last nineteenth-century American to write within the maritime tradition) tells us how one of his characters had—" 'Praise be unto God in Heaven' "— "become a sailor and a man." Sailing with a young friend from Norway to America on his uncle's ship, the boy had thought that "God must be something like Uncle Olaf [the captain], . . . immense and bearded and ageless, but ever so much more powerful." But when his friend falls from the weather yardarm and breaks his back, and then must be left helpless in the cabin when the remaining crew are forced to lash themselves high in the ship's rigging in order to survive a hurricane off the coast of Texas, the nineteenth-century boy becomes a twentieth-century man. Looking down to where his friend lay "drowning or drowned already," "he began screaming" into the wind, "There is no God!": "He moistened his split lips and shouted into the darkness, across the frenzied sea, above the grave of his shipmate, 'There is no God! There is no God!' " And when at last he was able to hear his own voice, it "gave him courage" (ch. 4).

The volume and variety of fiction published by American seamen over the last century is so extensive that only a brief survey of the most significant work is possible here. To proceed chronologically, then, one must begin with Morgan Robertson. By virtue of his ten years before the mast in both sail and steam, Robertson was one of the most experienced seamen in a tradition that—in Cooper, Dana, and Melville—had been established by young men who gained their brief sea experience during their late teens and early twenties. In a sense, Robertson fulfilled the tradition's democratic ideal, as established by Dana and Melville: from his early experience as a simple working seaman, representing one of the most exploited classes of American workers, he successfully entered the literary world, more or less as Jack London would a short time later. While one cannot pretend that Robertson was a great writer, he was a very prolific and well-known writer whose large readership included William Dean Howells, Joseph Conrad, and Jack London. To such readers, Robertson must have seemed, if not a very talented or profound author, certainly a fully representative figure of his time, the "Strenuous Age." If he is remembered at all today, it is as the author of *Futility; or, The Wreck of the Titan* (1898), in which, uncannily, he foretold the *Titanic* disaster by fourteen years. But *Futility* is also of interest as the earliest American sea novel to emphasize the hellish imagery of the boiler room (which would become most widely known through O'Neill's "*Hairy Ape*"), and the earliest to extend the Darwinist vision into a presentation of the great social struggle between steamships and sailing ships, ship owners and insurance brokers, and officers and men. In his several novels and many stories, he pursues these ideas as they figure in the Darwinian struggle among nations that Alfred Thayer Mahan assumed in his famous study *The Influence of Sea Power upon History* (1890). Robertson actually contributed to the development of this power by inventing and building the first periscope, which he sold to the U.S. Navy. But often criticizing the Darwinist underpinnings of Mahan's theory from a grimly comic point of view, in a kind of science fiction, Robertson wrote many sea stories of future warfare at sea in which the great battleships and submarines are instruments of global war in the evolutionary struggle among the races.

Thornton Jenkins Hains ("a licensed navigator of oceans for vessels of more than 700 tons," according to *Who Was Who in America*) wrote several novels and collections of sea stories that were very favorably reviewed. He is almost completely forgotten today, but at least two of his books deserve permanent places in the tradition of American sea fiction. One, *The Strife of the Sea* (1903), is a unique collection of stories about various sea creatures, from fishermen to sea turtles. Unified around the Darwinist theme suggested in its title, it is a kind of *The Call of the Wild* of the sea, in which Hains's response to the evolutionary theory is to imagine—while dramatizing and fully accepting the bloody order in which his creatures exist—that a more peaceful existence will evolve, one less possessed of the "killing spirit." In another collection, *The White Ghost of Disaster* (1912), he too (like Robertson) told of a great steamship's collision with an iceberg. And in his best novel, *The Voyage of the Arrow* (1906), he published one of the most readable novels of this period. Like his best stories in *The Strife of the Sea*, *The Voyage of the Arrow* is presented in a clear, plain style that befits its simple but interesting narrator, Mr. Gore, and the simplicity of the plot. It is another story of survival: Mr. Gore struggles in the competitive merchant marine, he survives a bloody conflict with renegade seamen, and he yields to the "necessity" of "propagating his species" in his relationship with the captain's daughter. But in presenting this Darwinian interpretation of Gore's life, Hains never allows his ideas to intrude on the narrative flow.

Among the major writers in the tradition of American sea fiction, Jack London is the crucial link between Melville (whose great influence on London Charles Watson studied in 1983) and Hemingway. London's influence on Hemingway is widely recognized, but also resisted, for London's significance in American literary history is still falsely disputed by many who see him only as the author of boys' books or stories of the north; or, mainly, it would seem, by those who are disturbed by the general turbulence of his time and the unsavoriness of the ideas that burst into view as twentieth-century man struggled toward self-discovery. In the fifty books he published during his sixteen-year career (at the strenuous pace of one thousand words a day), he took in a truly breathtaking expanse of the new American experience, but in his path to authorship he followed Dana and Melville.

From his first story ("Typhoon off the Coast of Japan") to his last ("The Water Baby"), he turned to the sea, finding there the most truthful basis he could imagine for understanding modern life. He was repelled (as Hemingway would be) by what he saw as the "imaginative orgies" in *Moby-Dick*, Melville's soaring celebrations of a coherent Christian universe in which London had no faith. But he saw twentieth-century possibilities for a great captain like Ahab, and he recognized the new relevance of Melville's exuberant interest in biology. "You lack biology," his autobiographical character Martin Eden charges in an argument with a university professor: "It has no place in your scheme of things. Oh, I mean the real interpretative biology, from the ground up, from the laboratory and the test-tube and the vitalized inorganic, right on up to the widest aesthetic and sociological generalizations" (ch. 27). Earlier in that novel he tells how he had learned from Herbert Spencer that "all things were related to all other things"—even, although it seemed "ridiculous and impossible," "a woman with hysterics and a schooner carrying a weather-helm or heaving to in a gale" (ch. 13). In this last idea exists the logic according to which a woman passenger was obliged to take part in so many literary voyages during these years—from Hains's *Voyage of the Arrow* to London's *Sea-Wolf* and *Mutiny of the Elsinore* to Frank Norris's *Moran of the Lady Letty:* evolutionary theory was incomplete without the woman's role in the evolution of the race, as Darwin had explained in *The Descent of Man, and Selection in Relation to Sex* (1871). In this effort to find a believable form for this larger plot of life, London wrecked his otherwise great novel, *The Sea-Wolf,* which was essentially an expression of faith in the idea that evolution does proceed positively, like the *Ghost,* with a "mighty rhythm"—"the lift and forward plunge of a ship on the sea." London *would* submerge the brutal Wolf Larsen in the stream of life and leave him in the wake of evolutionary development that was promised in the love of Maud Brewster and Humphrey Van Weyden—just as Melville had expressed his willed belief by extinguishing Ahab's fiery doubt in the waters of God.

By the time he wrote his other sea novel, *The Mutiny of the Elsinore* (based on his sailing trip with his wife Charmian around Cape Horn in 1912–13), London could see no other end to the evolutionary plot, but he had come to view it as darkly comic, absurd in a modern sense that is often misinterpreted (by what few readers the novel still has) as his thoughtless affirmation of racist brutality. Two years earlier, in the fine but ugly sea story, "Make Westing," he did not include a woman in the Cape Horn passage, but rather developed two other aspects of the Darwinian view of life that characterize the sea literature of this period—the brutal force of American commerce as exemplified in the competitive merchant marine, and the sense of evolutionary degeneration that is projected in the image of hairy, apelike seamen. "For seven weeks the *Mary Rogers* had been between 50 degrees south in the Atlantic and 50 degrees south in the Pacific . . . struggling to round Cape Horn," and her captain had become obsessed with the sailing directions for navigating those waters: "Whatever you do, make westing! make westing!" The captain, Dan Cullen, resembles Wolf Larsen only in his animal vitality. With none of Larsen's rhetorical grandeur (as in the language of Ecclesiastes), nor any of his "terrible beauty" (his "clean-shaven face" and skin "fair as the fairest woman's"), Cullen is, as his name suggests, an evolutionary cull. Working with Darwin's theory of sexual selection, London suggests that, without Wolf's beauty, Cullen is an even less highly evolved sea captain. "Hairy as an orangutan," he is doomed to be selected against—like a male bird who has failed to develop beautiful plumage. But the crucial result of Cullen's low evolutionary state is his lack of what Darwin called the "moral sense." Thus, in the navigational crisis, when a sailor falls from the rigging, Cullen drives on. The man had caught a life buoy in going over, and the ship could easily have come about; but Cullen had just begun to make his westing. After the ship's single passenger objects to his needless cruelty, "a sordid little drama in which the scales balanced an unknown sailor named Mops against a few miles of longitude," Cullen arranges for the passenger's "accidental" death, doctors the log, and "surge[s] along at his nine knots. A smile of satisfaction slowly dawned on his black and hairy face." London did not originate the apelike seaman. Morgan Robertson had worked with the idea, notably in a comic story called "The Mutiny," in which a crew of apes overpower the human officers and win the ship. But here London developed a variation on the traditional sense (that would culminate in O'Neill's *"Hairy Ape"*) by creating an apelike captain whose brutal authority is a just reflection of American maritime experience during those years, as documented by the International Seamen's Union of America in *The Red Record: A Brief Resume of Some of the Cruelties Perpetrated upon American Seamen at the Present Time* (1897).

Arthur Mason was born in Ireland in 1876, went to sea at age seventeen (beginning a career of more than twenty years), became a citizen of the United States in 1899, and began to publish his sea stories in 1920. Two of his books, *The Flying Bo'sun* (a novel, 1920) and *The Cook and the Captain Bold* (stories, 1924) are memorable contributions to the tradition for their realistic portrayal of American shipping at around the turn of the century and for the impress of his engaging plain style. As a reviewer for the *New York Times* remarked (regarding Mason's stories), "Mr. Mason writes with a humor that remains intensely human and that rubs shoulders time and time again with an unstressed and unmistakable pathos." While his stories project an awareness of the ideas of the time, Mason never allowed these ideas to overshadow his chief purpose, which was to portray the life he loved. His work is notable in particular for its balanced treatment of the conflict between captains (who are always mere men with no pretensions to the mythic stature of an Ahab or a Larsen) and the common seamen. He portrays the crew, for example, sheltering "themselves in the lee of the galley" as the ship weathers a gale: "Quiet and solemn they were, trusting their captain in that plain simplicity which is the everlasting bond of the sea" ("The Captain"). But he can also portray a captain's comic helplessness when he oversteps the limits of just authority. In one story, for example, a presumptuous captain disregards the advice of his experienced sailmaker; he wants a sail made according to *his* pattern, but the sailmaker "had his own way of doing things, and felt that he could point to the successes of years to uphold his resentment of interference." Having his own way, the captain produces a very poor sail, provoking muffled giggles among the crew, who gather about to watch, "for this was one of the moments when discipline makes for familiarity." Then, feeling "the sailors rub shoulders with him as he stooped over the deck chart," the captain "was reminded . . . that this was a terrible breach of discipline. Who was master of this ship anyway?" At the end of the story, after the tension has developed into a darker conflict in which the sailmaker is vindicated, the embarrassed captain is left with "nothing to say. He just stood wondering, with his hand on his Adam's apple" ("The Sailmaker").

Felix Riesenberg went to sea at age sixteen and worked himself up over a period of twelve years from ordinary seaman to captain. He accumulated an extraordinary amount of sea experience, from sailing around Cape Horn as a sailor before the mast to commanding the USS *Newport*, the New York State School Ship. He began his career as a writer with *Under Sail* (1918), an autobiographical account of his Cape Horn voyage on the *A. J. Fuller;* he followed this with two books that became standard reference works for seamen, *The Men on Deck* (1918) and *Standard Seamanship for the Merchant Service* (1922); and in addition to serving as associate editor and a regular writer for the *Nautical Gazette*, he wrote several other books of fiction (only some of which were of the sea) and informal maritime history. His chief contribution to American sea fiction is *Mother Sea* (1933), a well-written, engaging, and informative novel about the subject he had treated nonfictionally in *Under Sail:* "that phase of our sea life that formed and forged the link between the old and the new, between the last days of sail and the great new present of the America of steam and steel" (11–12). The novel traces the careers of two seamen, the old Captain Glade, whose long career as a whaler and wind seaman in the merchant marine link him to nineteenth-century maritime and literary experience; and his protégé, Captain Nicholson, who begins his career in sail, and, after losing his ship, makes a new career in steam. As the title suggests, it is a story of renewal and survival for those like Nicholson and Glade who are mothered by the sea—which "harden[s] those it [does] not kill." Riesenberg's fiction won high praise from a range of critics, who saw in his work both the kind of broad appeal that had led to one novel's being made into one of the earliest silent films (*East Side, West Side*, not a sea story), and the more purely literary appeal of an innovative stylist (his *Endless River* [1931] was compared with the work of Dos Passos and Hemingway). And William McFee, the tough-minded critic and author of sea fiction, gave *Mother Sea* the highest praise, claiming that it "would gratify Richard Henry Dana himself."

In 1883, Lincoln Ross Colcord was born at sea in a gale off Cape Horn. His father was captain of the *Charlotte A. Littlefield*, in the China trade, and Colcord sailed with him until he was fourteen. He grew up to become a noted critic and author of sea fiction whose work was admired by the maritime historian Samuel Eliot Morison and by literary critics as well. Colcord's "Notes on 'Moby Dick'" was one of the earliest pieces of criticism in the Melville revival, and it is still of interest as a unique comment on *Moby-Dick* by a critic with the rare credentials of an experienced seaman, a maritime historian, and a writer of serious sea fiction. Colcord wrote of maritime affairs and reviewed maritime literature for the *American Neptune* (which he was instrumental in founding), the *Nation*, the *Bookman*, the *American Mercury*, and the *New York Herald-Tribune*, *Books;* for this work and his own sea fiction he should be remembered as the unofficial dean of American sea literature during the 1920s and 1930s.

Colcord's one novel, *The Drifting Diamond* (1912), is a product of his apprentice years as a writer, when he was heavily influenced by Robert Louis Stevenson and Joseph Conrad. Its best feature is a long description of a hurricane, in which Colcord measured up to his own high standard of nautical verisimilitude. But he was a far better short story writer, and each of his collections (*The Game of Life and Death* [1914] and *An Instrument of the Gods* [1922]) contains stories that should be included in any collection of great American sea stories. His work is unusual in the tradition of American sea fiction in that it is presented from the quarterdeck point of view. Few writers—perhaps only Conrad—can surpass him in recreating what he called "the psychology of the quarter-deck, the psychology of handling a vessel," which he found lacking in Melville's work. But, unlike Conrad, Colcord portrays his sea captains in a way that often suggests an explicitly positive response to evolutionary thought and thereby reflects American sea fiction's tendency to affirm life. Colcord's most memorable captains (such as Armstrong and Gordon in "Thirst: An Incident of the Pacific") exercise their duties as "sea-fathers." And in another fine tale, "A Friend," Colcord develops the American tradition's commitment to the ideal of brotherhood by telling of two boyhood friends who become sea captains. When one's career goes on the rocks and he enters a period of dark and bitter depression, his friend keeps in touch and finally, in a scene recalling the "wedding" of Queequeg and Ishmael in "The Counterpane," "saves" the troubled man's life. The friend's physical presence is "a pure gift of human divinity" that enables the other to regain command of his "ship of life."

Richard Matthews Hallet went to sea at age twenty-five, older and more educated than Dana when he sailed on the *Pilgrim*. Hallet had earned both his bachelor's (1907) and law degrees (1910) from Harvard; but after a brief, disappointing experience in the legal world, and refusing to accept the "nauseating fact of mere existence," he signed on as an able seaman aboard the

Juteopolis, from Boston to Sydney. Some of his relatives had been seamen, and he knew of the literary possibilities that Dana had created for such voyages. Also, even in 1912, it was "fashionable . . . for young men fresh from college to embark in [steam-driven cattle ships] for their wanderyear" (*The Rolling World*, ch. 1); so when he found that the "Standard Oil Company still chartered square-rigged ships" such as the *Juteopolis*, he signed on, recognizing at once that the literary possibilities in sail were far greater than in steam. He did go on to make a notable career for himself as a sea writer, producing two novels, *The Lady Aft* (1915, based on the *Juteopolis* experience) and *Trial by Fire* (1916, based on his work as a fireman aboard the steam freighter *Orvieto* from Melbourne to London in 1913). He also published several short stories of the sea that were judged among the best in America—one, for example, "Making Port," was judged the best American short story of 1916 by Edward J. O'Brien, who dedicated his volume of that year to Hallet.

Hallet's novels are significant in the tradition of American sea fiction for a number of reasons. First, *The Lady Aft* is one of our last realistic accounts in fiction of the sea experience before the mast; and it was written by a highly educated and talented young man whose ironic style gives life and a meaningful perspective to the whole adventure. Constantly aware of the absurdity of his sailing before the mast in 1912, and of himself as the inexperienced "stiff" who would, according to the tradition, become a man—he found in this situation a fitting vehicle for presenting the absurdity of life in general, as it seemed to many early twentieth-century writers. He was influenced by Stephen Crane's view of "little men" confronting a meaningless universe and by the view of his teacher at Harvard, George Santayana, that "the material world carried the mind along as a ship carries a curious passenger. . . . 'Even what we call invention or fancy . . . is generated not by thought itself, but by the chance fertility of nebulous objects, floating and breeding in the primeval chaos' " (*Rolling World*, ch. 4). Thus, in becoming an able-bodied seaman, the young "stiff" suffers such experiences as the "soul-and-body lashing" of going aloft in a storm. As a result, he comes to see the irony of his existence in the stream of life as a mere organism, a "wrecked mechanism which would balance itself on its two legs, time after time, when given the least chance" (ch. 13). And in his violent sexual competition for the young lady aboard the ship (his involvement in the necessary "law of battle" in Darwin's theory of sexual selection), he is painfully aware of how "ridiculous" it all is.

In *Trial by Fire*, Hallet explores the forces of social Darwinism from the point of view of the exploited stokers aboard a Great Lakes ore ship, the *Yuly Yinks*. In his first story about his experience as a stoker aboard the *Orvieto* ("The Quest of London"), he had addressed the timely question of "the

heavy workers of the world" in a way that clearly prefigures O'Neill's treatment of the subject in *"The Hairy Ape."* In fact, *Trial by Fire,* which appeared
in 1916, six years before *"The Hairy Ape,"* is a major but unacknowledged
source for O'Neill's play. Hallet's leading character, the fireman Cagey, is
(like Yank) an apelike laborer trapped (as his name repeatedly suggests) by
the brute forces of American industry and by his own crippled imagination,
his sense that his labors are heroic: " 'Hah,' " he says, " 'some guy has got to
get the steam' " and "like a vast ape he [swings] himself down, down into that
whispering blackness," the fire room (chs. 17, 9). The unmistakable and extensive similarities between Yank and Cagey (and between other characters
and dramatic elements in the novel and the play) are clear, not only in their
language or the fiery imagery of their labor, but also in O'Neill's repeated
emphasis on Yank's efforts to "tink" in the posture of Rodin's *The Thinker.*
The source of O'Neill's "expressionistic" image is Hallet's far subtler image
of Cagey at the end of *Trial by Fire:* he sits, sunk "forward with his chin on his
fist and his elbow on his knee, staring with hard eyes into the hot blackness
between the dusty boilers" (ch. 17).

Archie Binns was the last American born in the nineteenth century who
went to sea as a working seaman and later wrote self-consciously within the
tradition. He co-authored one novel with Felix Riesenberg (*Maiden Voyage,*
1931), and produced a great work of American sea fiction in his novel *Lightship* (1934). *Lightship* won the highest possible praise from all who reviewed
it (including Lincoln Colcord and the English seaman-novelist David W.
Bone), but it has been sadly neglected since then. Like Hallet, Binns drew
on a family heritage of maritime experience and, writing from the point of
view of the simple sailor, extended Dana's perspective into the context of
the American labor movement early in the twentieth century. Among the
several characters in *Lightship* (whose life stories are sketched in the various
chapters), one had entered the American merchant marine by way of being
shanghaied—"drugged and robbed and sold into slavery in the year of grace
1905!" (ch. 8); and another (the one who as a boy had lost his friend and
screamed into the hurricane, "There is no God!") had come to put his faith
in the "powers for good in the hearts of people doing their duty," like Robert
La Follette, creator of the Seaman's Act (ch. 16).

As *Lightship 167*'s "first consideration is to give light to other ships," the
novel as a whole illuminates a time of "nautical darkness" in American fiction
by dramatizing how *Lightship 167,* in *its* navigational crisis, must use both sails
and steam power in order to avert disaster. As this drama suggests, Binns's
underlying purpose in *Lightship* is to draw on the traditional elements (e.g.,
the dream of the South Seas, the young man's entry into manhood by way

of the maritime experience, the affirmation of democratic possibilities in the brotherhood of shipmates, the crises of faith that occur in nautical extremities) in order to launch the tradition safely into the future. In this way his effort resembles Felix Riesenberg's in *Mother Sea*. Binns's remarkable success is due partly to the possible sailing directions he offers for future literary voyages: in the captain's faith in evolution, for example, his sense that "our blood is nothing but an ebbing and flowing tide of sea water," and that we are "too new to our present condition—fish just pulled out of the deep sea" to know what "we will be"; or in another seaman's faith in Crazy Horse's vision that "people are like water on the earth, driven in waves before the Great Wind" (chs. 15, 13). But, more deeply, Binns's success derives from his accompanying sense of doubt, the "haven of darkness inside" the lightship that provides the essential motive for new voyages of discovery (ch. 7).

Born the same year as Binns (1899), Ernest Hemingway was never a working seaman in the traditional sense, but during his years in Key West and Havana he worked self-consciously within the tradition, creating two memorable working seamen in Harry Morgan and Santiago. And during those years he created his own working relationship with the sea—as a seaman-writer, a devoted amateur marine biologist, and an unofficial sea captain in antisubmarine warfare. In his first sea novel, *To Have and Have Not* (1937), he wanted to tell a twentieth-century sea story; thus he referred only ironically to the tradition—as pictured in "a clipper ship running before a blow framed above [the] head" of a wealthy yachtsman. Hemingway's comment is, "It's great to be an Intrepid Voyager" like this representative of the "haves" (ch. 24). He is one of the many decadent Waste Landers in the novel in contrast to whom the brutal Harry Morgan's virtue is clear: Harry is "alive" (ch. 25). Hemingway's sympathetic portrayal is based on Harry's biological simplicity and his predicament in time: as he realizes, "There's no honest money going in boats any more" (ch. 18). With this awareness, with his simple motive to feed his family, and with his primitive sexual force (which is compared to that of a loggerhead turtle), he offers more promise than the impotent yachtsmen who only *seem* to have so much. Despite the Waste Land setting, Harry remains a traditional American voyager: in his biological vitality in the sea of life (a Wolf Larsen of the 1930s), and, most significant, in his sense that "You got to have confidence steering" (ch. 6).

In his two other men of the sea, Hemingway created one (Thomas Hudson in *Islands in the Stream*) in whom he intensified the deadly complexities of modern life, including the man's own awareness of his organic aggressiveness in love and war; and another (Santiago in *The Old Man and the Sea*) whose radical simplicity frees him from the complexities of modern life and enables him to make his peace with himself as a fated participant in the voracious order of life, "The Sea in Being," as Hemingway suggested in the book's original title. Guided not by the light of traditional Christian thought, but by its refraction in "the prisms in the deep dark water," Santiago comes to see that "the punishment of hunger . . . is everything" (84). And he endures by accepting himself as a brother of the fish, one who "was born to be a fisherman as the fish was born to be a fish" (116). Hemingway had projected the stories of Hudson and Santiago as a single four-part "sea-book." As he explained to Charles Scribner in 1951, "The first part is Idyllic until the Idyll is destroyed by violence. . . . The same people are in books 1, 2, 3. In the end there is only the old man and the boy" (*Selected Letters* 739). But he could find no coherent form for bringing these lives together in a single volume. In making Hudson "a painter of marine life for the Museum of Natural History" who vowed that "not even war must interfere with our studies" (316), he had sensed the way. But while, as a naturalist, Hudson can finally see and accept himself in the frightful biological order, he cannot undo time. This is Santiago's power. As an *old* man, he is free of the sexual conflict that had troubled Hudson. Thus, drawing on his own past as a turtle fisherman who had sailed before the mast, he can reduce life to its simplest terms by fishing from his wind-driven skiff.

Just two years after Hemingway published *The Old Man and the Sea*, Peter Matthiessen, who had spent three years as a commercial fisherman, began his career as a sea novelist by publishing *Race Rock* (1954). He went on to produce in *Far Tortuga* (1975) a masterpiece of sea fiction that, like *Moby-Dick*, *The Sea-Wolf*, and *The Old Man and the Sea*, feeds on the tradition and renews it. *Far Tortuga* is set in the Caribbean waters where Columbus first discovered the New World, and Matthiessen begins his novel by quoting an entry from Columbus's journal in 1503, when he sighted and named the islands called the Tortugas. The ensuing voyage, like Santiago's and many others in American sea fiction, is a story of biological survival: Captain Raib Avers is a turtle fisherman whose desperate economic circumstances have led to this last voyage. His ship, the former schooner *Lillias Eden*, is still rigged to sail but has new twin diesel engines that might make her more competitive in the doomed fishery. That the *Eden*'s engines are incompletely installed (the location of her wheel and controls make it impossible for the helmsman to see his way) suggests how much more desperate the circumstances of this voyage are than those of *Lightship 167*. In the "progress" from sail to steam to diesel, our ships proceed as does the tanker in the closing image of *To Have and Have Not*: "hugging the reef as she made to the westward to keep from wasting fuel against the stream."

In this metaphorical sense, we might envision the tradition of American sea fiction as a single vessel, "on its passage out, and not a voyage complete" (to use Melville's famous image of the world in "The Pulpit"). And the course that Matthiessen follows in *Far Tortuga* is that which Melville first charted in his cetological meditations on life. But of all the writers in the tradition of American sea fiction, Peter Matthiessen is by far the most highly accomplished naturalist. His book focuses on the same image of life feeding on life that drove Ahab under: his vision—immediately preceding "The Chase—First Day"—of the "smiling sky and the unsounded sea! Look! see yon albicore! who put it into him to chase and fang that flying-fish? Where do murderers go, man! Who's to doom when the judge himself is dragged to the bar?" With his voyaging mind, Ishmael survived this crisis, preserving his faith through his cetological meditations, Melville's natural theology. In *Far Tortuga,* Matthiessen relentlessly pursues the image of life feeding on life and the even darker implications of this violent order at its regenerative source—the aggressive sexuality of the great green turtles and the turtlers alike. And in a "modern time" of even greater "nautical darkness" than his predecessors had known, he ends the voyage in *Far Tortuga* with the vision of light that Captain Raib takes in at his death: he "looks straight up at the sun. Tears glisten in his eyes," and he utters his last words, "Oh, dat sun wild." Projecting this absolute reality of life on earth, the wild, searing simplicity of our existence, Matthiessen's seascape offers no transcendent promise like that which T. S. Eliot envisioned in "East Coker" in his seascape of "The wave cry, the wind cry, the vast waters / Of the petrel and the porpoise"— that "In my end is my beginning." But in his imagery of the "sun, coming hard around the world" again on the last day in *Far Tortuga*, and in his brilliant last word, "towarding" (without a period), Peter Matthiessen joins Melville in promising that the voyage is not complete.

BERT BENDER

Chapter Ten

GREAT LAKES MARITIME FICTION

The maritime literature of the Great Lakes, like the culture that produced it, did not grow from Protestant roots in the Atlantic colonies. Even as Governor Bradford chronicled the attempt to found a commonwealth according to Puritan doctrine, Father Marquette knelt before his crucifix on a wild and desolate Lake Michigan shore. He prayed to the Blessed Virgin Mary for protection and he prayed in French. For many years after, the only language those shores knew was the Québécois French of the voyageurs and the Algonquian of the Indians, interwoven occasionally with the clipped Scots of the *hivernauts* of the fur trade companies.

The first stories of storms on the lakes, tales of canoes that met the waves and the wind, were told in Indian lodges in winter when the spirits of the lakes were underground where they could not hear what was said about themselves and be offended. The next were French, told after meals of pemmican or beans and pork fat, as the voyageurs drew closer to the fire and spun their stories in song. There was no governed town with orderly streets and cultivated fields and solemnly dressed families bound for church on the Sabbath, no belief in the divine predestination in the founding of a settlement. Instead there was a string of isolated missions and trading posts, linked through the wilderness of rivers and portages and open lakes like beads on a rosary. The days were measured not by prayers, but by "pipes," a rest after three miles of paddling. This was the natural, not the heavenly world, despite the efforts of the black-robed priests. The colonists who settled America's first sea coast and bore her first mariners would not have recognized the Indians and voyageurs as seamen, nor believed their stories of the danger of the waters on which they traveled. Yet when their command of the lakes was lost to the English and the Americans, the first sailors of the Great Lakes left behind more than the names they gave to points and passages. From Death's Door to Cape Gargantua, the tales and legends they told have endured to be appropriated and transformed by the cultures that followed. It is with these narratives that Great Lakes fiction properly begins.

The Indians had navigated the lakes for millennia before Europeans appeared in the sixteenth century, and they treated the water with a fearful respect born of experience. Moreover, they believed the lakes were guarded by *manidog* who punished those who did not pay proper homage with fast and ceremony before embarking. Death's Door strait in Wisconsin takes its name from such a disaster, and there are stories of avenging waves and storms in the literatures of the Fox, Pottawatamie, and Chippewa. The Menomini tell of a party crossing Lake Michigan to receive bounty payments who, despite prayers and rituals before setting out, encountered a storm. Faced with imminent sinking, they made miniature canoes, filled them with lice, and set them upon the angry water with much weeping and mourning. Once the symbolic canoes had been swamped, the waves and wind grew quiet. Schoolcraft transcribes a story, "Peeta Quay; or, The Foam Woman," that tells how the lake spirits plot to punish a mother's overweening pride in her daughter's beauty by raising a storm that casts her adrift from the Lake Michigan shore and carries her through the Straits to Detroit. When the loss of her daughter has destroyed the mother's pride, the lake spirits raise another storm and bring her back, although now the daughter's beauty has been faded by age. These Native American themes, of malevolent *manidog* and of pride and over-confidence punished by storm, are reworked by writers as various as Mary Hartwell Catherwood (*Mackinac and Lakes Stories*, 1899), Charles Bert Reed (*Four Way Lodge*, 1924), MacHarg and Balmer (*The Indian Drum*, 1917), and Joan Skelton (*The Survivor of the Edmund Fitzgerald*, 1985) to become part of the fabric of lakes fiction.

The voyageurs took up the Indian trade routes, but not necessarily their respect for the water; the leading cause of death among them was drowning, usually because their canoes were swamped by storm waves before they could land. These men, perhaps the most hardy and brave mariners ever to ply the lakes, were also the least self-conscious of their accomplishments. Nor, unlike the Indians, did they consider storms and the concomitant drownings a sign of moral or spiritual failure. They simply were, like the rain on their faces, the hard beach on which they slept, and the packs they portaged. Although the songs associated with them were often old French folksongs such as "A la clair fontaine," they also composed what they termed *chansons de voyageur* about their experiences on the lakes. "Epouser le voyage" is characteristic in its elegiac complaint: "Dans le cours du voyage, / Exposé aux naufrages; / Le corps trempé dans l'eau, / Eveillé par les oiseaux, / Nous n'avons de repos / Ni le jour ni la nuit" (Archives, Université Laval). Historical novels written about them such as *Les Engagés du Grand Portage* (1938) often contain passages about storms on Lake Superior, but most fiction concerns their portages, the meetings at Mackinac Island, and their arrival at the lakehead. Their descendants still sail the lakes in novels such as Samuel Merwin's *His Little World; The Story of Hunch Badeau* (1903) and Henry Drago's *Where the Loon Calls* (1928).

French domination of the lakes was ended by the French and Indian War, and their antagonists, the British, continued to control the major fur trading posts until the War of 1812. Perry's victory on Lake Erie in 1813 is arguably the most important naval engagement of the war, and a number of historical novels treat the battle. Robert Harper's *Trumpet in the Wilderness* (1940), Carl Lane's *Fleet in the Forest* (1943), Samuel Woodworth's *Champions of Freedom* (1816), and N. C. Iron's *Double Hero* (1861) are typical in their focus on an Easterner who comes west, becomes involved with the building of Perry's fleet, fights at his side, and accompanies him when he crosses from the *Lawrence* to the *Niagara*. C. H. J. Snider's *Story of the "Nancy" and Other Eighteen-Twelvers* (1926) and Fred Swayze's *Rowboat War on the Great Lakes* (1965) describe the conflict from a British/Canadian perspective. The fiction set in the war is not typical of the lakes, however, for its themes are particular to that conflict. After the Treaty of Ghent and after the opening of the Erie Canal in 1825 came the rise of the first large-scale commerce on the lakes, built to carry freight to the East and immigrants to the West. It is during these years before the Civil War that modern Great Lakes fiction properly begins.

Scenes on Lake Huron (1836) shows its saltwater origins. Unlike Cooper's *Pathfinder* (1840), *Scenes* takes place almost exclusively on the water. It was written, the anonymous author tells us, to

> place the Lake Seamen upon an equal footing; or redeem if possible, a race of the most hardy and skillful men from the imputation, which has been often cast upon them by their Atlantic brethren [*sic*], old in the profession; that they were, in fact, "no seamen at all;" when in truth, after they themselves had tried the Lakes for some one or two seasons, and taken lectures from our fresh water gentlemen; the more candid have universally acknowledged them to be as efficient as any other, and that they ought to be ranked among the most able on the globe. (ch. 1)

In a grueling, month-long passage from Mackinac Island to Detroit in 1822, the skill and mettle of the crew are well displayed. But the writer was not an artist of Cooper's rank; the characters remain types and the action owes much to the patterns of voyaging literature popular for a century. Save for the geographic references, *Scenes* could have been set on salt water equally well. That is understandable, since there were no models for what lakes fiction could or should be except the saltwater novels appearing at this time, and besides, what better way to show the skills of lake mariners than put them in a storm that any saltwater man would recognize instantly?

With Cooper's historical novel *The Pathfinder*, the themes and antagonists of lakes fiction take the shape they will bear for the rest of the nineteenth century and into the twentieth. Few writers after him who describe windship sailing on the lakes can do so without setting up a conflict between a quiet, competent lakes captain and a contemptuous, condescending, incompetent saltwater man who, like Cooper's Uncle Cap, does not respect the lakes until he gets his comeuppance during a gale. This frontier theme of the local triumphing over the uppity Easterner, a staple of westerns, endures with the same tenacity on the lakes. Norman Reilly Raine uses it for "The Deep Water Mate" (1928); Robert Carse alludes to it in *The Beckoning Waters* (1953); several lesser-known writers use it for stories; and it is a staple of dime novels. Those who do not appropriate Cooper's theme outright take the tack of *Scenes* and include at least one, and usually several, digressions about the undeserved scorn lake mariners endure.

Melville, too, uses this theme in *Moby-Dick*. With perhaps the most frequently quoted description of the lakes, "they are swept by Borean and dismasting blasts as direful as any that lash the salted wave; they know what shipwrecks are, for out of sight of land, however inland, they have drowned full many a midnight ship with all its shrieking crew" (ch. 54), he introduces Steelkilt, who, unlike anyone on the crew of the *Pequod*, defies his unreasonable captain and triumphs. "The Town-Ho's Story" appears to suggest that the Great Lakes men are a hardier and more courageous breed than those of the Atlantic, but of all the writers who use this theme, Melville is the most equable in his characterizations of the lakes sailor and the Nantucketers, suggesting that neither is above reproach. What one must remember is that Steelkilt survives while his Nantucket antagonist does not.

At the turn of the century, Morgan Robertson took up the idea of Steelkilt and his ability to terrorize saltwater captains by force and cunning and developed the tragicomic Sinful Peck who, with his cohorts from Buffalo and Cleveland, creates mayhem on salt water in *Where Angels Fear to Tread* (1899) and *Sinful Peck* (1903). One need not seek far for the historical reasons underlying the persistence of this theme. Whenever times were bad on the lakes or on the coast, sailors left their usual berths looking for work. Saltwater men came west to Buffalo, where they competed for jobs with lakemen, and they could hardly have been welcome. For every Steelkilt or Sinful Peck who made life miserable for a deepwater man, there were saltwater men who were, in the words of lakeman Captain Thomas Murray, "The toughest lot of pirates who ever walked the decks of a ship. An officer never got them on watch without having to go down and drag them out . . . and he would find every man had a knife at his side" (*Inland Seas* 2 [January 1946]). Animosity and sailors flowed both ways in history and fiction, but few were the characters or captains who were as mild-mannered and patient as Cooper's Jasper Eau douce.

The Pathfinder sets in place a number of other motifs that persist in lakes fiction. Because he had sailed the lakes as a professional seaman and knew his subject well, Cooper makes much of the coastal piloting and ship-handling skills of lakes sailors. Eau douce sails without chart or compass or sextant, and according to Cap, makes decisions that fly in the face of reason. Sailing by dead reckoning alone and superb ship-handling are still characteristic of lakes pilots, and these skills appear in almost every work of fiction written about the lakes, most notably in *The Marked Man* by Karl Detzer (1927), "Out of the Trough" by Ralph Emberg (1948), and *Spindrift* by Harold Titus (1925). Cooper also structures his novel so that the action takes place as much on land as on water. Certainly this is partly because the Pathfinder's natural home is the wilderness, but it also reflects the geography of the lakes, where land, as lee shore or settlement, is never far away. Most Great Lakes fiction, even in the late twentieth century when lakes merchantmen may not get home for six months at a stretch, is set on land and water equally. The lines that hold ship and men to shore are not easily severed here; there are no Conradian voyages with a small group of men cut adrift from the constraints of civilization to work out their destiny. The shore, with all its concomitant problems, intrudes continually. With the change from the small schooner owned by the men who sailed her, to company steamships with funnels darkening the sky, to the thousand-foot diesels of the transportation conglomerates, the conflict between salt water and fresh becomes less pronounced and the theme of a sailor's place in the societies of ship and shore assumes prominence.

One can run away to sea on the lakes, and many do, but soon must come the confrontation with all that was left behind. Richard Matthews Hallet's *Trial by Fire* (1916) predates O'Neill's "*Hairy Ape*" by five years. Hallet's novel, although employing much the same imagery as O'Neill's play, reaches a different conclusion. Hallet limns a clear contrast between poor and rich on the Great Lakes, between the shacks of the "sailors flats" section of Cleveland and the wealthy neighborhoods of Chicago, but interestingly, Hallet's protagonist does not die an ape. He returns to his stokehold defiant and convinced of his worth to the end because he has confronted the life of the shore that made him and has triumphed over it. Unlike the cities, the lakes he sails are still a frontier, and all the sailors who have sailed that frontier before him—Eau douce, Steelkilt, Sinful Peck—have absorbed those lessons the frontier teaches: freedom, self-respect, and individuality. Myth-based though those lessons may have been, particularly by the turn of the century, they convinced men that they were not caged, but free and responsible to no master. This is Melville's theme in "The Town-Ho's Story," which is, significantly, the chapter he chose to publish separately in *Harper's* before *Moby-Dick* was printed; it is Cooper's theme as well, as it is Robertson's.

But frontier freedom and self-respect in the nineteenth century can easily become unrestrained license and self-aggrandizement in the twentieth. Mary Frances Doner's *Glass Mountain* (1942) and Jay McCormick's *November Storm* (1943), perhaps the two best novels written about the Great Lakes merchant marine, are both concerned with the impossibility of escape from the shore into complete freedom aboard ship. *Glass Mountain* tells the story of an orphan's drive to become master of a laker because he covets the prestige he will have in his hometown. His competitiveness, and the foolhardy wish for recognition that fuels it, cause the death of his son, the loss of his ship, the ruin of his career, and nearly end his marriage. The shore he tried to flee and yet to impress is his place at last. *November Storm* is more subtle but no less concerned with lives left behind and the impossibility of escape from them. In thoughtfully delineated portraits of the crew of the *Blackfoot*, McCormick works out anew the old conflict between an aging, alcoholic master who has nothing but his ship and an ambitious first mate. The tension between them provokes the captain to prove he is still capable of command, and so he pits his ship, one of the best on the lakes, against a fall gale. When the ship is wrecked, McCormick implies that the skipper who has isolated himself in his power until he believes he is invincible, even against the lakes, is not the best man to command; there must be a life ashore interconnected with that afloat. To refuse to acknowledge these connections is death for the soul and perhaps the ship as well. Walter Havighurst also considers this theme in *Signature of Time* (1949); Constance Fenimore Woolson used it earlier in *Castle Nowhere* (1875); and George Vukelich describes it in the grim context of winter sailing in the 1960 story "The Bosun's Chair."

This rhythm between shore and lake informs the fiction of the fishing fleets as well—not surprisingly, since most fishermen work only with their families or a small hired crew, and seldom, if ever, pass the night on the water. But there is a difference in the land-water theme of the fishing novels. For the fisherman, the shore (aside from his family and other fishermen) represents the law and the enforcers of the law who know nothing of fish but would take away the fisherman's livelihood. In Great Lakes fiction, the fishermen occasionally fight against each other, as they do in countless saltwater novels, but they also always fight against the conservation officers and police who have tried to regulate catches of the declining stocks of fish for decades. The fisherman may confront the regulators with anger and defiance, as in the fiction of James Oliver Curwood (*Falkner of the Inland Seas,* 1931) and George Vukelich (*Fisherman's Beach,* 1962), or with passive trust in God as in Louis Kintziger's *Bay Mild* (1945), but for all fishermen the lawbound society of the shore is hostile.

Yet despite the vagaries of time and genre, as sail gave way to steam and steam to diesel, the one archetypal theme of lakes fiction is the sudden and deadly storm; there are few works of lakes fiction that do not reach their climax in a gale. *November Storm* is a classic title in this respect, since fall storms as testers of mettle and winnowers of character begin with *Scenes on Lake Huron.* These storms in fiction are no coincidence, since in fall, with the waters still warm from summer, the cold wind sweeps down on them from Canada and breeds gales of hurricane force that last for days. The shipping history of the lakes, one of the world's most wreck-strewn shores, catalogs its greatest losses in November, most recently the 729-foot *Edmund Fitzgerald,* which sank with all hands on Lake Superior on November 10, 1975, faster than anyone in the pilot house could radio Mayday. For a maritime culture that had begun to think itself immune to gales by virtue of technology, the loss of the *Fitz* was a stunning blow.

Once the flurry of new journalism subsided, a slender Canadian feminist novel by Joan Skelton, *The Survivor of the Edmund Fitzgerald* (1985), began to ask more profound questions. *Survivor* is one of few Canadian novels about the lakes, most of which are lighthearted entertainment. In contrast, this is the story of a reclusive Toronto woman awaiting death on the north shore of Lake Superior from a massive, untreatable infection. She meets or imagines another hermit, a young Canadian artist who has stowed away aboard the *Edmund Fitzgerald* on her last voyage. After trying and failing to warn the complacent crew of water in the hold, he crawls into a lifeboat that survives the sinking. Aside from the ingenuousness of Skelton's implication that Americans and American culture are not healthy for women, ships, and other living things, the novel reworks a theme first inscribed by *Scenes on Lake Huron* and Cooper. Now, a century and a half after Uncle Cap mocked the lakes, it is the Americans who have become the arrogant sailors. They have proved their place as mariners on the lakes and assumed their power, but with that power has come a complacency that wrecks ships and destroys lives. They must now be instructed by another seeming novice, a Canadian who knows nothing about ships but who will, by simple common sense and honesty, question what they believe. The telling difference now is that unlike the saltwater sailors of the nineteenth century, the Americans do not live to repent of their mistakes.

In one of the last of her visions before dying, the narrator of *Survivor* imagines that she sees Missipeshu, the great spined water-cat manitou of the Chippewa whose lashing tail raised storms on Lake Superior. She sees the waves rising like animals to come crashing over the ship, punishment, the Indians of centuries ago would have said, for those who set out arrogantly or unthinkingly upon the waters. The land surrounding the lakes has been explored, conquered, mined, and settled until it is a wilderness no longer; the ships float serenely in a bubble of modern technology. But the unpredictable ferocity of the waters has not changed since the glacier that gave them birth retreated growling to the Arctic. This Janus-faced character of the lakes, such bounded and seemingly placid waters that can become so quickly terrifying, informs nearly everything written about them. Kipling, traveling to Lake Superior in the nineteenth century describes it well:

> There is a quiet horror about the Great Lakes which grows as one revisits them. Fresh water has no right or call to dip over the horizon, pulling down and pushing up hulls of big steamers, no right to tread the slow, deep sea dance-step between wrinkled cliffs; nor to roar in on weed and sand beaches between vast headlands that run out for leagues into bays and sea fog. Lake Superior is all the same stuff of what towns pay taxes for (fresh water), but it engulfs and wrecks and drives ashore like a fully accredited ocean—a hideous thing to find in the heart of a continent. ("Letters to the Family")

His uneasiness mirrors that of many other writers who feel compelled to offer some explanation for waters that long since should have yielded to civilization as has the surrounding land, but which have not. Thus the fiction of the lakes seldom characterizes them as uncaringly destructive like the ocean in Melville or Conrad or Crane; rather they are most often personified as willfully malevolent or haunted by spirits who will punish those who dare too much, who brag too freely of their omnipotence on a summer's day. In fiction that consciously appropriates Indian legends for cheap dramatic effects, such personification is to be expected, but the characterization of the lakes as a presence bent on destruction occurs in nearly all fiction, poetry, and folksong across a long span of time.

In *November Storm*, a realistic novel that makes no attempt to draw on legends and folklore, a deckhand clings to the engine room grating in the gale, afraid to move because it would be "inviting [the] storm's attention and a greater wrath . . . to dare action [would] deliberately infuriate the destroying force by setting puny deeds, machines in its path" (ch. 12). His realization that technology is no protection from the uncontrollable force of the storm, that instead it encourages men toward untenable risks they might not attempt otherwise, is a recurring theme in twentieth-century lakes fiction. In this novel and others, the Great Lakes give the lie to the North American dream of conquest by settlement and technology, even as they are surrounded by it and their ships help make its success possible; this irony forces us to confront far older and less comforting truths of existence we usually avoid acknowledging. Steven Dietz, author of the recent play *Ten November* (1987) about the sinking of the *Edmund Fitzgerald*, describes the focus of his drama as "the myth of invincibility in our culture. . . . We get cocky with nature. Technology has enabled us, we think, to reinvent it in our own image. [But Nature] still melts our wings when it needs to" (Playwright's Note). In a bleak drama set on a stark stage, he examines all the old lakes myths that purport to explain what causes ships to sink, and concludes that the greatest myth of all is technology.

Despite what some would criticize as a narrowly regional focus, lakes fiction is valuable because it partakes of two traditions—the recent one of the conquest of a continent, and the ancient one of the sea that never changes. The presence of an oceanic body of water in the midst of the land is unique in North America, and the stories that have grown from men's experiences sailing there question our myths of control over nature in a way that saltwater fiction does not. We do not expect to control the forces of the ocean; shipwreck and destruction there are accepted traditions in literature and life. Yet we have never ceased to believe we can control the waters of the Great Lakes as we have controlled their shores, and we have consistently failed—a lesson the Indians taught in their stories generations ago. Is it so surprising then that in our stories we have tried to attribute our failure to a force outside ourselves? If one can blame a mythical presence—a spined water-cat reincarnated anew, a nameless presence bent on destruction—the failure does not lie within us. Carl Jung and D. H. Lawrence might suggest that the personification of nature in lakes fiction is an illustration of how the spirits of place endure despite differences in time and culture; others would say that any ship in danger on any large body of water prompts an atavistic urge to personify what cannot be understood. But what all the storytellers of the lakes, from a Chippewa in his winter lodge calling up the specter of Missiphesu to actors on a stage in Chicago recreating the *Fitz*, really illustrate is our own frailty in the face of nature, and our fear. This is a lesson we do not usually encounter in North American fiction, and it is one we would do well to heed.

VICTORIA BREHM

Chapter Eleven

AFRICAN-AMERICAN LITERATURE

Historically and culturally, the African-American experience has been an inland one. Documentation describing the Middle Passage, the shipboard journey across the Atlantic from Africa to America, proves it to have been debilitating, degrading, dehumanizing, destructive of human life and civilization. Black Americans, consequently, it appears, have not generally turned seaward in their literature. They have instead concentrated on creating a culture in this strange and terrible new land. In their literature, oral and written, the dominant elemental images call on swamps, floods, and rivers rather than ocean; dominant travel images invoke railroads and wagons rather than ships. This contrasts with the literature of white Americans who crossed the sea with expectations of glory and who subsequently worked the sea with expectations of profit. In the wondrous African-American folk story of slaves who, on discovering conditions in this strange new land to be unbearable, simply and magically flew to Africa and freedom, there is no reference to the sea; they left the earth, saturated with their blood and sweat, for the benign and blessed sky, but the sea, not represented as help or hindrance, does not appear at all.

However, the sea is not absent from African-American literature. Although the Middle Passage is seldom the subject for African-American writers until the twentieth century, from the time of eighteenth-century slave narratives, the sea, as image and element, is represented by black Americans as touching their lives and extending their imaginations. Indeed, the changing responses to the sea reveal their changing responses to historical circumstances, to their identity as Americans, African Americans, individual human beings. Two separate traditions dominate nineteenth-century African-American literature: on the one hand, that created orally by the folk, and, on the other, that created in writing by known individuals. The former, consisting of songs, stories, sayings, and sermons, tends toward myth, metaphor, symbol, humor, while the latter, slave narratives and novels, leans toward history, realistic description, allegory, didacticism. In the richly complex twentieth-century works, the two traditions join. Appearing sporadically in both traditions in the nineteenth century, the sea is represented with increasing diversity in the twentieth century—and increasing frequency. In contemporary African-American literature, writers combine oral and written traditions and thus, in representing the sea, draw simultaneously on both myth and history, evoking both metaphysical and physical possibilities, proving it to be a barrier and, often ironically, a catalyst to life. Consequently, contemporary black writers have been able to confront the full implications of the Middle Passage, African Americans' primal experience of the sea, their inaugural experience of America. To the grand vision of the sea as the unconscious and as a manifestation of infinity, as expressed by white Americans from the nineteenth century on, the contemporary African-American writer adds an emphatic historical dimension.

Spirituals referring to the River Jordan or that river which is deep and wide are well known, these references being recognized as allusions to actual rivers—the Ohio, the Mississippi—which marked the slave community's geographical boundaries, but not the limits of their spiritual aspirations. References to the sea in the spirituals and sermons of the oral tradition are directly derived from the Bible. They refer to God's creation of the land and the sea, to Noah and the Ark, to Jonah and the whale, to Moses at the Red Sea, and to Jesus on the Sea of Galilee. For the slave community, for whom the sea-crossing would have been an appalling memory, that some songs should identify the sea as an uncontrollable element is probably not surprising. What is noteworthy, however, is the existence of spirituals and sermons in which an individual either prevails against the sea or harnesses it to his use. Thus Noah maneuvers the Ark after rocking and reeling, to snug harbor on the mountain top; Moses smites the Red Sea while "Pharaoh, Pharaoh, and his host got drownded," and Jesus calms Galilee.

The sea also becomes in certain songs, as it did historically for slaves on the Atlantic coast, a means of escape to freedom. Thus in the spiritual "Didn't My Lord Deliver Daniel?," the community celebrates the fact that as Daniel was delivered from the lion's den and Jonah from the whale's belly, all will be freed; it concludes, "I set my foot on de Gospel ship / An' de ship begin to sail / It landed me over on Canaan's shore / An' I'll never come back no mo'." Another spiritual triumphantly exclaims, "My ship is on de ocean. / . . . Po sinner, fare you well." Sermons about Jonah also have overtones of wish-fulfillment, implying that the sea might lead to liberation.

Thus, although W. E. B. Du Bois called the spirituals "sorrow songs," those spirituals and sermons invoking the sea inspired and encouraged the slave community, strengthening both their collective hopes for a savior in this world and individuals' hopes for saving themselves.

Among blues songs, those often ironic, always personal expressions of experience that evolved from a rich tradition of musical forms, including the spirituals, the sea figures scarcely at all. Here the singer confronts the elemental earth, ploughs that furrow, lifts that hammer. Here backwaters, not tides, rise; levees, not leeboards, fail, and the singer drinks muddy, not salty, waters. Only occasionally is love described as being like "the deep blue sea." In a rare blues song describing a sudden storm at sea, the singer, a woman, feels overwhelmed by the sea's intensity. Though personal incompatibility and betrayal are the more conventional reasons for the loss of love and the subject of countless blues lyrics, here the sea, uncontrollable, unpredictable, reflecting the insecurity of coastal fishermen, is responsible for the singer's separation from her lover. In her despair, she imagines her psychological state on land as analogous to her lover's physical situation on the sea:

> It's rainin' and it's stormin' on the sea,
> It's rainin', it's stormin' on the sea.
> I feel like someone has shipwrecked poor me.

In response to the 1912 sinking of the *Titanic*, from whose maiden voyage all persons of color, servants as well as Jack Johnson, the heavyweight boxing champion of the world, were excluded, a number of folk songs were created. All suggest, however, that justice ultimately prevails as it did when Pharaoh's host was drowned. The excesses of capitalism and racism, and the absurdity of greed and white supremacy are proven, although "it was sad when the great ship went down." One song proclaims that Jack Johnson was emphatically vindicated by staying on shore.

As Langston Hughes noted and recorded in *The Book of Negro Folklore*, however, "folk versifiers insist that there was one Negro aboard" the ill-fated vessel (367). The story of Shine suggests not only the triumph of justice, but also the triumph of a black man over oceanic forces and humor over despair. Shine, forced to work in the ship's great boiler room, ultimately has the last laugh on the imperious, cock-sure captain. As the *Titanic* sinks, the captain, a banker, and his seductive daughter all try to bargain with him to save their lives, offering him money, stock, and sex. Knowing that there is more of each on land than at sea, Shine denies their pleas, and at the end of each stanza, "he swimmed on," until the poem's final ironic couplet:

> When all them white folks went to heaven,
> Shine was in Sugar Ray's Bar drinking Seagrams Seven.

Shine, back home in Harlem, alive and well on the land, is far better off than he would be either in heaven or on the sea.

All of the elements—earth, air, fire, water—appear with regularity in African-American folk tales, and like Shine, the representative character in the tales usually proves able to put these elements and the natural world to use. In tales involving the sea, this character, despite his small size, uses the sea to demonstrate his moral superiority to his larger antagonist. Adapted from West-African tales, these stories, whether of man or beast, have been interpreted as a means by which the slave community was empowered. Thus, in tales such as "Wolf and Birds and the Fish-Horse" and "The Two Johns," the Fish-Horse (the manatee) and Little John, abused by a stronger but stupid and greedy opponent, succeed in drowning him in the sea. Another story, giving us an African-American version of the Deluge, has lowly crawfishes drowning all the mighty of the world:

> Dey bo'd inter de groun' en kep' on bo'in twel dey onloost de fountains er de earf; en de waters squirt out, en riz higher en higher twel de hills wuz kivvered, en de creeturs wuz all drowned; en all bekaze dey let on 'mong deyselves dat dey wuz bigger dan de Crawfishes. (*A Treasury of Afro-American Folklore* 499)

Although such tales, with their apocalyptic dimension, also project wishful thinking by allying the powerless with the forces both of virtue and the sea, they would have strengthened the African-American community. (Several contemporary African-American works, notably Richard Wright's "Man Who Lived Underground" and Ralph Ellison's *Invisible Man*, describe their protagonists as burrowing, like the crawfish, into the earth's bowels. Such a vantage point reflects not only the lives many blacks are forced to live, but also a position chosen for security and sabotage. However, neither Wright's nor Ellison's alienated black men have the collective support of a community to get down far enough to loose revolution on the world by fire or by water.)

In a particular category of folk tales, a black man, cast from his ship into the sea, does not seek retribution but instead discovers the sea's mysteriousness astonishingly personified by a mermaid. Unlike Pip, the black cabin boy of *Moby-Dick*, who, abandoned at sea, is "carried down alive to wondrous depths, where . . . the miser-merman . . . revealed his hoarded heaps" (ch. 93), these black men, wrapped in the mermaid's hair, are carried down to rather ordinary domestic scenes. Pip, rescued, becomes insane while the castaways of African-American folk tales return to land hardly the worse for wear. In only one of the mermaid tales cited by Richard Dorson in *American Negro Folktales* does the sailor undergo a sea change as a result of his oceanic tenure; here, Sarah "Aunt Jim" Jackson recalls: "The people didn't know him 'cause he was so hairy. He didn't have on no clothes. But he knew them. He shaved up and got natural; he favored himself. But he looked like some kind of animal before." Although the stories vary, consistently the mermaid has magical powers over the ships and the men who sail them. Seemingly omniscient, she can call their names and their bluff; she can wreck their ships and control their lives. Her "home is on the bottom of the sea," "in the wall of the sea like the alligator," or in a hole "dark as midnight," and it is thus tempting to consider that she and her sisters are like the crawfish, creating an undersea existence as an alternative to life on land. The mermaid's tail, however, renders her helpless on land, to which the men eventually return to live mundane lives and to tell their tales of the wondrous sea. If the mermaids are threatening to men, as women and the sea may be, the men who survive have their revenge, not on the ship captains who cast them overboard, but on these female denizens of the deep, who having become dependent on men, are doomed to perish without them.

Before the Civil War, the sea figured in African-American lives in several respects. Although the United States outlawed the slave trade in 1808, condemning it as piracy in 1820, Africans continued to be transported across the Atlantic until the Civil War. Slaves living along the Atlantic seaboard, however, often sought escape from slavery by the sea. Escaped slaves as well as free blacks also found that the sea provided various means of employment. By 1859, of the twenty-five thousand native-born American seamen working out of New Bedford, more than half were blacks, with twenty-nine hundred serving in the whale fishery, the others in the navy or the merchant service. (New Bedford, home of the black Paul Cuffee, prosperous shipowner and merchant, civil rights advocate and proponent of African resettlement plans, home of tolerant Quaker families, became an important terminal on the Underwater Railroad and the subsequent home of many black families.) In contrast to works in the oral tradition, which represent the sea figuratively, pre–Civil War literature written by blacks draws largely on these historical conditions and usually gives us the literal sea.

The Interesting Narrative of the Life of Olaudah Equiano; or, Gustavus Vassa, the African, Written by Himself (1789) is one of the earliest slave narratives and one of the very few describing the agonies of the Middle Passage. Equiano, enslaved as a boy in the interior of Africa, had never seen "any water larger than a pond or rivulet," and when he was brought, in the course of a six- or seven-month journey, to the sea coast, astonishment turned to terror as he believed he "had gotten into a world of bad spirits" (ch. 2).

Equiano's sense of reality continues to be undermined during his voyage by both the apparent technological mastery of the whites over the sea through sail and quadrant and their tyranny over himself and his fellow Africans. Beneath decks, suffocating from pestilential smells and lack of air, chafed by chains, surrounded by the shrieks of the dying, he describes a "scene of horror almost inconceivable." On a day of "smooth sea and moderate wind," two of Equiano's fellow sufferers choose the freedom of the sea over the confinement of the ship and leap to their deaths. These initial experiences at sea prove so disorienting that Equiano feels "more persuaded than ever, that I was in another world, and that every thing about me was magic" (ch. 2).

Equiano lived to tell his story of the Middle Passage and to sail the sea again. Through his acquaintance with young white sailors he rapidly lost all superstitious dread of the sea, although he remained acutely aware of its unconscious cruelty and of the deliberate cruelty of human beings. He could never take the sea for granted, just as he never took freedom for granted.

Sold to a series of owners until he was eventually able to purchase his freedom from his American owner, he made repeated trips across various seas—to the West Indies, the United States, Great Britain, France, Spain, Turkey, and even to Greenland. At sea, he became a jack-of-all-trades, learning not only the sailor's basic ropes but also how to navigate, stand a naval siege, purify salt water, market sugar and rum—and participate himself in transporting slaves. Chapters 3 through 12 of Equiano's narrative detail his sea-crossings; yet he never romanticizes the sea. Quite simply, life at sea offers him the respect of his fellows and the dignity of work. Although he sojourns briefly on land and at times yearns for the formal education and social security the land would appear to offer, he seems invariably drawn again to the sea. Late in his life, for example, he writes: "I thought of visiting old ocean again" (ch. 12). Yet he never ceases to value the freedom he lost on the Middle Passage and that he witnessed his fellow slaves desperately seeking to retrieve by throwing themselves overboard, committing himself at the end of his life to recolonization efforts in Sierra Leone.

For several slaves, writing in the nineteenth century of their escape from plantations on the Atlantic coast, the sea becomes literally and figuratively the means of attaining freedom. In these narratives, memory of the Middle Passage has been supplanted by more immediate horrors of the Peculiar Institution; the sea, by contrast, prompts hope, suggests possibilities. Thus, James W. C. Pennington, having been enslaved in Maryland, can draw imaginatively on his perceptions of the sea in his 1849 narrative as he describes his hopes on the first night of his escape overland:

> I felt like a mariner who has gotten his ship outside of the harbor and had spread his sails to the breeze. The cargo is on board—the ship is cleared—and the voyage I must make; besides this being my first night, almost everything will depend upon my clearing the coast before the day dawns. (ch. 2)

For Frederick Douglass, writing his narrative in 1845, the sea and the ships on it, which he views from the Maryland coast, torment him by their evocation of liberation. Contrasting with his condition, they exacerbate his conscious desire to be free, and precipitate the most self-conscious, but perhaps the most eloquent rhetoric in the narrative:

> "You are loosed from your moorings, and are free; I am fast in my chains, and am a slave! You move merrily before the gentle gale, and I sadly before the bloody whip! You are freedom's swift-winged angels, that fly round the world; I am confined in the bands of iron. O that I were free! O, that I were on one of your gallant decks, and under your protecting wing! . . . The glad ship is gone, she hides in the dim distance. I am left in the hottest hell of unending slavery. . . . It cannot be that I shall live and die a slave. I will take to the water. This very bay shall bear me yet to freedom." (ch. 10)

Douglass actually did not take to water to effect his liberty; however, working as a caulker on the Baltimore wharves, he became acquainted with ships and sailors. Such knowledge provided him with the means of escape. "Rigged out in sailor style" with "a red shirt and a tarpaulin hat and black cravat, tied in sailor fashion, carelessly and loosely about [his] neck" and carrying a free black sailor's "seaman's protection papers," Douglass succeeded in taking a train out of slavery into safety. He explains, "My knowledge of ships and sailor's talk came much to my assistance, for I know a ship from stem to stern, and from keelson to crosstrees, and could talk sailor like an 'old salt.'" In showing the conductor his papers, Douglass proclaimed, "'I have a paper with the American eagle on it, that will carry me round the world,'" thereby identifying himself with a particularly American promise of freedom on the sea (*Life and Times*, pt. 2, ch. 1). (One thinks of Ishmael's expectation, as he embarks on the *Pequod*, of being able to "circumnavigate the globe.") Arriving in Philadelphia, Douglass made his way through New York to New Bedford, where he found work again among seafaring men. There, as a free man, working for his own wages, he saw "the nearest approach to freedom and equality that I had ever seen" (ch. 11).

For Harriet Jacobs, who actually fled North by sea after seven years of confinement in an attic, the sea evoked the exhilaration of freedom. Although her arrival in Philadelphia was fraught with anxiety, Jacobs suggests in her 1861 narrative, *Incidents in the Life of a Slave Girl*, through her description of the birth of the sun out of the sea, her own sense of rebirth.

As Douglass and Jacobs turned their rhetorical skills to the physical liberation of African-Americans, John Thompson in his 1856 narrative, *The Life of John Thompson: A Fugitive Slave*, sought the spiritual liberation of humankind at large. Thompson, following his escape by land, continued to fear pursuers and so took to the sea; like Douglass, he fled from Maryland to Philadelphia to New York, and then to New Bedford, where he eventually shipped as a cook on a whaling vessel. In a manner comparable to Melville's cetological chapters, he describes in precise and concrete detail the tools and techniques of the whaling enterprise and his experience on shipboard; in his last chapter, however, he shifts dramatically to an elaborate metaphysical conceit equating "the passage of a Christian from earth to glory [to] a gallant ship under full sail for some distant port." But:

> if Christ is not the sole foundation, and his righteousness the grand security, then on the slightest trial, the seams open, the vessel bilges, and every soul on board is lost. . . . nor must the fairest gales entice us to sea without the heavenly pilot: . . . we dare trust no other at the helm, because no other can safely steer us past the rocks and quicksands.

In the years leading up to the Civil War, during the height of abolitionist activity and the publication of slave narratives, African Americans had also started to fictionalize the experiences of slavery. The first to do so was Frederick Douglass. His "Heroic Slave" (1853), whose language and details show Douglass's familiarity with nautical matters, is based on the true story of Madison Washington, a slave who led a successful mutiny aboard the ship *Creole*, bound from Virginia for the slave market of New Orleans with a cargo of 134 human beings. Sailing instead for Nassau, Washington brought them all to freedom. In two novels, William Wells Brown's *Clotel; or, The President's Daughter* (1853) and Martin Delany's *Blake; or, The Huts of America* (1859, 1861–62), the literary possibilities for the remote and horrific history of the Middle Passage as well as the recent and hopeful history of blacks escaping by sea become apparent. Brown, who himself as a former slave had written an account of his escape by land and river in 1847, derives much of his novel from abolitionist rhetoric and his own experience; however, his brief editorial, contrasting the significance of the Mayflower's journey to the New World in 1620 with that of the Jamestown slave trader's in 1619, indicates the stretch of his historical consciousness and his imagination. In *Clotel*'s narrative, nineteenth-century ships continue to represent these moral polarities. Clotel, the daughter of Thomas Jefferson, and the tragic mulatto after whom the novel is named, is captured as a fugitive slave and destined to be transported by sea from Richmond, Virginia, to the slave markets of New Orleans, prolonging the experiences of the Middle Passage. Escaping again, she chooses suicide, drowning herself not in the sea, but in the Potomac. Water, however, proves more merciful to Clotel's daughter and her husband-to-be, for both use the sea to escape from America to Europe, reversing the historical experience of the Middle Passage.

With the exceptions of its representation in Equiano's and Thompson's accounts, the sea is treated with greater complexity in Delany's *Blake* than in any other nineteenth-century African-American work. Writing at the beginning of the Civil War, Delany examines history in his reflection on the interrelationship between African Americans and the sea: in particular, America's continued involvement in the slave trade; shipboard mutinies by slaves, including the 1839 *Amistad* incident and the 1817 slave revolt that led to Melville's "Benito Cereno"; and the increasing appearance of free and fugitive slaves in maritime service. In his novel, Delany describes the attempts of Henrico Blacus, later known as Blake, a free black, who is sold into American slavery in Cuba, to escape and return to Cuba to organize an international insurrection against slavery. The scene of most extended dramatic intensity and complexity in the novel occurs at sea. Blake, having hired out as the sailing master on a slave trader bound for Africa, determines to use the return trip as the basis for generating unrest among the slaves. Thus Delany envisions the possibility of the horrors of the Middle Passage being transformed into triumphant revolution.

Blake's employer regards him, superstitiously, as one able to tame the sea. Little does he realize, however, that while Blake may be harnessing the ocean's storms, he will be unleashing the storms of rebellion. As the slaver, appropriately named the *Vulture*, makes its way back to Africa, the largely black crew unnerve the white officers by singing a song that celebrates both theirs and the sea's unfettered condition. Evidently they take inspiration from the sea, and when he is reprimanded for not controlling the black sailors, Blake proclaims their membership in the democratic fraternity of sailors. After the ship has taken on its cargo of human lives and begins its return across the Atlantic, however, joyful song ceases. In a sequence of three Middle Passage chapters Delany dramatizes the despair of the enslaved Africans and the justification for all blacks to revolt—the turbulent conditions on the ship mirroring the turbulence of the sea. Thus Delany's description of the agonies of the Middle Passage reflects Equiano's and anticipates Alex Haley's more than one hundred years later. Although Blake fails to commandeer the ship, the seeds of rebellion are sown among the enslaved Africans who are greeted on arrival in Cuba by a rainbow promising their future success.

Historians observe that after the Civil War, several factors contributed to a decline in black seafaring. Although they remained in commercial shipping and in the navy, blacks were finding work in industrial jobs in the North and in the West. The general decline in American shipping saw a commensurate loss in numbers of jobs at sea, with many of them going to white immigrants. It has been claimed that following the war an increased confidence among black sailors resulted not only in their demanding more rights at sea but also in whites reacting with "institutional racism" in the forecastle. Shine's job on the *Titanic* was, after all, apocryphal. Langston Hughes, recalling his experiences on a freighter in the 1920s, tells of his third mate, "a fine old New Englander of abolitionist stock," who, after twenty-nine African voyages with black seamen, considered them "splendid sailors," and who regretted their general absence on modern ships (*The Big Sea*, pt. 2, "Voyage Home"). One can conclude that such racist and economic factors interrupted the continuity of living memory among blacks who had experienced the sea. After the Civil War, the memory of the sea in African-American literature as leading to enslavement, or inspiring freedom, as prompting magic or providing work, also receded. Until the resurgence of African-American writing in the 1960s, references to the sea are sparse. Survival occurs in the fields, in the cities; liberation must be won on land; a culture cannot be created at sea where women and children are habitually absent. While the oral tradition may have sustained an awareness in the black community of the sea as the source of metaphor and myth, the slave narratives and the novels *Clotel* and *Blake* fell out of print until the 1960s.

When the sound of the sea *is* heard in African-American literature of the early twentieth century, it seems removed, distanced by literary convention and by the fading of historical memory as black writers tried to shape their personal, aesthetic, and political consciousnesses by recognizing dual kinship to African-American and Western traditions. Thus Paul Laurence Dunbar wrote poems in black dialect conjuring up a never-never land for African Americans, reflecting, in Melville's terms, "all that's kind to our mortalities," as well as poems in standard English, evincing a deep dissatisfaction with life. While the former poems focus on the cosy cottage and the pleasant meadow, the latter ones often center on the sea, always represented as antithetical to the poet's life on land. Thus, the sea that had helped Moses hinders Dunbar when he plants his feet "In Life's Red Sea," and "No voice yet bids th' opposing waves divide!" In "On the Sea Wall," the poet observes from his vantage point on shore, the wind and the sea contending, with the sea invariably raging and rising up to betray the wind and to defeat the ships sailing across it. If the sea seems to be all that is dangerous, unpredictable, inexplicable, irrational, in such a poem as "Ships That Pass in the Night," Dunbar in addition sees it as embodying unattainable promise; however, his expression of heartfelt yearning seems to pale beside Douglass's paean to freedom. Finally, in a poem entitled "Anchored," Dunbar opts for the land over the sea, as difficult as he knows such an option is.

In contrast to Dunbar's romantic despair, the writers of the Harlem Renaissance view the sea through a lens of romantic possibilities of exotic lands and loves. As their consciousness of America's racial and economic inequities evolved, and Africa and the West Indies came to suggest desirable alternatives to this country, the horrors of the Middle Passage seem to be forgotten while the seas surrounding the homelands are remembered with nostalgia. Thus Countee Cullen, tormented by his three-centuries-long alienation in America, poignantly asks in his poem "Heritage," "What is Africa to me: / Copper sun or scarlet sea[?]" Claude McKay, born in Jamaica to become an American citizen and a world traveler, always remembers the sea-ringed island of his youth as a pastoral Eden. In "When Dawn Comes to the City," the poet watches "Dark figures" "sadly shuffle" from city tenements, "cold as stone"; yet in his mind's eye, he sees life waking, waters rushing "on the island of the sea, / In the heart of the island of the sea / . . . [and] There would I be at dawn." Another poem, "Subway Wind," compares the city's "captive wind," with the fresh, free sea island winds (*Selected Poems*). For Hughes, the sea evokes less-specified dreams. He may be the first African-American writer for whom the sea is simply mysterious. In "Long Trip," it is "a wilderness of waves"; in "Sea Calm," it is preternaturally tranquil. Hughes also discovers that magically, for some, the sea makes dreams come true; but in his short poem "Catch," he partially dissipates this sense of wonder by creating a comic fish story and reversing the tales of sailors caught by mermaids: "Big Boy came / Carrying a mermaid / On his shoulders." Like Big Boy, the sailors of Hughes's poems come home from the sea (*Selected Poems*). Although the "Young Sailor" in his poem of that title may gain strength from the green sea, he acquires laughter from the brown land; and in the "Port Town," a whore is able to entice the young sailor with romantic promises of another sort.

In their prose, much of which is autobiographical, McKay and Hughes continue to contrast sea with land. Although the sea offers them and their characters romantic options for renewal and rebirth—as opposed to the more realistic options of work or escape—invariably they return to the land. World travelers both, McKay and Hughes realized that the racial tensions ashore were frequently transferred to the ship where the owners and authorities were always white and where close quarters intensified hostilities. Their prose also reflects their realization that return to Africa, via either Marcus Garvey's Black Star Line or an ordinary cargo ship, or to the Caribbean's "Salty-warm blue bays where black boys dive down deep into the deep waters, where the ships shear on foamy waves . . . and black pilots bring them in to anchor" (*Banjo*, ch. 23) does not resolve the African American's problem of cultural and racial identity.

In his two early novels, *Home to Harlem* (1928) and *Banjo* (1929), McKay pairs Ray, educated, intellectual, yearning for a sensual, devil-may-care existence, with Jake/Banjo who lives it. Both men are sailors—vagabonds; yet both novels are set in port. Despite the poverty and the oppression of their lives on land, Jake chooses it for the sake of his Harlem sweetie, and Banjo and his friends choose it over "'wrestling with salt water'" (ch. 3). In the conclusion of *Home to Harlem*, tormented by questions of identity, Ray ships out to sea as a mess-boy. His appreciation for Banjo, however, in McKay's second novel leads him to reverse the decision of the earlier book. Intellectually recognizing the sea's romantic lure, he finally rejects the sea for the land's warmth, for its variety of colors and odors, for its bawdy, brawling, boisterous human possibilities. He also rejects the condition of homelessness, which McKay claims is the lot of "Colored seamen who had lived their lives in the great careless tradition, and had lost their papers," leading them to be classified as "Nationally Doubtful" (ch. 25). In his 1937 autobiography, *A Long Way from Home*, McKay describes his own stay in the lusty port of Marseilles as well as his odyssey through the United States, Russia, Europe, and North Africa; yet he mentions the sea only in passing as his means of transport.

Like McKay and his characters, Hughes knew the seaman's life. His two autobiographies, *The Big Sea* (1940) and *I Wonder as I Wander* (1956), document his journeys across continents and seas. For the young Hughes of *The Big Sea*, his journey to sea liberated him from the social insecurities and constraints and the intellectual pretensions he felt himself prey to on land; going to sea, he is replenished, almost ritualistically reborn. Yet for the older Hughes of *I Wonder as I Wander*, the experiences at sea are to be casually enjoyed or merely endured. *The Big Sea* begins with Hughes at the pivotal age of twenty-one, leaning over the rail of a freighter, throwing his collection of books into the sea. In the course of his voyage, he crosses the equator and is "baptised in salt water and shaved with crude oil by Father Neptune." But a more significant rebirth occurs because of his determination, like other youths before him, "to see the world" by going to sea rather than by reading books; freed of his books, Hughes believes he is freed not only of the falsifications of life, but also of:

> everything unpleasant and miserable out of my past: the memory of my
> father, the poverty and uncertainties of my mother's life, the stupidities
> of color-prejudice, black in a white world, the fear of not finding a job,
> the bewilderment of no one to talk to about things that trouble you, the
> feeling of always being controlled by others—by parents, by employers,
> by some outer necessity not your own. (Pt. 1, "Time to Leave")

On this first voyage and on his subsequent ones in *The Big Sea*, as deckhand and mess-boy, Hughes is initiated into a life of diversity and change. As he describes his varied experiences, he evolves an attitude of humor, optimism, and tolerance that is as capacious and unsullied as the sea seems to be on his first day out of New York. Thus, as the sea harmonizes, in his narration Hughes balances incidents of racial inequity on shipboard with ones of racial ease, descriptions of the sea turbulent with depictions of the sea tranquil, the tragic tale of a mulatto African boy with the comic antics of monkeys loose in the rigging; his own seasickness, brought on by his drinking the night before, which he describes as parodying jazz rhythms—"I rocked! And the sea rocked! And the boat rocked! And the world went round and round!" (Pt. 2, "Winter Seas to Rotterdam")—is juxtaposed with the courage of his shipmates in a storm.

The experiences of the Middle Passage seem far removed from Hughes's references to the sea in both his poetry and his prose. More than any previous African-American writer, however, he responds to the sea as a source of the imagination. Thus the memories of his own experiences are fused with his interest in folklore, expressing the sea's symbolic possibilities. If John Thompson regarded the voyage, the ship, and the sea allegorically, Hughes came to see it symbolically, drawing on it to illuminate his experiences ashore. In both of his autobiographies the sea is also an analogue for black music. In *I Wonder as I Wander*, it is wavelike—"fluid intense"—and has a rhythm "beneath the melody like the soft undertow of sea waves" (9). In *The Big Sea*, the repetition of refrains may come in "like the boom of waves" (23); like "the waves of the sea coming one after another ... like the beat of the human heart" (209). Through the sea's association with music, Hughes affirms his own commitment to literature. In *The Big Sea,* Hughes had expressed his desire and determination to "write poems like the songs they sang on Seventh Steet," songs that had the pulse of people who, like the sea, "kept on going" (209). In the closing lines of this autobiography, he proclaims that "Literature is a big sea full of many fish. I let down my nets and pulled." Thus from discarding his books in the sea, Hughes has turned to creating his own books. He realizes the sea is finally within himself.

Zora Neale Hurston's representation of the sea in her 1937 novel *Their Eyes Were Watching God* is entirely symbolic. Like both McKay and Hughes, Hurston had crossed seas, traveling to Jamaica and Haiti to do anthropological research. Although her writings, all of which draw on her extensive knowledge of black folklore, are rich in allusions to lakes, swamps, rivers, ponds, and waterfalls—those bodies of water in proximity to which her people settled and about which they developed a range of stories and rituals, the sea appears to have little place in the communal imagination. The primary settings for *Their Eyes Were Watching God* are a small black community, such as the one on the Florida peninsula in which Hurston herself grew up, and the Everglades; nevertheless, despite the fact that Janie and her beloved Tea Cake—the novel's principal characters—make trips to the coastal cities of Palm Beach, Fort Myers, and Fort Lauderdale, and despite the fact that a hurricane sweeps in from the sea to destroy their connubial bliss and their community, the sea—neither historically nor literally—figures in their story. It does, however, figure in Hurston's literary imagination, providing a symbolic frame for the novel, and becoming part of Janie's imagination.

Hurston begins her novel, which focuses on Janie's expanding consciousness, by contrasting men's and women's perceptions in terms of the sea. Her opening paragraph explains that the sea, with its distant and infinite horizon, holds a future of infinite possibilities; in Hurston's image it is men who focus on the sea's horizon and the future. Her second paragraph describes the dreams of women, implying that for them there is neither sea nor horizon. "Now, women forget all those things they don't want to remember, and remember everything they don't want to forget. The dream is the truth. Then they act and do things accordingly." Men, whose gaze is only seaward toward the future, are doomed to disappointment, but women's perceptions may be circumscribed by the past if it includes no vision of the sea's horizon. Thus Hurston describes Janie's search for a horizon; increasingly Janie determines that her vision of herself and her possibilities must not be landlocked. She realizes that she had allowed herself to be limited by images associated with land creatures: by her grandmother's sense that " 'de nigger woman is de mule uh de world,' " by her husband's sense that " 'women and chillun and chickens and cows . . . don't think none theirselves,'" and by her having been "whipped like a cur dog, and run off down a back road after *things.*" She realizes, midway in her story, that "She had been getting ready for her great journey to the horizons. . . . It was all according to the way you see things. Some people could look at a mud-puddle and see an ocean with ships" (ch. 9). By the novel's conclusion, she can assert in her own voice, "'Ah done been tuh de horizon and back and now Ah kin set heah in mah house and live by comparisons.'" Thus she has transcended the gender-defined parameters set forth in the novel's opening paragraphs; she has experienced possibilities as represented by the sea's horizon and by the future and converted them to memory, placing past and future in a continuum of time.

She has also learned that "'Love is lak de sea. It's uh movin' thing, but still and all, it takes its shape from de shore it meets, and it's different with every shore'" (ch. 20). For Hurston and for Janie, finally, the most significant of the possibilities the sea can represent and can cause us to dream of is love; yet as Janie herself has come to experience, love is neither immutable nor immortal. The novel concludes with Janie's remembering Tea Cake, but her memories are not confined to the things women do not want to forget, for she has been to the sea of possibilities, and, as Hughes had seen himself pulling in his net, so Hurston explains in her final paragraph how Janie "pulled in her horizon like a great fish-net. Pulled it from around the waist of the world and draped it over her shoulder. So much of life in its meshes!"

With few exceptions, African-American literature written between the Harlem Renaissance and the Black Power and Black Aesthetic movements of the 1960s did not look seaward. In the novels of the major black fiction writers of these middle decades—Richard Wright, Ralph Ellison, Ann Petry, Chester Himes, James Baldwin—characters are trapped in apartments or ideologies; they run across rooftops, through streets, down into subways to take up residence underground; although water may gush from sewers and rush up to meet the suicide leaping from a river bridge, neither the authors nor their characters focus on water; for them it is the fire next time, if not right now. The successes of the civil rights movement in the 1960s, however, precipitated not only renewed political and historical awareness but also a reinvigorated cultural and aesthetic consciousness. During this time, blacks were reexamining their history, necessitating, on the one hand, their recognition of the horrors of the Middle Passage, and, on the other, such triumphs as that of the *Amistad.* History, however, for the black artist remembering, becomes myth, endlessly relived, repeated, reevaluated; from the 1960s on, written and oral traditions are joined.

On the brink of the changes wrought by the civil rights movement, two poets, Melvin Tolson and Robert Hayden, were writing carefully crafted poetry that resonates with allusion and reference to Western, African, and African-American culture. Several of Tolson's and Hayden's poems, through their complex historical references, remind us of the costs of the Africans' first voyages across the sea to America.

Tolson's *Libretto for the Republic of Liberia* (1953), an epic poem that focuses on the pilgrimage of blacks—from America to resettle in Africa—does not describe the experience at sea as a fortuitous return. It is as if the memories of the Middle Passage are too horrific to be repressed. They still conjure up a living hell:

> This is the Middle Passage: here
> Gehenna hatchways vomit up
> The debits of pounds of flesh.
>
> This is the Middle Passage: here
> The sharks wax fattest and the stench
> Goads God to hold His nose!
> ("Sol" 149–54)

One character from Tolson's diverse *Gallery of Harlem Portraits* (1935), Marzimmu Heffner, remembers the inferno of the Middle Passage. Tolson's portrait of Marzimmu, however, mirrors the historical experience of black Americans. He wanders the seven seas, but neither time nor sea releases him from the slavery into which he was plunged on the Middle Passage. Tolson concludes the poem ironically with Marzimmu reaching Harlem to die, "happy . . . at last . . . in the land of his servitude." In his later, twenty-four-part *Harlem Gallery, Book I* (1965), Tolson continues to associate the sea with disaster and destruction for blacks; his penultimate section concludes with a nightmare vision comparing attempts by blacks at self or group identity to a shipwreck. Earlier in the ode, however, in imitation of an African-American animal tale, Tolson suggests the possibility that the turtle, like the weaker but more wily African American, may survive the treacherous sharks, the dangerous shoals, and the abysses of the white American sea; the turtle may even succeed in destroying the conditions of oppression and proclaim its freedom. For Tolson, the sea is also associated with the Flood, as well as with the Middle Passage. "And every ark awaits its raven," he writes in the *Libretto for the Republic of Liberia* ("Sol" 224), and even though at the end of the *Harlem Gallery*, he claims to have "no Noah's ark, / no peak of Ararat," he sees that "the binnacle of imagination / steers the work of art aright"; here in the last lines of his ode he asserts that his own work of art has amply chronicled "a people's New World odyssey / from chattel to Esquire!" ("Omega"), that the sea voyage, mythic, historical, and aesthetic has been successful.

Hayden's account of the Middle Passage in his long poem of that title, written in 1962, compresses centuries and complexities of the slave trade on the seas between Africa and America. The effect of the poem is not only to indict the trade but also to reveal the consequent and irrefutable inter-relationships between blacks and whites resulting from the horrors of the sea voyage. Hayden achieves this effect by synchronizing his own brilliant images of the voyage's sharks and disease with the "bright ironical names" of the slave ships, the equally ironic passages from Protestant hymns ("Jesus Saviour Pilot Me / Over Life's Tempestuous Sea"), reminiscences, journal entries, and legal transcriptions from slave traders and sailors—all white. Concluding this powerful poem is an account of the *Amistad* mutiny, in which the black leader Cinquez was defended by the white lawyer John Quincy Adams. Hayden's poem finally makes the experience on the *Amistad* seem paradigmatic of the American experience, with the Middle Passage be-coming, in the last lines of the poem, a "Voyage through death / to life upon these shores."

Although Hayden continues to assert the specific role of the sea in African-American history, he also universalizes it through symbol—always, how-ever, envisioning it as libidinously treacherous. Thus, in "The Diver" and in "Veracruz" (1962), we must learn to resist the sea's easeful seduction, struggling as did Cinquez and the slaves to survive the Middle Passage with a sense of social and personal responsibilities. A later poem, "Aunt Jemima of the Ocean Waves" (1971), describes the poet's encounter with a matronly black woman on the beach. As she tells him the stories of her life—a series of attempts to discover her identity—he imagines another identity for her, recalling "An antique etching":

> "The Sable Venus" naked on
> a baroque Cellini shell—voluptuous
> imago floating in the wake
> of slave ships on fantastic seas.

Again alluding to the historic circumstances of the Middle Passage, Hayden implies that because of the violations against African peoples, of their cul-tural and personal identities, African Americans continue to struggle against the fantastic identities superimposed on them even as they attempt to fanta-size new ones. Aunt Jemima, looking at the waves rushing onto the shore, had speculated on the ocean's destructive intentions. Certainly the poet recognizes the cost of voyages over those "fantastic seas" to both himself and Aunt Jemima and, consequently, might be inclined to join her in desiring "them big old waves" to destroy "this no-good world." Nevertheless, in the poem's conclusion, he can see "in the surf, / adagios of sun and flashing foam," the possibility for continuing to find form and meaning in the seas of life.

As Hayden synthesizes the particular, historical experiences of the Middle Passage with contemporary black lives, so does Michael Harper in a short 1977 poem entitled "American History":

> Those four black girls blown up
> in that Alabama church
> remind me of five hundred
> middle passage blacks
> in a net, under water
> in Charleston Harbor
> so *redcoats* wouldn't find them.
> Can't find what you can't see
> can you?

Drowning five hundred blacks in Charleston Harbor in order to save the country from tyranny is analogous to the murder of four black children in order to save the country for white Christians. And tragically, Harper's poem implies that the deaths of blacks—by fire or by water—continue to be ignored by those who write American history and control American politics. Thus for the contemporary African-American writer, whose memory stretches back to the Middle Passage, although water may signal baptism, it may also designate death.

In America's bicentennial year, 1976, Alex Haley's *Roots: The Saga of an American Family* irrevocably restored the Middle Passage to America's historical memory. Although only six of Haley's 120 chapters describe his ancestor Kunta Kinte crossing the Atlantic, this experience becomes the basis for his life in the New World. In these chapters, Haley elaborates on Equiano's litany of excesses. The sea itself is only casually mentioned in Haley's reenactment of the crossing. In contrast to the immorality of the white sailors, it is a neutral, natural phenomenon, but in the face of both these factors, Kinte and the other uprooted Africans must forge a new culture. Thus, to complement the destruction wrought by the Middle Passage, Haley also demonstrates that a community centered on trust, a common memory of the past, shared experiences of suffering, a determination to rebel, but above all to survive and to communicate across all barriers was also created in the hell-hold of the ship. Even as Kinte remembers the pastoral African land, he is preparing for the life in a new land and a new community, which are the subject of Haley's twelve years of research into his family's history and the subject of his novel. His six introductory chapters, however, personalize and dramatize Hayden's words: "Voyage through death / to life upon these shores."

At the conclusion of *Roots*, Haley steps forward to tell, in his own voice, how he came to write his family history as "symbolic saga of all African-descent people." He maintains that his twenty years at sea with the U.S. Coast Guard, during which time he taught himself to write, were instrumental in his creation of *Roots*: "And because I had come to love the sea, my early writing was about dramatic sea adventures gleaned out of yellowing old maritime records in the U.S. Coast Guard Archives. I couldn't have acquired a much better preparation to meet the maritime research challenges that this book would bring." *Roots*, however, cannot be considered by any stretch of the imagination to be a "dramatic sea adventure." More significant than Haley's "love of the sea" in the creation of his saga was his personal reenactment of the Middle Passage. By shipping aboard a freighter sailing directly from Africa to the United States and by spending his nights in the cargo hold, he attempted to relive "the ghastly ordeal endured by Kunta Kinte, his companions, and all those other millions who lay chained and shackled in terror and their own filth for an average of eighty or ninety days, at the end of which awaited new physical and psychic horrors. . . . finally I wrote of the ocean crossing—from the perspective of the human cargo." His facility with maritime research led him to discover the name of the precise ship on which his ancestor was transported to America, the dates of its departure from the Gambia River and its arrival in Annapolis, and the fact that of the 140 slaves shipped, only 98 survived the voyage. Again through his historical data, he attempts to recreate the experience. Thus, two hundred years to the day after the ship bearing Kinte had docked, Haley stands on a pier in Annapolis, and "staring out to seaward across those waters over which my great-great-great-great-grandfather had been brought," weeps.

Although Paule Marshall does not write historical fiction, her characters, like Haley himself, seek to understand their present condition by extending their consciousnesses to imagine the past. Marshall is aware of the dangers of romanticizing the sea, and only those of her characters for whom the sea is associated with the Middle Passage grow. Deighton Boyce in Marshall's 1959 novel *Brown Girl, Brownstones*, trapped in an American city, in marriage, and in poverty, which stifle his fantasies, remembers his youth in Barbados with pleasure—" 'Stay in the water all day shooting the waves, mahn' " (ch. 1). When Deighton, returning to his youth and his island home, drowns within sight of the coast of Barbados, a pathetic life ends. Yet Deighton's daughter subsequently struggles to understand what the islands meant to her father. In her short fiction "Barbados" (1961), "To Da-duh, In Memoriam" (1967), and "Merle," adapted in 1983 from her 1969 novel *The Chosen Place, the Timeless People,* all of which are set in the Caribbean, the sea is a backdrop for human history, a test and a mirror of human character in the present.

Throughout "Merle," the sea echoes the inner life of the novella's central character, Merle Kinbona, whom Marshall calls, in her introduction to this story, "the most passionate and political of [her] heroines" (*Reena and Other Stories*). Until the penultimate chapter the sea in this story is described variously as "keening," "perennially grieving," "enraged," reflecting Merle's anger and sorrow at the exploitation of the island and people of Bournehills. Merle, in whose bedroom hangs a diagram of a Bristol slaver showing its human cargo, is in tune with the sea's unconscious memory of the tragedy of the Middle Passage and its response of rage, lamentation, and commemoration. Her receptivity to the sea is in direct antithesis to the response of Lyle Hutson who seeks approval for a big promotional campaign in the United States and Canada to sell the island as the newest vacation paradise. When Merle's anguish at the island's developers and landowners reaches its peak, Marshall synchronizes her actions and emotions with those of the Africans betrayed by slave traders. In a vast barn, as gloomy as "a ship's hold," she rages against injustices, and as her furor turns apocalyptic, she becomes insane, screaming "like one drowning" (ch. 13). But instead of Merle, a white woman associated with the commercial developers drowns—because of which, atonement for the sins of the past and present seems possible; toward the novella's end the sea becomes cleansed, its "ritual keening" diminished, its "spume-like tears" less hysterical (ch. 17). Released from her own despair, Merle leaves for Africa in the final paragraph to be reunited with her husband and son and implicitly with all Africans who had gone before her. She does not go by sea, but flies, and in a final cartographic image Marshall seems to eliminate the sea altogether, sending Merle "from Recife, where the great arm of the hemisphere reaches out toward the massive shoulder of Africa as though yearning to be joined to it again" (ch. 17).

In her 1983 novel *Praisesong for the Widow,* Marshall more emphatically associates the sea with a historical unconscious, distinguishing her work from that of those white writers for whom the sea is a Jungian reference and from those black male writers for whom it is a historical locale. As in "Merle," her protagonist comes to be able to relate her personal history to the communal history of African peoples. Thus in the course of Marshall's novel, Avey Johnson, a black matron from North White Plains, is nudged back through her own history, back through the history of her people, with the sea as the setting for each stage of her own psychic rebirth. At the beginning of the novel, Avey disembarks in Grenada from a luxury cruise ship. Her voyage appears incomprehensible to her upper-middle-class friends and to herself; she knows only that she has had a strange dream and that she wants to "go home." For Avey, at this stage in her return, the sea and its islands can be experienced only in terms of her materialistic life; thus each elegant table in the ship's dining room appears as "an island separated from another on a sea of Persian carpeting" (pt. 1, ch. 4), while the sea itself seems a "wide, silvertoned sheet of plate metal" (pt. 1, ch. 1). Symptomatic of changes to come, Avey suffers from a headache and a queasy stomach and feels as if the stable luxury liner is foundering under her.

A dream, however, suggests a different level of consciousness, in which the sea is neutral. She dreams of being summoned by her great-aunt to journey, as she had as a child, down to Tatem Island, South Carolina, where the slave ships once docked. There, her aunt had told Avey the story, long since repressed, which her grandmother had told her—of the newly arrived slaves who turned around and, singing, walked back across the Atlantic to Africa. The child Avey had resisted the logic of the story, although she had accepted the fact that Jesus could walk on water; the adult Avey continues to resist not only the miracle of the Ibos' return to Africa but also the culture of her great-aunt, with its roots in African communal life.

When Avey decides to join a group of "out-islanders" on their annual excursion home to the small Caribbean island of Carriacou, she calls up another memory from childhood associated with water. She remembers that she and her family joined multitudes of Harlemites every summer for an excursion on the SS *Robert Fulton*. Waiting for the boat, she recognized not only her parents' love for each other but also her own connection with the crowds; she felt that "the moment she began to founder those on shore would simply . . . haul her in" (pt. 3, ch. 4). This deeper memory of "a huge wide confraternity," nurturing her, sustaining her, rescuing her from drowning in forbidding waters, seems a reversal of her Tatem Island memory. There she had resisted communion with her people; here she is buoyed by it, despite the sea. But Marshall subsequently suggests that before Avey can transcend the sea, the terrifying unknown in her personal and communal past, and be restored to the Ibos and to Harlem, she must experience that Middle Passage, the dark night of her personal life and her people's life. En route to Carriacou, she joins the throng of men, women, and children aboard a battered schooner, the *Emanuel C.* Midway in the passage, at the rough confluence of two currents, other memories "swept her . . . darker and more powerful than any wave the sea could have thrown up" (pt. 3, ch. 6). As she recalls moments of humiliation when, as a child, she had vomited and urinated uncontrollably in public, Avey's bowels and stomach betray her urbane exterior. Her ignominy is ironically praised by the people on board the *Emanuel C,* who, Marshall implies, intuitively identify her experience as one that purges and purifies. If it releases her from the "waste and pretense" of her life as imitation white suburban matron, it simultaneously associates her with the profound humiliation of African people during the Middle Passage; vicariously she relives their history, her own history.

On the island, Avey participates in traditional African and African-Caribbean rituals; she is restored to her past and to herself. As she returns to North White Plains, Marshall links her with the Ancient Mariner. "Like the obsessed old sailor she has read about in high school," she leaves the sea to haunt suburban street corners and front lawns, metropolitan shopping malls and train stations, determined to tell the story of her rebirth (pt. 4, ch. 3). If she has paid her dues to the albatross of whiteness, she has also gained a sense of her own and her people's rich darkness; if she has survived her sea voyage like the Ancient Mariner, thus becoming, Marshall implies, one with her Western heritage, she has also become an African-American *griot,* like her great-aunt telling the history of her people, spanning the sea through myth and uniting the past and present and the peoples of the Diaspora. The sea, however, conduit to understanding, is left behind (as it is in most African-American literature in which it figures) as Avey returns to life in the community.

In two of Toni Morrison's novels, *Song of Solomon* (1977) and *Tar Baby* (1981), and Ntozake Shange's 1982 novel *Sassafrass, Cypress & Indigo,* the sea, although tangentially associated with the memory of the historical past, seems more emphatically associated with myth, and like Hurston's sea, with symbol, with dream and the unconscious. In these works by black women, the sea calls to black men, luring them as it did historically to find freedom from slavery, from poverty, or from authority. Thus in *Song of Solomon,* Milkman Dead, having grown up near the Great Lakes, which have given him a vision of the sea, is not one of the truly landlocked; "dream-bitten," he has come to believe, Morrison explains, as coastal people do, that flight, "final exit and total escape are the only journeys left" (ch. 7). Son, in *Tar Baby,* seeks solitude at sea—his own space where he may be his own authority and find his own power. At sea, like McKay's Banjo, he would know only the company of "other solitary people," having joined

> that great underclass of undocumented men . . . an international legion
> of day laborers and muscle-men, gamblers, sidewalk merchants, mi-

grants, unlicensed crewmen on ships with volatile cargo, part-time mer-
cenaries, full-time gigolos, or curbside musicians. What distinguished
them . . . was their refusal to equate work with life and an inability to
stay anywhere for long. (ch. 5)

The father of the three sisters Sassafrass, Cypress, and Indigo chooses to go
to sea for more explicit reasons; as the girls' mother explains, "*It's real clear
there wasn't much work for a skilled Negro carpenter in Charleston, but I figure he
never felt at home here on the mainland anyhow. Geechees from the islands don't
take to being called foreigners . . . I suppose he went to sea so he could get away from
all that*" (190).

In each of these novels, however, the sea's promise of freedom results in
compromise with the land, with the past, and with others. Having left Michigan, Milkman, in a dream sequence, comes to feel that he flies "over the dark
sea, but it didn't frighten him because he knew he could not fall" (ch. 12).
An extension of this exhilaration occurs when he exclaims to his lover that
he needs the sea, "'The whole goddam sea! . . . The sea! I have to swim
in the sea. Don't give me no itty bitty teeny tiny tub, girl. I need the whole
entire complete deep blue sea!'" (ch. 15). Although Milkman, discovering
his kinship with the flying Africans, learns to fly, his lover does not learn
to swim, and he, by the novel's conclusion, understands the responsibilities of loving. Thus he realizes that his primary mentor, his aunt, who had
shown him the meaning of love, could fly "Without ever leaving the ground"
(ch. 15). Like the persona in Hayden's poem "The Diver," he must make a
"shoreward turn."

Tar Baby begins with Son's immersion in the sea to escape shipboard confinement, and concludes with his emergence on "an island that, three hundred years ago, had struck slaves blind the moment they saw it" (ch. 1). During his attempt at the beginning of the novel to direct his motions in water, he
struggles with the current, "the water-lady," who was forcing him terrifyingly
"toward the horizon in a pitch-black sea" (ch. 1). As the novel progresses,
Son's associations with the sea continue to relate to women and to the difficulties of perception. Thus in New York City, as he anxiously awaits the
arrival of his citified, sophisticated lover, he draws water for a bath, "smiling
to think of what the leaden waves of the Atlantic had become in the hands
of civilization" (ch. 7). And Son, whom his lover now associates with "blue-
sky water," is so obsessed with her, he has become domesticated—and blind.
At the conclusion of the novel, the blind Therese (Tiresias?)—herself a de-
scendent of the blind slaves—who, like Avey Johnson's great-aunt and her
own ancestors, can see "'with the eye of the mind,'" guides him across the

treacherous channel to the island where he had first wooed his lover. Therese leaves him alone on the shore to which the slaves, having survived the vicissitudes of the sea, naked and blind, had fled. There, Son relinquishes all his possessions, including his lover, and recognizes his relationship with a people and the past. As he leaves Therese in the novel's last paragraph, Son crawls, like a child, out on the rocks of the island and begins to run with his ghostly ancestors, leaving "the nursing sound of the sea . . . behind him." His immersion has somehow shown him how to see more clearly his limitations as well as his strengths; yet he is mythicized in the conclusion: a man on the brink of maturity and civilization, a man on the shore of an island, between sea and land.

In *Sassafrass, Cypress & Indigo*, the father of the three women in Shange's title drowns off Zanzibar, again suggesting the sea's mastery over human affairs. His daughters, however, believing in the sea's restorative powers, "*toss nickels & food & wine in the sea down the coast so daddy wd have all he need to live a good life in the other world*" (108). The sea provides Sassafrass and Indigo with all they need to live a good life in this world. Sassafrass seeks, through the sea, to know her father and other African cultures; she visits the wharves, learning, as her father had, to mend nets, to sing "*like the sailors & dance like the west indians / who was crew on a lotta boats . . . & listen to the tales of other colored folks lives in the islands & as far off as new guinea*" (108). Through association with the sea, she enriches her knowledge of her personal and cultural past and develops her plans for the future. As explained in the conclusion of the novel, however, Indigo, mystic and musician, conjures mythic associations from the sea and thus creates her future. With her music, Indigo not only heals women and children but also moves tides. The old people know that not since *Blue Sunday*, the slave girl who turned crocodile in order to remain free, had a black woman on Daufuskie Island been able to call on the sea for assistance. *Blue Sunday*, protecting herself from her lascivious master, could churn the sea into "fuming, swinging whips of salt water round the house where the white folks lived" (222). Inheriting *Blue Sunday*'s magic, Indigo turns it to creative purposes, encouraging the rough seas to bring free children into the world. Thus by linking it with African-American folklore, Shange enriches the traditional concept of the sea as the source of life. *Sassafrass, Cypress & Indigo* concludes with Indigo's departure from Daufuskie to join her sister and her mother in helping with the birth of Sassafrass's child; as she leaves, "2,000 *Blue Sundays*" come out dancing and "the aqua-blue men . . . the slaves who were ourselves" come striding out of the sea (224).

"Many sea stories and histories of ships from many cultures went into research" for *Middle Passage*, according to Charles Johnson's acknowledgments, hinting at the intertextuality of his 1990 National Book Award–winning novel. Among those works that Johnson most obviously drew on are the *Odyssey*, *Moby-Dick*, "Benito Cereno," *The Sea-Wolf*, *The Narrative of Arthur Gordon Pym*, slave narratives describing the Middle Passage, and Hayden's "Middle Passage." The Ishmael of *Middle Passage*, Rutherford Calhoun, "a newly freed bondman," has also swum through libraries in search of scientific, philosophical, theological, and anthropological truths. It is his experiences during the Middle Passage in the year 1830 rather than his book-learning, however, that bring about his sea change.

To keep his recently attained freedom, which has become endangered by the prospect of marriage, Calhoun lights out for the sea. Quickly, he discovers that the oppression of wedlock seems mild by comparison with the conditions of bondage aboard the ship on which he has stowed away, the allegorically named *Republic*, a slaver bound for Africa to collect the last members of a legendary tribe, the Allmuseri, as well as their mysterious god. From a distance, the *Republic* appears beautiful, but in reality, she is "unstable. She was perpetually flying apart and re-forming during the voyage . . . , from stem to stern, a process" (35–36). Her crew is motley, "all refugees from responsibility and, like social misfits ever pushing westward to escape citified life, [who] took to the sea as the last frontier that welcomed miscreants, dreamers, and fools" (40). Her captain, who carries traces of Ahab and Wolf Larsen, is Ebenezer Falcon, megalomanic empire builder, inventor, Puritanical perfectionist, child-abuser, "a specialist in survival"; warped in body and soul, he epitomizes the American structure of power.

Calhoun, like Ishmael, is at first attracted to the sea as a route of escape from vanity, mediocrity, selfishness; for him, "the formless, omnific sea" (4) was the land's antithesis: an analogue for and the means for exploring all experiences and extremes. Aboard the *Republic*, however, Calhoun discovers that he is "literally at sea" and that the sea could also be an analogue for death. As he witnesses repeated human violence, and as he participates in this violence by assisting in the suicide of the captain and by cannibalizing his friend, the first mate, Johnson's narrator learns the truth of Falcon's words: " 'The sea does things to your head, Calhoun, terrible unravelings of belief that aren't in a cultured man's metaphysic.' " In the course of the Middle Passage, the sea's formlessness (it is in one instance "a thrashing Void," in another a "chaosmos," and in another "this theater of transformations") appears increasingly as a metaphor for the confusion of Calhoun's values and beliefs; the Middle Passage on that sea assumes psychological and philosophical as well as political and historical implications. If it refers to the horrendous historical experience at the center of the lives of those black and white Americans engaged in the transatlantic slave trade, it also implies the life crisis—a personal middle passage—in the life of a single black American man.

Against the wild upheaval and uproar of the sea and of the anarchy aboard ship and in his own mind, Calhoun gradually comes to recognize the significance of well-defined democratic and humanistic values and beliefs. He rejects "the masculine imperative" prevailing on the ship as well as the capitalistic desires and philosophical Manichaeism on which Falcon bases his power. From a drunk old white cook and an innocent young black girl, he learns to assume responsibility for others, and from the Allmuseri he learns the necessity for recognizing a "unity of Being," empathetic connections among all forms of life. In a moment of calm prior to the mutiny, Calhoun perceives the common humanity and equality of all aboard the *Republic*.

Following a slave revolt, the death of the captain, and the loss of leadership among the blacks, his compassion extends to the *Republic* itself, "a crippled Ark" (152). On the brink of death from starvation and disease, he expresses the insignificance of his own life and his concern for others. He does survive, of course, along with the cook and three Allmuseri children, attributing his survival to his belief in a democratic idealism, in its way analogous to Melville's and Crane's in *Moby-Dick* and "The Open Boat": "if you hoped to see shore, you must devote yourself to the welfare of everyone" (187).

In the concluding chapter of *Middle Passage*, Johnson's postmodern book shifts from being a revision of a nineteenth-century sea narrative to a revision of a nineteenth-century domestic romance. On board the rescuing ship is Isadora, his intended, who, like Penelope, has been loyal to her absent sailor despite the seductive entreaties of others; and, like Odysseus, Calhoun has come to yearn for home. Calhoun's personal and psychological middle passage brings an eager acceptance of all that Ishmael associates with the land: "the wife, the heart, the bed, the table, the saddle, the fire-side, the country" (ch. 23). In the novel's last lines, as he and Isadora lie in bed, they envision a new life that will transcend the terrors of the Middle Passage and the cold ocean. If home thus has personal implications for Calhoun, it also resonates with political ones. Although the *Republic* goes down, Calhoun "desperately dreamed" of America, of "home":

> the States were hardly the sort of place a Negro would pine for, but pine for them I did. Even for *that* I was ready now after months at sea, for the strangeness and mystery of black life . . . , for there were indeed triumphs, I remembered, that balanced the suffering on shore. (179)

Thus the contemporary African-American writer, acknowledging the sea's historical association with the horror, grief, rage, and loss caused by slavery, can also recognize its philosophical and psychological implications and its mythic and symbolic possibilities for creation and transformation. The Middle Passage as a "voyage through death" becomes a passage into wisdom and life; Equiano's bad spirits are metamorphosed into Shange's "aqua-blue men," and Johnson's Calhoun, having crossed "countless seas of suffering," knows how to live in this land.

ELIZABETH SCHULTZ

Chapter Twelve

TWENTIETH-CENTURY POETRY

In twentieth-century American poetry, the sea, as inspiration and metaphor, is more ubiquitous and richer in its evocations than might be expected in the literature of a nation that has generally been regarded as continental in outlook rather than maritime like the British; but it is captious to categorize poets into the seafaring and the landlocked. Nevertheless, few British poets could say with Emily Dickinson "I never saw the sea," and not many American poets have regular access to or extensive experience of it. Reed Whittemore, in the title poem of his first volume, *The Self-Made Man* (1959), highlights this latter point:

> I spend my winters a thousand miles from the sea.
> Images of the beaches and the long grass,
> And pools left in the rocks by tides, and flats
> Scattered with driftwood, and shells—all of these
> Come to me ringing as out of caves, as I
> Struggle to rise out of sleep, climbing, descending,
> And look out my cold window at the snow.

"Images . . . Come to me ringing": it is as remembered images that the sea comes most often into American poetry, seldom as subject matter. With the twentieth-century decline of a tradition of narrative verse, it is in imagery especially that one seeks the sea, where it still has an important role. The paradox of Whittemore's images coming to him "ringing" is also suggestive, for it is often the sound of the sea that survives in the memories of those distanced from it. Wallace Stevens, in a poem discussed below, spoke evocatively of "the dark voice of the sea," and it is often as a voice, rather than as a physical presence, that the sea impinges on the American poetic imagination. That sound may be exhilarating or menacing, though seldom the "melancholy, long withdrawing roar" that Matthew Arnold hears in "Dover

Beach." "Sophocles long ago / Heard it on the Aegean," Arnold recalled; in this century, it can be heard in the work of poets influenced by the classics such as H.D., Ezra Pound, T. S. Eliot, and Allen Tate. Other modern poets are as attentive as Arnold to the forlorn vastness of what he calls elsewhere "the unplumbed, salt, estranging sea"; they seize eagerly on its adaptability to imagery of separation and death. To others, its tidal motion will suggest images of cleansing, purification, and renewal, with intimations of transcendence and eternity, while more recent poetry is characterized by a marked difference of tone and by its greater interest in sea creatures. Throughout the century, as an elemental, natural force that may represent energy, change, eternity, or destruction, the sea has been poetically valuable as a universal symbol of much that preoccupies the age. These are the themes that this chapter must try to bring together in a survey that is primarily, though not slavishly, chronological. Confronted with so kaleidoscopically changing a cluster of images, one sees why Homer and legend regard Proteus as the archetypal sea figure.

I.

In the second decade of this century, what T. E. Hulme identified as "*hard, dry* classical verse" became a major influence on American poetry even more than on English. Flowering as Imagism in the work of H.D., Pound, and the young Eliot, it was less austere and forbidding than Hulme's phrase threatens, but the more closely it adhered to his principle of "*accurate, precise and definite description*" (again the italics are his), the more it risked subordinating content and power of evocation to technique. Moreover, despite its manifesto's bold insistence on "absolute freedom in the choice of subject" and its claim to "believe passionately in the artistic value of modern life," Imagism's more restrictive aesthetic aims necessarily excluded whole areas of contemporary experience. Of World War I, especially, only Hulme himself had the firsthand knowledge on which to base "accurate, precise and definite description," but he was killed in action in 1917 leaving only one war poem of disputed authorship.

Consideration of the role of the sea, however, illuminates the intrinsic merit of the poetry of the other purest and most consistent Imagist, H.D. Characteristically she visualizes it in a context of classical mythology; less predictable is the interchangeability of image in her juxtaposition of sea and land. In "At Ithica," for example, she thinks of "The sea-blue coast of home," while the six-line "Oread" exhorts the sea to "splash your great pines / on our rocks . . . cover us with your pools of fir." Her first collection of poems

(1916) is significantly entitled *Sea Garden* and includes five poems with such titles as "Sea Rose," "Sea Lily," and "Sea Violet." One commentator has observed that for her, "sea and shore refer to much the same thing—the world of flux and suffering." In such a world of pain and loss, the coastal flowers can be emblems of the self. The sea violet, for example, tucked precariously into the sand, is all the more beautiful for its capacity to withstand the predations of the sea. The realistic recognition that neither sea nor shore offers immunity from suffering or violence is "accurate, precise and definite" in rejecting a sentimental idealization of either as a haven, and is entirely consistent with the robust anti-Romanticism Hulme wished Imagism to exhibit. In "The Islands," the security of the sea garden and the beauty of its sea flowers are destroyed by the violence of wind and wave, and in "Lethe" (which, it has been said, echoes tidal rhythm), it is not mere oblivion that engulfs humanity so much as physical drowning. Drowning and shipwreck are again brought about by the "Crash on crash of the sea" with which "Sea Heroes" opens; but against the crash, "the line of heroes stands, god-like . . . they live beyond the wrack and death of cities." Other poems in the collection also offer cause for hope in the midst of painful extremity, as in "Sea Gods," where the poet sings to the gods who seem to have been battered and destroyed by the combined effects of waves and rocks. She knows the gods will return and offers gifts of violets of every kind, from inland, as homage:

> For you will come,
> you will come,
> you will answer our taut hearts,
> you will break the lie of men's thoughts,
> and cherish and shelter us.

Here and elsewhere, H.D. identifies heroic attributes by the use of incrementally repetitive compounds: "lover-of-the-sea, lover-of-the-sea-ebb, lover-of-the-swift-sea." The device, as characteristic of Anglo-Saxon epic as of Homeric, may even be indebted to Ezra Pound's adaptation of "The Seafarer" from the Anglo-Saxon in *Ripostes* (1912). Labored and self-conscious though Pound's version may be, hovering at moments dangerously on the brink of the later cheerful pastiche of "Ancient Music," his recognition that there is no hero "But shall have his sorrow for sea-fare," and must "the salt-wavy tumult traverse alone" leads him to reject the belief "That any earthweal eternal standeth." Yet there is a force that "Whets for the whale-path the heart irresistibly," and like H.D., he is attracted by the ambivalent symbolism of the sea, destroyer and purifier. Elsewhere he is content with a more generalized evocation of the ocean, like the artist in "The Age Demanded," who is left "delighted with the imaginary / Audition of sea-surge." The concordance to *The Cantos* lists more than 110 allusions to "sea" or "seas" and almost thirty separate compounds in which "sea" is the first element, yet disappointingly few merit comment, though the first Canto opens, appropriately, with a sea-voyage of adventurous exploration and discovery. Pound's own explorations were notoriously preoccupied with economics: hence the unusual frequency of references to the sea as trade route and the fascination with the law of the sea. (The quotation attributed to Antoninus, "Law rules the sea," occurs in six different Cantos.)

T. S. Eliot, to whom Pound was friend and mentor, uses the sea, in his 1917 volume, as a key image in "The Love Song of J. Alfred Prufrock." The soliloquy of a man on his apprehensive way to a London salon where he is anxious to impress one woman, most of it is essentially urban in its atmosphere and its imagery conveying his vacillation, self-questioning, and self-distrust. At midpoint of the poem, however, one thought of unprecedented certainty is suddenly isolated in a couplet:

> I should have been a pair of ragged claws,
> Scuttling across the floors of silent seas.
>
> (71–72)

After this moment of truth, the earlier mood is resumed, forcing Prufrock to recognize that he is "At times, indeed, almost ridiculous," reduced to vacationing on a beach in "white flannel trousers"; but, he affirms, he will now wear the bottoms of those trousers rolled. Elderly men, at least at British coastal resorts, can still be seen in this mode, walking barefoot in the shallows of the water's edge, perhaps recalling more youthful days when swimming was still a possibility for them. It is the perfect image for the prematurely aged Prufrock, but also for his longing to immerse himself in an element that he fears will prove destructive to him. That this is a trope for sexual commitment, and that Prufrock is to be pitied rather than mocked, is underlined by his nostalgia for the mermaids, whose song he has heard but which, he knows, will not be sung to him. The lyrical vision of them "riding seaward on the waves" and wreathing with seaweed those who have "lingered in the chambers of the sea" recalls Keats's vision, in "Ode to a Nightingale," through "magic casements opening on the foam / Of perilous seas, in faery lands forlorn" (69–70). Both Keats and Prufrock are recalled from reverie by sound, and for both, that recall to the world immediately about them is a form of death, or annihilation. Keats conveys this by the image of a funeral bell where Eliot dramatically sustains the sea imagery: "Till human voices wake us, and we drown." Significantly, the *I* of the soliloquy only becomes *we* in those final

three lines: Prufrock is not alone in his disappointment that Romanticism has
to give way to present reality, nor are Eliot and Keats alone in finding a sexual
meaning in imagery of immersion and drowning in a destructive element.

More Imagist than Romantic in being "accurate, precise and definite," the
crab image recurs, more menacing and more fully developed, in "Rhapsody
on a Windy Night." In "Mr. Apollinax" the eponymous figure disturbs the
tea party by his laughter,

> submarine and profound
> Like the old man of the sea's
> Hidden under coral islands
> Where worried bodies of drowned men drift down in the green silence,
> Dropping from fingers of surf.
> I looked for the head of Mr. Apollinax rolling under a chair
>
> Or grinning over a screen
> With seaweed in its hair.

Grotesqueness, incongruousness, and the obsession with drowning undermine and give a sinister twist to the romanticism of the imagery, just as, in
Poems 1920, the expansive literary allusions to a more classical seascape in the
epigraph and opening stanzas of "Sweeney Erect" will be quickly modified by
the brutal coarseness of the simian Sweeney and his entourage. In the same
volume, the drowned sailor, Phlebas the Phoenician, made his first appearance in "Dans le Restaurant." Section 4 of *The Waste Land* (1922), "Death by
Water," is, of course, also devoted to him and to the sea's destructive powers.

For Marianne Moore also (another poet with origins in Imagism), though
she is not blind to its beauty, the sea is primarily "A Grave":

> It is human nature to stand in the middle of a thing,
> but you cannot stand in the middle of this;
> the sea has nothing to give but a well excavated grave.

Like Eliot's, this ocean is one "in which dropped things are bound to sink—/
in which, if they turn and twist, it is neither with volition nor consciousness."

At the end of *The Waste Land* (lines 418–22), however, the sea becomes,
for the first time in Eliot, something of a regenerative agent. This concept he
would develop richly in the haunting "Marina" (1930), the theme of which
F. O. Matthiessen rightly defined as "regaining the purified vision in later
life." Unfortunately, Matthiessen did not draw attention to the importance,
in that process and that poem, of sea imagery generating ideas of creativity—
ideas to which the final section of this chapter will return in the context of

two contemporaries of Eliot, Hart Crane and Wallace Stevens. In "The Dry Salvages," Eliot was later to remind us that "The sea has many voices, / Many gods and many voices," that "We cannot think of a time that is oceanless," but above all that "the sound of the sea-bell's / Perpetual angelus" is God's message to voyagers to fare forward. The purified vision can encompass and transcend drowning because it can also encompass the continuity of created life and "the stillness / Between two waves of the sea" with which, in "Little Gidding," the *Four Quartets* are concluded.

An American contemporary of Eliot, consistently neglected in his own country, is John Peale Bishop. If his "Colloquy with a King-Crab" ("I consent like him to go on claws") suggests affinities with Eliot, "The Submarine Bed" goes much further. Here, his successfully sustained use of the sea and drowning as images for sexual coupling and its implications is more adventurous in its complexity than Eliot's in "Prufrock." Its six stanzas constitute a fascinating twentieth-century metaphysical poem using wit in the seventeenth-century sense. That the sea is even more prominent in Bishop's background than in Eliot's is overt in "Beyond Connecticut, Beyond the Sea" but implicit throughout his work. Elegizing Scott Fitzgerald in "The Hours," he turns, as Robert Lowell does in "The Quaker Graveyard in Nantucket," to "Lycidas" for a model and to the sea as theme, but for Bishop the sea is a personally experienced symbol of both flux and timelessness, evoking that sense of mutability set against continuity that bereavement accentuates so painfully. "A Subject of Sea Change" in 1942 relates the theme to war and to history:

> The oldest contours
> of these refluent shores
> Are subject always to the sea's revenge.

Yet, though "The sea is old, / Severe and cold, secret as antiquity," the poet remembers

> how Love came on the immediate swell
> Reflecting in a shiver of resplendent spray
> The dawn enclosed within her secret shell.

Admitting that the sea "rants," he also responds to its profound poetry. In a fascinating study of three of Bishop's sea poems, Paul Grant Stanwood demonstrates how the sea reconciles the opposing strains of Southern aristocrat and Yankee trader in the poet's personal background. In adding that, for Bishop, the sea also "brings European and American, classical and pragmatic civilization together," he might have drawn a useful comparison with the way in which Bishop's friend Allen Tate used "The Mediterranean" for a similar purpose.

From these main currents of poetry between the wars, Robinson Jeffers stands interestingly isolated. His long narrative poems, often more melodramatic than truly tragic, are hardly more to the modern taste than are his prosody or his often-tortuous language and syntax. What is important to him, however, is not the sea but one particular sea, the Pacific Ocean; that is absolutely central to his work. Living for most of his life on its spectacular coast in the area of Carmel and Big Sur, Jeffers uses the Pacific, vast, violent, mysterious, inhuman, to give much of his poetry its specific location and atmosphere. For Jeffers, to know the sea is, in large part, to come to terms with the "bitter earnestness / That makes beauty," that is one of his subjects in "Boats in a Fog" (1924). As in William Carlos Williams's "Yachts," boats themselves, lovely and powerful, are the actors in the poem, rather than the humans who build and use them. Like Williams (and many prose writers too), Jeffers alludes to human experience in describing the boats and their behavior.

> At length, a stone's throw out, between the rocks and the vapor,
> One by one moved shadows
> Out of the mystery, shadows, fishing-boats, trailing each other,
> Following the cliff for guidance,
> Holding a difficult path between the peril of the sea-fog
> And the foam on the shore granite.

Beauty and danger have roughly equal parts in this "essential reality / Of creatures going about their business among the equally / Earnest elements of nature," a balance that also shapes the sardine fishermen's life as Jeffers describes it in "The Purse-Seine" (1937). The men fish from their boats with nets, using an ancient method that strikes the inner eye as a combination of dance and religious rite:

> They close the circle
> And purse the bottom of the net, then with great labor haul it in.
> I cannot tell you
> How beautiful the scene is, and a little terrible, then, when the
>   crowded fish
> Know they are caught, and wildly beat from one wall to the other
>   of their closing destiny the phosphorescent
> Water to a pool of flame, each beautiful slender body sheeted
>   with flame, like a live rocket

The scene occurs to the speaker again when he is standing on a mountaintop, looking at a plain of city lights: "How could I help but recall the seine-net / Gathering the luminous fish?" To be at home on the sea, according to Jeffers's vision in this poem, is to be forced to make an inevitable and none-too-heartening comparison between the fish who are trapped and helpless and human beings at the mercy of governments, laws, and "Progress." That no apocalyptic disaster has yet come is no comfort, since we know it *will* come: "we must watch the net draw narrower."

The Pacific is also the background, spiritual as well as geographical, to poems of human passion such as "Tamar" and "Thurso's Landing," set on what, in "Apology for Bad Dreams," he calls "This coast crying out for tragedy like all beautiful places." Land and sea are for him, in his own way, as inextricably interwoven as for H.D. The elemental force of the ocean beating on the beauty of the coast is matched by the elemental violence of the human passions unleashed in the physical beauty of his characters, especially such women as Tamar Cauldwell, Helen Thurso, and Fayne Fraser. The sea is literally and symbolically essential to all these narratives: in "Tamar," for example, by reviving Lee Cauldwell after his potentially fatal fall, it sets in train the whole tragic sequence of events. Tamar's dance on the seashore while "the sea moved, on the obscure bed of her eternity," is as much a Conradian ritual immersion in the destructive element as, in "Thurso's Landing," is Helen's dragging of Hester into the sea. Tamar's bathing with her brother, which begins their incestuous relationship, takes place in an inland pool but is linked with the sea in Jeffers's question:

> Was it the wild coast
> Of her breeding, and the reckless wind
> In the beaten trees and the gaunt booming crashes
> Of breakers under the rocks, or rather the amplitude
> And wing-subduing immense earth-ending water
> That moves all the west taught her this freedom?

If the sea in these poems typifies freedom, humankind is destined to distort freedom into the unbridled license of self-destructive passion. Just as, in *Desire Under the Elms*, the gaunt isolation of the barren New England farm makes more credible the dreadful burden of in-turned lust, hate, and guilt of its inhabitants, so here a similar intensity of emotion and violence is linked directly to the wildness of the inhospitable landscape and motivated by the sea itself. In another poem, Jeffers himself titles this "Haunted Country"; there, his imagery of parturition and struggle contrasts with the destructive force of the sea to suggest the end-of-the-world vision that haunts his poetry.

It is reinforced by the reference to "the last migration" that evokes simultaneously the American historical experience and the insecurity of human tenure on the globe. This is echoed in "The Loving Shepherdess," where he sees man as

> Walking with numbed and cut feet
> Along the last ridge of migration
> On the last coast above the not-to-be-colonized
> Ocean.

Jeffers later makes that ocean the setting for one of the best American poems of World War II, "The Eye." Characterizing the Atlantic as a "stormy moat" and the Mediterranean as having for more than five thousand years "drunk sacrifice / Of ships and blood," he distinguishes them from the more mysterious Pacific, to which "Our ships, planes, wars are perfectly irrelevant":

> Here from this mountain shore, headland beyond stormy headland
>   plunging like dolphins through the blue sea-smoke
> Into pale sea—look west at the hill of water: it is half
>   the planet: this dome, this half-globe, this bulging
> Eyeball of water, arched over to Asia,
> Australia and white Antarctica: those are the eyelids that
>   never close; this is the staring, unsleeping
> Eye of the earth; and what it watches is not our wars.

Realistic description modulates into surrealistic image to express the poet's awe at the mystery of this wholly alien element. Humanizing the sea with this gift of sight, which it uses in this inhuman way, Jeffers manages for once to validate powerfully and memorably the vision of a hostile and desolate universe that, in so much of his poetry, can become oppressive in its ultimate inhumanity.

By comparison, Yvor Winters's "Slow Pacific Swell" is less memorable, though it encapsulates many of the poetic attitudes to the sea already defined. The first of its three sections recalls the sea as seen in childhood, from great distance. The apparent softness, steadiness, and calm, the speaker now knows, are "illusion." Its second part recalls a contrasting, immediate experience of the ocean, a voyage in which the sea

> Hove its huge weight like sand to tangle me
> Upon the washing deck, to crush the hull;

But that same voyage also revealed "gray whales for miles," and the wonder of a breach, something like the verbal equivalent of a Rockwell Kent illustration for *Moby-Dick:* "The whale stood shining, and then sank in spray." The third section strikes a balance. Opening with the admission "A landsman, I," it nevertheless regrets that "I have lived inland long. The land is numb." The speaker finally seems to seek or prefer a seaside position, near the "limber margin" where "precision ends" but not on that "chaos of commingling power." In Winters's poem, the sea is sought for thought, to be watched as it "gathers seaward, ebbing out of mind."

Even more of a landsman than Winters, Robert Frost seldom wrote about the sea. When he did, the result was different from the work of any other poet so far discussed. In "Once by the Pacific," his speaker thinks, like Jeffers, of the sea as a living thing that "thought of doing something to the shore / That water never did to land before."

As in the case of H.D., W. S. Merwin, Washington Irving ("The Voyage"), Herman Melville (*Moby-Dick*), Stephen Crane ("The Open Boat"), and other authors, as well as of America's earliest writers, readers can recognize "Once by the Pacific" as being in what might be called the American sea-fear mode—in which the sea is fearfully and unpreventably destructive. At times that fear becomes apocalyptic, as in Richard Steere's seventeenth-century poem "On a Sea-Storm Nigh the Coast." In terms remarkably similar to Frost's, Steere's observer voices his fear that the sea may have decided to reduce the world to chaos. Frost's viewer, on the shore of the ironically named Pacific, faces an approaching sea storm that apparently intends to wipe out our earth once and for all, and not merely cover it for a while as Noah's flood did. It looked, says the speaker, "as if" the shore was lucky to have a continent behind it; it looked "as if" a night of dark intent / Was coming, and not only a night, an age." But the speaker soon distances himself from these apparent implications of the sea's appearance by saying that "someone" "had better be prepared for rage."

"Once by the Pacific" might be best understood as saying that those who read the sea in this apocalyptic way are thereby obliged by the implications of their reading. It is for them, and them alone, to get ready for the rage they think the seascape prophesies. Frost, then, unlike any other American writer who comes to mind, seems to be saying that seeing the angry sea as a harbinger and symbol of the end is foolish because, though the ocean evokes in the water-gazer biblical analogies and symbolic meanings, those meanings are in the gazer, not in the ocean itself. In this poem, as elsewhere in Frost, what seems to be said (often in an apparently traditional context) is denied at the same time, as the author has it both ways at once.

In other poems (such as "Neither Out Far nor In Deep" and "Sand

Dunes"), Frost seems to distrust, if not deny, the ability of the sea to mean much to us at all. In the former, the "wetter ground" actually "reflects" more than does the sea. "Does No One at All Ever Feel This Way in the Least?"— a late poem—presents an ironic, American version of Tiresias's instruction to Odysseus to put an oar on his shoulder and go inland to where the sea is unknown. And what may be Frost's only allusion to Melville comes as a possibly meaningful glimmer of something white—in a well ("For Once, Then, Something"). Despite New England's long and continuing intimacy with the sea, Frost's New England pastoralism, however bleak at times, does not accord with the sea that is "unharvested" and implicitly unharvestable. Homeliness does not harmonize with an ocean that is "not to be colonized."

If Frost and Jeffers were untouched by the sort of poetry Eliot advocated, William Carlos Williams actively resisted it, establishing his own equally influential line. In its reliance on imagery and irony (characteristic modes of modernism), "The Yachts" is more complex than Williams's usual idiom, yet Randall Jarrell was right to praise it as "a poem that is a paradigm of all the unjust beauty, the necessary and unnecessary injustice in the world." It is a paradigm apprehended sensorily through the pictorial reality of its evocation of the sea, and it opens deceptively with a sea surface full of ships and a dominant impression of color, grace, and beauty. The first five lines, seeming to echo the sea-fear of the "Brit" chapter of *Moby-Dick*, admittedly introduce

> an ungoverned ocean which when it chooses
> tortures the biggest hulls, the best man knows
> to pit against its beatings, and sinks them pitilessly.

From this open sea, however, the yachts are shielded by the partially enclosing land while they contend in their races. Their sea is a human playground. The yachts' beauty and movement are stressed as they operate in brilliant light. Williams's sea imagery is sufficiently pictorial to recall the seascapes of Winslow Homer and, in the hard, brilliant purity of its light, the paintings of the Immaculates, such as Sheeler and Demuth, both of whom Williams knew personally, and whose pictures he always went to see.

The first epithet applied to the yachts is of transience: they are "Mothlike in mists, scintillant in the minute"—at least until the reader realizes that the sense runs over to the next stanza, as in all but three of the poem's eleven. The mothlike yachts are "scintillant in the minute / brilliance of cloudless days." Does "minute" have an unusual adjectival connotation here of a unit of time, or its more conventional one of precision and smallness? The ambiguity is unresolved, but it is deliberate, constructive, and entirely symptomatic of the period that saw William Empson identifying *Seven Types of Ambiguity* in 1930, only five years before the first book-publication of "The Yachts."

The "mothlike" yachts are man-made and man-tended with "ant like industry," "solicitously groomed" in "a well-guarded arena of open water," surrounded sycophantically by other vessels. However, the youthful spontaneity they embody is immediately threatened (we might have expected it) by the sea, which

> is moody, lapping their glossy sides, as if feeling
> for some slightest flaw but fails completely.

The race is postponed, but on its resumption in a rising wind the sea renews the attack, again unsuccessfully because "they are too / well made." Now the poem takes an unexpected and sinister turn as the sea becomes "a sea of faces about them in agony" and "The skillful yachts pass over" the struggling "entanglement of watery bodies" with a cruelty that constitutes an implicit criticism of their grace. The familiar metaphor of a sea of faces or a sea of humanity acquires a macabre literalness of application, so that the sea, no longer the destroyer, shares defeat with the human victims. Thus the line "the horror of the race dawns staggering the mind" alludes both to the yacht race and the human race, doomed to be ridden down by its own creations of youthful beauty and uncaring energy. The linking of beauty and horror recalls Poe, but as in Jeffers, the natural, normal and real sea is being evoked in such a way that the poet's art enhances the credibility of the more surrealistic aspects of the scene.

In a much later poem, "The Sea Farer," Williams reverts to the more familiar identification of humanity with the rocks on which the sea pounds, but he replaces horror with tragic exultation in his insistence that humankind "invites the storm, [it] / lives by it!" This greater reliance on the deceptively simple stating of an observed analogy makes "The Sea Farer" more characteristic of Williams than "The Yachts." Man is stimulated, not frightened, by the predicament:

> so that the rocks
> seem rather to leap
> at the sea than the sea
> to envelope them. They strain
> forward to grasp ships
> or even the sky itself that
> bends down to be torn

upon them. To which he says,
It is I! I who am the rocks!
Without me nothing laughs.

The rocks straining at the ships are as doomed to defeat as the hands grasping at the yachts, but here, in this ability to laugh at an ever-present danger, the speaker achieves a measured power over the sea. It may be compared to the position to which, in "Lapis Lazuli," W. B. Yeats arrives by a different route—a vision, in the midst of human tragedy, of "Gaiety transfiguring all that dread."

II.

The laughter of Williams and Yeats is heroic, akin to the epic mood in H.D., but that mood has become less prevalent. Loneliness and mortality continue to be the poet's primary associations with the sea, but since World War II, the tone becomes antiheroic, witty, and conversational, as in William Meredith's "Open Sea":

> Oh, there are people, all right, settled in the sea;
> It is as populous as Maine today,
> But no one who will give you the time of day.

Meredith refers more literally to the drowned than does Poe in "The City in the Sea," but for each the sea's shutting-off effect emphasizes his own isolation:

> We say the sea is lonely: better say
> Ourselves are lonesome creatures whom the sea
> Gives neither yes nor no for company.

Generally, to earlier poets, the creatures of the sea were alien and often menacing, but over the past forty years, poets have come to empathize, even identify, with them and their condition (as has happened in fiction, and as, in the theater, Edward Albee tried less successfully to do in *Seascape*).

Daniel G. Hoffman's "Armada of Thirty Whales" (title poem, in 1954, of his first volume) is a parable of the sea creatures who, dissatisfied with their element, "nudge the beach with their noses / eager for hedgerows and roses." Hoffman's poems meet Frost's requirement of beginning in delight and ending in wisdom: here the moral is simple, foreseeable, but appropriate:

But they who won't swim and can't stand

lie mired in mud and in sand,
And the sea and the wind and the worms

will contest the last will of the Sperms.

Reversing the roles in "Natural History," Hoffman envisages the whales, this time wiser and in their element, reporting, "in whale calligraphy on harpoon handle inked," on man: they

Say but, 'This unfinned sport with little spine
Was fierce as froth asnarl on the crest of the surf is.
Habitat: found only on the surface;
Tried beyond its depth to spout; believed extinct.'

"The Seals in Penobscot Bay" "hadn't heard of the atom bomb / so I shouted a warning to them." The warning is unheeded by the animals, secure and seemingly happy in their natural element. It is from the deck of a destroyer that Hoffman observes them with delight. For him, implicitly, it is not the sea that is hostile but his own world "when the boom / of guns punched dark holes in the sky" and where man can no longer move with the confident ease and grace that he envies in the seals. Many years later, Robert Lowell would carry a similar idea even further in his "Seals," the penultimate poem in *For Lizzie and Harriet.*

If we must live again, not us; we might
go into seals, we'd handle ourselves better:

"all too at home in our three elements / . . . creature could face creator in this suit."

Hoffman, in sea poems much less confessional than Lowell's, derives from the otherness of the sea more of serenity and strength (as well as less fear) than many poets because its denizens prompt him to contemplative metaphysical speculations. Richard Eberhart in "Seals, Terns, Time" sees:

the seals at play off Western Isle
In the loose flowing of the summer tide
And burden of our strange estate.

"Our" quietly stresses the morality those "blurred kind forms / That rise and peer from elemental water" share with humanity, a sentiment that recurs frequently, as does the sea itself, in the poetry of Elizabeth Bishop. (She was one of Lowell's close friends, and the two captured parts of their lifelong friendship by writing poems for each other set in favorite coastal spots.) In "At the Fishhouses" (1955), for example,

> One seal particularly
> I have seen here evening after evening.
> He was curious about me. He was interested in music;
> like me he believed in total immersion,
> so I used to sing him Baptist hymns.

The element of whimsy in Bishop does not distract or detract from her fundamental seriousness, however. Darkness mutes the scene in "At the Fishhouses" and gives everything an unearthly beauty. The setting has enormous sensual appeal; Bishop depicts a vibrant stasis.

From this world, we look toward another; there is an edge between the land and water, a horizontal line intersected by a ramp suggesting that movement from one world to another is possible. Bishop's speaker turns that seal she sees in the water into a kindred spirit, curious about the other world "against his better judgement." Her whimsical and affectionate picture of the seal shares space in her vision with a much more frightening set of images. The final third of the poem establishes a beautifully rhythmic sound for her description of the seal's element, "bearable to no mortal." The water is "icily free" and swings indifferently over the stones below. It can burn, "as if the water were a transmutation of fire."

> If you tasted it, it would first taste bitter,
> then briny, then surely burn your tongue.
> It is like what we imagine knowledge to be:
> dark, salt, clear, moving, utterly free,
> drawn from the cold hard mouth
> of the world, derived from the rocky breasts
> forever, flowing and drawn, and since
> our knowledge is historical, flowing, and flown.

The poet's effort to record the scene and time's effects on humans and their artifacts takes on added poignancy in light of the fact that so much of knowledge comes to us by way of suffering and loss. Bishop offers the transmutation of fire as an affirmation of something like Hopkins's concept of inscape: if you were to go into this dangerous world you would know water for what it really is. Her identification of this trope with knowledge suggests that she approaches the latter with a combination of desire and fear, shaped by her awareness of the cost of vision. The poem explores a connection between human and natural worlds and a link between the visible and the invisible.

In examining that connection, Bishop suggests that the mystery in that other world is at least potentially threatening—the speaker asserts no more than it is *like* what we *imagine*—and suggests we are lucky to come no closer to it than imagination can bring us. Such a judgment is also evident in "Seascape" or in "The Unbeliever." For her, "This celestial seascape . . . does look like heaven," but "the spangled sea" "is hard as diamonds; it wants to destroy us all."

For Howard Nemerov, in "The Salt Garden," a gull, "image of the wild / Wave where it beats the air," brings him a message, "brutal, mysterious," of this mortality. In "The Lives of Gulls and Children," sympathetic youngsters come upon a dying gull on a deserted shore and attempt to help him in his apparent loneliness. As the bird rejects them, they recognize their difference from "the Atlantic kind he was," though they too are "Bearing the lonely pride of those who die"; and in "The Gulls," Nemerov again celebrates these "White wanderers, sky-bearers from the wide / Rage of the waters" who sing "of mortal courage."

Hoffman, Eberhart, Bishop, and Nemerov all use vivid, accurate observation to reinforce the larger assertion or implication that human beings share with their fellows in the sea what Nemerov, in "The Lives of Gulls and Children," calls the "lonely pride of those who die." But in "I Only Am Escaped Alone to Tell Thee," Nemerov has recourse to a quite different use for the sea in poetry. The title is from the Book of Job, and its Ishmael-association prepares us for a whale world like that in *Moby-Dick,* but in fact, the poem might well have used Pound's title "Portrait d'une Femme," for a woman is its theme.

Pound, too, draws on sea imagery in his 1912 poem. His opening line, "Your mind and you are our Sargasso Sea," strikes the ambivalent keynote in this portrait of a lovely woman by extended analogy. She is essentially the repository for "oddments"—stories, gossip, ideas—that have come to her over the years. Pound's images add up to a sad portrait, finally; despite all the riches that she has accumulated, the woman remains "second always" to the needs and desires of those who come to her looking for "strange spars of knowledge and dimmed wares of price."

Pound's speaker believes that the woman has found some recompense in her unconventional situation, that she prefers this way of being "to the usual thing: / One dull man, dulling and uxorious, / One average mind—with one thought less, each year." Nevertheless, it is clear that her only real success has been to make the best of a bad bargain. The sea metaphor contributes particular images of the woman as lovely, mysterious, difficult to fathom, but it also points to the inescapable loneliness in a life devoted to her suitors rather than herself.

> These are your riches, your great store; and yet
> For all this sea-hoard of deciduous things,
> Strange woods half sodden, and new brighter stuff:
> In the slow float of differing light and deep,
> No! there is nothing! In the whole and all,
> Nothing that's quite your own.
> Yet this is you.
>
> (24–30)

Nemerov's approach is more indirect. The woman is seen in a hall doorway, in a dim light; "an immense shadow" on the floor between them is symbolized as "A giant crab readying to walk." In the next stanza, the dignity of her bearing is conveyed by the simile of a stately ship—not imitatively, though other poets have used it. He is then led by a mention of corsets to the reflection that "all that whalebone came from whales." His final vision of her is in the hall

> Where the mirror's lashed to blood and foam.
> And the black flukes of agony
> Beat at the air till the light blows out.

If Nemerov's agony of parting invites association with the final dark isolation of Ishmael, the "black flukes of agony" also lead us to "The Fish" (1946), in which Elizabeth Bishop describes how she "caught a tremendous fish / and held him beside the boat / half out of the water." Admiring him as a living organism, visualizing his internal as well as his external structure, she notices "five big hooks / grown firmly in his mouth":

> Like medals with their ribbons
> frayed and wavering
> a five-haired beard of wisdom
> trailing from his aching jaw.

Her sympathy for his suffering seems to have a saving power for both woman and fish: "victory filled up / the little rented boat / . . . until everything / was rainbow, rainbow, rainbow! / And I let the fish go." A few years later, in *The Old Man and the Sea*, Hemingway's Santiago is equally respectful of his great fish but, refusing to release it, has to see it eaten by sharks. Yet as in Bishop's poem, a kind of victory is attributable to both the catcher and the caught.

A struggle as epic as Santiago's and a victory equally Pyrrhic are celebrated in James Dickey's "The Shark's Parlor" in his *Buckdancer's Choice* (1965). The first significant American narrative poem of the sea since Robinson Jeffers, Dickey's belongs also in the category of initiation literature that has so strong an American resonance. Two boys, by inference teenagers, are eagerly watching the sea from the porch of a beach house that stands on wooden pilings in the water. To one of these, they have attached a rope that runs three hundred yards out to sea with a baited "drop-forged shark hook" and the contents of "a bucket of entrails and blood" in the water around. At length, their patience is rewarded by the rope leaping out of the water so violently as to threaten the whole edifice. The narrator, his companion having run for help, tries vainly to steady the rope, is pulled briefly into the water, and manages to return to the porch as help arrives. As the hammerhead they have hooked comes into view, the boy tries in vain to let up on the straining rope:

> But the rope still strained behind me   the town had gone
> Pulling-mad in our house: far away in a field of sand they struggled
> They had turned their backs on the sea

By now the shark is thrashing dangerously around beneath the porch:

> The screen door banged and tore off   he scrambled
>   on his tail slid
> Curved   did a thing from another world   and was out of his
>   element and in
> Our vacation paradise   cutting all four legs from under
>   the dinner table

The use of the vernacular, the laconic vividness of narration, the fascination with physical process and prowess, all recall Hemingway. At last, when the shark's prolonged death-agony is over, he is seen to have left a prominent blood stain on the wall of the room he has wrecked. For what that mark symbolizes, the narrator, when older, has no choice but to buy, and restore, the house.

In "Leviathan," though, W. S. Merwin introduces no note of human victory. Here, the vastness of the sea is matched by the bulk of the huge "black sea-brute bulling through wave-wrack, / Ancient as ocean's shifting hills . . . past bone-wreck of vessels, / Tide ruin, wash of lost bodies," his size "frightening to foolhardiest / Mariners." Yet this monster can also lie, like "a lost angel / On the waste's unease, no eye of man moving":

> The sea curling,
> Star-climbed, wind-combed, cumbered with itself still
> As at first it was, is the hand not yet contented
> Of the Creator. And he waits for the world to begin.

This hint of creation, of Genesis, is not easily sustained in a postwar world that seems rather to be lapsing back into chaos than emerging from it, and Merwin wisely undercuts optimism with his non-Christian stress on the huge whale and the vast ocean as clumsy forces of destruction. Merwin's twenty additional poems about the sea, eight of them closing *Green with Beasts* (1956) and twelve opening his next volume, *The Drunk in the Furnace* (1960), assert the unbreachable human ignorance of the sea that makes it dangerous. The sequence attempts, despite the admitted human fear of full knowledge, to penetrate the mystery, but ends in images of wreckage, tones of warning and despair.

The problem of living in a world of destructive forces recurs in Denise Levertov's "Sharks." They arouse in her, despite her conversational under-statement of it, a reaction reminiscent of Melville's to the "Pale ravener of horrible meat" in "The Maldive Shark" and anticipatory of, though more profound than, the popular frissons of the *Jaws* films. The sharks arrive as the speaker has dared for the first time "to swim out of my depth," and in a setting between sundown and moonrise, childlike perception is caught, with tact and delicacy, between innocence and experience. For the modern poet, it is the same sea as Melville's with its predators, but yet, in a new sense, it can never be the same sea again. Albany, in *King Lear*, foresaw that

> It will come,
> Humanity must perforce prey upon itself,
> Like monsters of the deep.

The seals in Penobscot Bay may be unaware of the atom bomb but are no less vulnerable than the poet to the force, more fearfully far-reaching than any monster of the deep, released by humanity in preying on itself in modern warfare.

The impact of World War II on American poetry of the sea may be unexpected but is undeniable. As outstanding as "The Eye" by Jeffers is "The Quaker Graveyard in Nantucket" by Robert Lowell, with its echoes of Thoreau and Melville. That his cousin, for whom it is a threnody, was serving on a naval vessel that sank with all hands necessarily establishes a war con-text for this 1946 poem, and Lowell's fiercely held pacifism determines his attitude, as it does to his retired-naval-officer father in other poems, but here the savage and hypocritical cruelty of the Quakers as whale fishers also intensifies Lowell's antagonism to violence. Yet the sea itself seems to have been contaminated and its very nature changed by the ferocity of war, "Man's inhumanity to man" as well as to whales. Not only is the familiar theme of death by drowning treated with an unusually relentless and macabre literalism, until the Atlantic is "fouled with the blue sailors," but dreadnoughts confess the sea's "hell-bent deity," "the guns of the steeled fleet" salute "the earthshaker" and, a few lines later, "blast the eelgrass" and "roil the salt and sand." While the high tide "mutters to its hurt self," the whole ocean becomes a Quaker graveyard into which "We are poured out like water." The final reminder that "The Lord God formed man from the sea's slime" brings the argument full circle: in preying on the sea, humanity is preying on itself, a self created from the sea. However, the implication that, in so doing, we are also collaborating with the sea's own destructive power seems at odds with the poem's religious faith, and makes its final assertion that "The Lord survives the rainbow of His will" particularly enigmatic. As a source of imagery, the sea off New England recurs throughout Lowell's poetry; its stormy violence is particularly apposite to the moods of *The Dolphin* (published in 1973 simultaneously with *For Lizzie and Harriet*).

Levertov, however, reflecting on a later war in "An Interim" (*Relearning the Alphabet*, 1970), counterpoints violence with the beauty and eternity of the sea:

> To repossess our souls we fly
> to the sea. To be reminded
> of its immensity.

From her first volume, *The Double Image*, published in 1946 in Britain, two years before her emigration to the States, the sea plays, throughout her work, a part more central, rich, and varied than can be traced fully here. Most characteristically, it has for her a disturbing, mysterious, and essentially sundering influence on human destiny. It separates lovers, it can quite literally separate the soul from the body by drowning or, as metaphor, separate the known from the unknown, the new from the familiar. In the six lines that comprise "Leaving Forever" (*O Taste and See!*, 1964), the movement of a ship away from the shore is a skillful trope for the distancing perspective given to life by experience; its aphoristic simplicity merits comparison with Robert Graves's eight-line "Quayside," written many years earlier in a more ironically detached vein.

The same volume contains one of Levertov's most original and again deceptively simple sea metaphors in "September 1961":

> This is the year the old ones,
> the old great ones
> leave us alone on the road.
>
> The road leads to the sea.
> We have the words in our pockets,
> obscure directions.

These "old great ones" are Ezra Pound, William Carlos Williams, and H.D., all at the end of their creative lives, "not dying" but "withdrawn / into a painful privacy / learning to live without words." Their inheritors, uncertain, left alone to "count the / words in our pockets," have to follow as best they can "this endless / road to the sea," a sea that symbolizes extinction and yet consummation also. The poem's plangent lyricism is evocatively haunting but, in the lonely sadness of its acceptance of the burden of creativity and the human condition, significantly different in its hesitant outlook and tone from the attitude more characteristic of the "old great ones" and their generation. Yet in *The Jacob's Ladder*, published in 1961, the dedicatory "To the Reader" makes clear that for Levertov, the sea is always a timeless, inescapable presence:

> and as you read
> the sea is turning its dark pages,
> turning
> its dark pages.

Richard Wilbur, however, does not find the sea an open book. To penetrate its darkness (compare Elizabeth Bishop's "Seascape"), he needs the strong beam of "The Beacon" (*Things of This World*, 1956), "with cutlass gaze / Solving the Gordian waters":

>   Rail
> At the deaf unbeatable sea, my soul, and weep
> Your Alexandrine tears, but look:
> The beacon-blaze unsheathing turns
> The face of darkness pale.

Wilbur is yet another poet who can contain the sea only by relating it to humanity. Peopling the waves with "the buxom, lavish / Romp of the ocean-daughters," he sustains the classical allusions until, "with one grand chop," the frightening, lonely darkness is illuminated, and experience unified, by the dual light of the poetic imagination and the lighthouse beam focusing on a passing vessel:

> It is the Nereid's kick endears
> The tossing spray; a sighted ship
> Assembles all the sea.

Of the same generation as Levertov and Wilbur, Adrienne Rich is also painfully aware of living uneasily in a world illuminated not at all by "The old masters, the old sources." Like Levertov, acutely conscious of the problems of womanhood, she seeks alternative visions, not the myths that seem to help Wilbur. The sea as access to another world is the central image in her "Diving into the Wreck" (1973). Its speaker recounts the story of preparing for and then risking her first descent into an unfamiliar element. Immersion is a premeditated act: the speaker is a diver who not only wears and carries protective gear but also has first "read the book of myths" about the dangerous sea. The poem might almost stand as a revision of Bishop's "At the Fishhouses," in that the female speaker has decided to risk unknown dangers, perhaps death, perhaps the loss of her own purpose—"it is easy to forget what I came for"—in her attempt to gain knowledge. Her dive is solitary, "not like Cousteau with his / assiduous team." And yet she is not entirely without guidance: she places some trust in accounts by divers who have returned from the wreck.

In Rich's poem, as in many others, the sea is inhospitable, but it nevertheless has a weird loveliness, especially in and around the wreck, silent testament to "the damage that was done / and the treasures that prevail." The poem offers a number of ways to understand its symbols, but internal evidence suggests that the *I* who carries "a book of myths / in which / our names do not appear" represents female experience in particular. The available myths provide only limited help for the female diver, now determined to explore a history that shaped her experience but did not preserve her story. She brings a light to examine the thing she came for:

> the wreck and not the story of the wreck
> the thing itself and not the myth.

The diver enters a reality she has only heard about before, and claims the past as her own. The act of claiming is possible because the *I* has temporarily moved outside of her human nature; with her streaming hair and armored body, she is simultaneously mermaid and merman. Rich's diver leaves Prufrock far behind, and her audacity is of another sort than that of the singer in Wallace Stevens's "The Idea of Order at Key West," but Rich's poem develops tropes that many of her predecessors have explored: familiar images of the sea as threat, as mystery, as repository of the past give Rich a vocabulary for describing her own new experiences.

III.

One American poet of this century exceptionally transcends many of the generalizations in this chapter: that, of course, is Charles Olson. Other poets draw on the sea for illustrative metaphor, for personal anecdote, or for symbolism. For none of them is the sea as central and integral to their entire work as it is to Olson's nor as elusive in its susceptibility to definition. This is due not so much to the difficulty of Olson's idiosyncratic verse as to the difficulty of identifying precisely why his concept of the sea seems so unusually and peculiarly American.

The New England seacoast and the town of Gloucester, Massachusetts, are at the heart of all his writing, but it is much more than that. A British anthologist has recently suggested that "the sea was the beginning of English journeys, it was the end of American ones." Certainly it was the sea by which, through the centuries when Britain was exploring and colonizing so much of the known world, settlers from Britain and Europe were traveling to new homes on America's eastern seaboard. This immigration is a major theme in Olson's *Maximus* poems especially. However, Gloucester was a fishing port and the often-hazardous voyages of Massachusetts fishermen and whalers form another of Olson's concerns that again he explores in a broadly historical dimension. Ships, he emphasizes,

> on waters which are tides . . . ,
> are not gods
>
> on waves (and waves
> are not the same as deep water)

As he puts it elsewhere, "Ships / have always represented a large capital investment," and his presentation of Gloucester insists on economic fact as relentlessly as does Pound in the *Cantos*. Sex and money also he sees as "facts, to be dealt with, as the sea is."

That he enjoyed sailing and other firsthand forms of contact with the sea is another fact, but he admits equally factually, in "Maximus, to himself,"

> I have had to learn the simplest things
> last. Which made for difficulties.
> Even at sea I was slow, to get the hand out, or to cross
> a wet deck.
> > The sea was not, finally, my trade.

He is learning, but "It is undone business / I speak of this morning." However, it must be one's own business for, as he asks in another poem,

> who
> can tell another how
> to manage the swimming?

This does not, though, prevent his describing, in "Tyrian Business" and elsewhere, the technical processes of sailing and fishing with a grasp of detail as authoritative and graphic as Dickey's.

The sea for Olson is many things and no summary of them in this scope can do their intricacy justice. First, the sea is space, and in *Call Me Ishmael: A Study of Melville*, he declares unequivocally, "I take SPACE to be the central fact to man born in America." Don Byrd has observed in *Charles Olson's Maximus* that, because of the will to change, "space is atomized in an endless succession of discrete events." This is a possible explanation of Olson's constant citing of sea experiences in the lives of others in earlier centuries as a means of appropriating them into his own and reunifying them into a continuing relationship with the sea in the American consciousness.

However, as Byrd also observes, "Like all American writers before him, Olson was obsessed with beginnings." In "Mareoceanum," this obsession combines with the impulse to unification, enabling the poet to resolve the problems inherent in the loss of a world containing

> Newfoundland a large peninsula
> joined almost to
> Biscay, Cape Race and
> Finisterre almost
> stuck together

He replaces Matthew Arnold's vision of "the salt, estranging sea" with one of the sea as a linking agent of communication rather than as disintegrative. This can take an even more mystical form, which Byrd associates with Olson's interest in Pytheas and this extract from a modern translation of the *Geographia* of Strabo: "There is no longer either land proper so-called, or sea, or air, but a kind of substance concreted from all these elements." Certainly Olson is never averse to laying the ancient Greeks or anybody else under contribution to his cosmography. It is this oneness of vision that produces the Frost-like realism of the advice to the reader to "Love the World—and stay inside it."

This kind of vision allows Olson, unlike other poets, to conceive of the sea and the city not as antitheses but as parts of a whole. Thus "Maximus in Dogtown II" begins with a peremptory injunction:

The Sea—turn yr back on
the Sea, go in, to
Dogtown.

In "Maximus, to Gloucester: LETTER 2," approaching the town from the sea
has an expectedly literal and topographical importance:

The point is
the light does go one way toward the post office,
and quite another way down to Main Street. Nor is that all:

coming from the sea, up Middle, it is more white, very white
as it passes the grey of the Unitarian church. But at Pleasant Street,
it is abruptly
black.

Nor is it straining interpretation to suggest that, for Olson, the whiteness
of the light is directly attributable to its coming from the sea. His frequent
invocations of Fitz Hugh Lane (1804–65) are relevant in this connection.
Not only was Lane born in Gloucester, built a house, and died there, but
his marine paintings of the sea and the ships off that coast are celebrated
for their remarkable rendition of light, characterized by the title "Lumin-
ism" now applied to the work of Lane and his associates. George Butterick
identifies nine references to Lane in the *Maximus* poems alone, and in the
later collection, *Archeologist of Morning,* there are others even more detailed.
The very painterly nature of these passages is of interest in itself and lifts
them well above the merely perfunctory. Indeed, the two poems " 'light sits
under one's eyes' " and "An 'Enthusiasm' " in the latter volume do not merely
evoke the quality of Lane's painting techniques superbly but, by linking him
with Hawthorne, Melville, Parkman, Prescott, Noah Webster, Whitman, and
Charles Peirce, epitomize much of the spirit of Americanness that informs
Olson's own attitude toward the sea and toward so much else—a sensitive
and erudite perceptiveness.

Olson's friend Charles Boer once reminded the poet of his habit of staring
at things outdoors "and of course the sea, the omnipresent Gloucester sea. . . .
And you described what you saw with a painter's precision rather than a
poet's license." This is true, but Olson could also see the sea through eyes
other than a painter's. In "Mareoceanum," for example, the sea, in one of
its manifestations as a beginning, is regarded biologically as the primal chaos
from which all life originally emerged. In "The Ring of" (*Archeologist of Morning*), this idea is developed with more of "a poet's license" when it describes
how Venus "rose/ from the genital/ wave,"

> this new thing born
> of the ring of the sea pink
> and naked, this girl.

These ideas, and many more, are brought conveniently together in the closing lines of another of the *Maximus* poems:

> The sea does
> contain the beauty I had looked at
> until the sweat
> stood out in my eyes. The wonder is
> limitless, of my own term, the compound
> to compound until the beast rises from the sea.

The sea as space and as timelessness, as change and stasis, as economic fact and force of nature, friend to man and enemy, source of beauty and beasts, strangeness and familiarity, yet above all a source of unending wonder— these paradoxes run through all twentieth-century American sea poetry as surely as they run through Olson's work, but seldom more challengingly.

IV.

For the American poets of our century, the sea has stood for so much as almost to defy summary. For most, perhaps, its essential otherness has constituted a challenge—not merely in how to assimilate it into poetry, and not merely as the physical challenge it presents to those involved with it in actuality. Its violence can give it an apocalyptic dimension threatening life and civilization as we know them. It may seem to separate human beings and yet sometimes to bring them together in a shared, sea-engendered experience. Its strength and timelessness can remind poets of the frailty of our mortality or encourage them to a fellow-feeling with the sea creatures, similarly vulnerable to the cruelty of sea and of humankind alike. Alternatively, that strength and timelessness may present themselves to poets as a model for emulation, a challenge to acquire a parallel, life-enhancing, if stoic, endurance. Yet again, mindful of evolutionary theory, poets may see the ocean as a life-giving force, a symbol of energy, and with an easy transition, associate it, positively or negatively, with the forces of love and sex. Protean as the characteristics of the sea may seem, it has for many American poets an irresistible fascination, something with which, from their several viewpoints, they are forced to come to terms in some way. Nowhere, perhaps, is this more clearly exemplified than in the work of Hart Crane and Wallace Stevens.

Venus rising from the foam is the key culminating image of the six magnificent lyrics that form Hart Crane's "Voyages" (in *White Buildings,* 1926). "Voyages I" is the most straightforward: the sight of "kids" playing on a beach arouses Crane's desire to warn them that "The bottom of the sea is cruel," much as Nemerov was to warn a later generation of youngsters. The more complex "Voyages II" was praised by Yvor Winters as "one of the most powerful and one of the most nearly perfect poems of the past two hundred years." Opening with a Jeffers-like allusion to the sea as "this great wink of eternity," each of the first two stanzas devotes four lines to evoking the sea's majestic, incomprehensible vastness and then, in a fifth line, contrasts that with the intimate warmth of human love. Personification of the sea as a woman and emphasis on its physical beauty lead to the fourth stanza's contrast between the eternity of the sea and human mortality.

Although Allen Tate once said that he did not think the association was consciously in Crane's mind, the image of currency is crucial here: it begins in the oxymoron of "penniless rich," is sustained in "pass," and then qualified in "superscription." The sea is passing to us some of its riches, but since its palms are penniless, this is clearly not material wealth. In the Authorized Version (Matthew 22), Jesus cites the "superscription" on the penny as proof of its validity, just as in the poem the word validates the gift, though only by the ephemeral "bent foam and wave." This benefaction, coupled with the measuring of time by the turning shoulders of the sea, makes the more urgent our acceptance of our own mortality. The theme is resolved in the final invocatory stanza, when the familiar image of man as a voyager on the sea of life is invested with a new intensity and force:

Bind us in time, O Seasons clear, and awe.
O minstrel galleons of Carib fire,
Bequeath us to no earthly shore until
Is answered in the vortex of our grave
The seal's wide spindrift gaze toward paradise.

The sequence ends, in "Voyages VI," with a connection between Venus rising from the sea and "Creation's blithe and petalled word," "the imaged Word" that constitutes "the unbetrayable reply." Yet in "the Bathers," Crane speaks of Venus as nursing "In Silence, beauty blessed and beauty cursed": beauty, like the sea, is ambivalent—like poetic creativity, as the haunted genius of Hart Crane knew all too well. Tate's acute observation that Crane is at his best "when he writes from sensation" is particularly applicable when it is the sea whose beauty he is celebrating; but it was, of course, in the sea that Crane sought, and found, his own death.

"Death is the mother of beauty." That formulation is Wallace Stevens's. "The Idea of Order at Key West," like so many of the sea poems already discussed, exhibits the poet at his best, but it also epitomizes felicitously several points made above. Here is the salvation Prufrock longs for, in a reality transformed and a beauty made durable by imagination. Here is both "the dark voice of the sea" and the song the poet can make of that voice. There are two makers in the poem, the poet telling the story and the woman who "sang beyond the genius of the sea." The poem is a celebration as well as a narrative, its subject the poet's power to move beyond the chaos inherent in human experience to the order and value that poetry makes possible. The poet recalls what is clearly a treasured memory, of the singer creating one of the supreme fictions we need in order to live. He eschews any facile identification of the sea with the human being. The constant cry it utters is a cry "That was not ours, although we understood, / Inhuman, of the veritable ocean." The mysterious sea itself, although a rich setting for the poet's song, remains less important than the creative power that transforms it, "For she was the maker of the song she sang."

> The sea was not a mask. No more was she.
> The song and water were not medleyed sound,
> Even if what she sang was what she heard,
> Since what she sang she uttered word by word.
> It may be that in all her phrases stirred
> The grinding water and the grasping wind,
> But it was she and not the sea we heard.

Literature, obliged by its nature to adopt a linear motion (proceeding word by word), cannot reproduce the simultaneity of our apprehension of and response to the sea in actuality, nor should it wish to. Its strength lies rather in its power for order, its ability, by what Coleridge called the "shaping spirit of Imagination," to make us understand more fully the significance the sea has for us, for "it was she and not the sea we heard." The point is not new, but the poem makes it memorable by its perfect harmony between meditative toughness (it is almost an essay in criticism) and the beauty of its rhetoric. Imaginatively, the poet and the singer have apprehended a natural order that they have then shaped in their art. Stevens's epithets for the sea exemplify the attributes usually associated with it: "inhuman," "the grinding water," "ever-hooded, tragic-gestured," "the sunken coral water-walled," "meaningless plungings," "bronze shadows heaped on high horizons," "dimly-starred." Stevens is not the only American poet for whom the sea is reality, but for him it is also the tragic beauty of reality. Others since Stevens have also recognized that it is not "only the dark voice of the sea / That rose." Audible in that voice they also hear "the still, sad music of humanity." If, however, the more pressing reality of war and its aftermath has indeed blunted our sensibility to the beauty that Stevens and his contemporaries, exulting in "Creation's blithe and petalled word," could find and express in their poems as they "Mastered the night and portioned out the sea," we are the poorer for it.

DENNIS WELLAND

Chapter Thirteen

MODERNIST PROSE AND
ITS ANTECEDENTS

In American literature, modernism generally means the kind of early twentieth-century writing that made a conscious break with the past and experimented with literary form in order to accommodate contemporary thought. As a serious movement, it followed the introduction to America of modern painting at the New York Armory Show in 1913. New techniques for fusing idea and emotion were adopted by writers who, like the painters, sought new ways to express the concerns of those who felt themselves increasingly "lost" after American soldiers had experienced the first world war of the modern age. In sea fiction, the advent of modernism also meant the reformulation of long-used paradigms; in particular, it meant that the journey-quest was no longer governed by the mystical and mythic models of previous ages.

American sea symbolism has its origin in such classical models as Homer's *Odyssey* and Virgil's *Aeneid*, where typically the sea represents aspects of cosmos and prehistory, and from which the sea-journey motif emerged as a powerful expression of human endeavor. Later reworkings preserved the meanings of the mythological conventions of the originals, while addressing the human experience of the intervening centuries, as, for example, Defoe focused on shipwreck imagery in order to make telling comments on eighteenth-century British culture in *Robinson Crusoe*. In twentieth-century psychoanalytic terms, sea journeys often represent a quest for selfhood, with the sea as an archetype of the unconscious and analogue of the womb. In this view, the sea threatens drowning and extinction, but it also provides the opportunity for ritual immersion and rebirth.

Before modernism, in the later nineteenth century and early in the realistic movement, when the influence of the Romantic tradition waned and strong Victorian doubts anticipated Darwinian negativism, literary journeys of learning reflected the radically different sensibility of a new age. Mark Twain is an example with his satiric ocean and land journey of pilgrims to the Holy Land in *Innocents Abroad* (1869) and in the account of his own overland journey as a tenderfoot to the far West in *Roughing It* (1872).

Comic as they were, such presentations preserved the journey motif for serious fictive expression of human endeavors, as Mark Twain himself illustrates in the archetypal journey to learning and selfhood taken by Huckleberry Finn on the Mississippi River in *Adventures of Huckleberry Finn*. Despite a seemingly facile romance context (shipwrecks, piratical villains, damsels in distress), Twain's realistic ironies advanced serious themes by furnishing a surrogate sea that supports a "sailor" as the hero-quester. The ironic subversion of romance in the novel made it an influential model for American writers of fiction who followed, even into the twentieth century when Ernest Hemingway would claim at the beginning of *Green Hills of Africa* (1935) that all modern American literature came from this early realistic model.

As psychological shifts occurred in American culture under the pressures of changing social, technological, political, and economic forces, including urbanization, writers who appeared after Twain in the later nineteenth century responded with new and darker views, often embodied in land journeys. Stephen Crane's *Maggie: A Girl of the Streets* (1893) and Frank Norris's *McTeague* (1899) are instances of the turn from the citation of the waters as a nurturing medium toward a dryland mockery of meaningful journeys. Maggie journeys through a nightmarish urban world to hopeless death in a New York City river. McTeague is a prototypical Darwinian creature more animal than man. Unable to compete effectively in the San Francisco environment at the end of the nineteenth century, and impelled by biological and environmental forces he cannot understand, he commits murder, and flees the authorities in a journey that ends in the desert with his impending death.

Both Norris and Crane also wrote tales in which sea symbolism moderates the despairing views of their land fictions. Norris's *Moran of the Lady Letty* (1898), which ends as the murdered body of Moran, a strong "Viking" woman, floats off on a derelict ship out of San Francisco Bay toward the open ocean, employs sea imagery to create a symbolic ground for primal psychic encounters that go beyond the book's overt Darwinian philosophizing, as Ross, a "modern" and "civilized" San Francisco gentleman, responds to the primitivism of his Viking. Crane's masterpiece, "The Open Boat" (1897), is a unique compound of Darwinian empiricism and profound psychic resonances generated by his artful handling of sea imagery. The struggle of four men in a very small boat on a ferociously dangerous though indifferent sea is an epitome of the human condition. In its main consciousness, the correspondent, Crane registers the responses of the alienated self as it encounters that power and indifference. Crane's presentation of the plight of a self-conscious, alienated self in "The Open Boat" is the first clear modernist treatment of that self in American sea fiction. In its sea symbolism, the story creates a union of image and theme that those American writers of the realist-naturalist tradition who followed in the twentieth century would emulate many times over: the only apparent hope for the lonely self is brotherhood, social solidarity.

The differing perspectives revealed through land and sea journeys are explained by W. H. Auden in his seminal comments on Romantic consciousness in *The Enchafèd Flood* (1950). In his illuminating exposition of the sea journey's primary meanings from Romantic to post-Romantic, modernist literature, Auden elucidates the psychic predicament and the changing consciousness of each era. He points to the primacy of sea and desert images throughout literary history, and notes that, especially for the Romantic sensibility, the desert and the sea were major symbols of opposition that governed artistic vision and expression (ch. 1).

Both desert and sea represent psychic dimensions, says Auden, that lie outside the community and historical change, and thus the modern voyaging individual may escape from communal responsibilities; but the very nature of this condition leaves the individual alienated, isolated, and lonely. The desert represents "actualized triviality" and "lifeless decadence" in this view, but there is also the sea, "the symbol of primitive power" and "living barbarism." Possessed of a sufficiently heroic soul (and most of the great romantic questers are), the voyager escapes the "trivial" desert waste land and plunges into the symbolic sea (ch. 1). Through immersion comes reintegration: the quester discovers the uniqueness of his own soul and reenters the community he has fled. There he discovers that he participates spiritually in the community of all souls.

Auden believed that the modern age, his own age, had lost the capacity for meaningful engagement in such a spiritual dialectic, let alone the hope of transcendence or reintegration that had been available to the Romantics. He sums up the implications of his observations for the modern era when he observes that the heroic image is no longer that of a "nomad wanderer" through the desert or over the ocean, but that of "the builder," who rebuilds the ruined walls of the False City by prostituting himself (ch. 3).

If subject matters of many works seem remote from sea concerns and renewals, often there is a sea consciousness that testifies to the continuing attraction of the sea as an empowering image for the self in the modern False

City and desert that Auden describes (ch. 3). The mimetic task makes all writers of all ages necessarily impose some sort of order on the experience they imitate; in this sense, as a modern image the sea substitutes for the myths that in the past had provided the "informing structures" of literature, in Northrop Frye's phrase.

Such is the case for Henry James, who in the last decades of the nineteenth century and early into the twentieth wrote fictions about a declining Victorian society by employing subject matters seemingly far removed from the sea fictions of the past. But James's sea consciousness makes the ocean a symbol of the cultural forces that alternately link and separate America and Europe for the characters in his "International" stories and novels. James was acutely conscious of the relatively rapid transit over the ocean that was possible in the age of steam, and he exploited this "modern" phenomenon in the cultural clashes occasioned by the "migration" of expatriated Americans in such works as "Daisy Miller" (1878) and *The Portrait of a Lady* (1881).

In "The Patagonia" (1888), a short story of social manners and mores modeled on the traditional "ship of fools" concept, most of the action takes place aboard a steamer sailing from America to Europe. James employs sea and ship imagery metaphorically in a way that Auden relates in *The Enchafèd Flood*, when he observes that ships are employed as metaphors for society in danger from within or without. Here, the danger is from within, for in the narrator's probing the motives of various characters, he finally is as culpable morally for the drowning-suicide of the heroine as is her lover, who presumably has led her into an act of unfaithfulness to her betrothed.

If James's art seems barely to acknowledge his peers and immediate predecessors, Jack London, who died the same year James did, brought a unique romantic realism to the new century that was solidly anchored in the earlier naturalism of Crane and Norris. In *The Sea-Wolf* (1904), London blends biological determinism with romantic hopefulness in the story of a tyro who journeys over a symbolic sea of existence under the tutelage of Wolf Larsen, who, himself, leads a hunted and "haunted" existence on a sea that represents a Darwinian universe.

The immense popular success of *The Sea-Wolf* gives London almost sole credit for preserving sea-romance elements for the age of steam and the succeeding diesel and nuclear ages. Most of the great stories of the sea in the age of sail derive their power over the imagination from the representation of direct and dramatic physical encounters of individuals with the forces of the sea. Despite the passing of that era when men struggled against nature under canvas, the thematic appropriateness of that kind of struggle did not entirely disappear.

Though the machine had removed much of the burden, the danger of the sailor's occupation persisted. The sea, with its power to destroy even the strongest iron or steel ships, was still an attractive subject matter with a meaningful resemblance to human struggle with nature or cosmos. In America, the sea tradition that began with nautical narratives in colonial times or earlier continued unabated into the twentieth century in both popular and serious works, including such mainstays of the canon as Eugene O'Neill's sea plays and Ernest Hemingway's *Old Man and the Sea.*

Crane, Norris, and London had responded in meaningful ways to their epoch by committing themselves to a naturalistic viewpoint, with metaphors such as "red badge," "octopus," and "northland." These figurations served well in an epoch that saw the universe in bleak Darwinian terms, but the American writers of the twentieth century who inherited these metaphors, along with the scientific positivism and religious skepticism out of which they emerged, found their own tropes to express the effects of that blighted inheritance on a new age.

Most notable of these was T. S. Eliot, who published *The Waste Land* in 1922 and framed for his age so powerful a metaphor of moral and spiritual aridity that it virtually dominated the literary scene until well after World War II. Eliot's poem is more than a social criticism of the modern age, and its methodology conjures a way of looking at human life so potent that for a long time it seemed there was no other way of expressing the nature of existence. *The Waste Land* signaled another "revolutionary change in style and sensibility" of which Auden writes. In terms of Auden's desert-sea opposition, what Eliot had done to the literary imagination was sever the desert from the sea and remove the illusion of a garden-island sanctuary.

The domination of the desert metaphor seemed to deny the possibility of an opposing image that would provide an affirmative expression. The sea, however, was just such an affecting image and it had been available in twentieth-century America for some time before Eliot's poem. In the highly unstable realms of belief and cultural practices from which the artist draws his metaphors, such images undergo transformations, and in the work of Eugene O'Neill, the process was already underway. In advance of other major American writers of the early part of the century, O'Neill chose the sea and sea life as subject matter for several dramas and for a group of virtuoso one-act pieces. Like the seaman-writers before him, O'Neill, who had lived on the waterfronts and served as a seaman, found early in his career that his personal experiences of the sea were the materials through which he could best present his fictional critique of life in the new age; and in a series of dramas, he employed the sea voyage as a metaphor of the life voyage. In *Bound East for Cardiff* (1916), for instance, in which Yank, an American sailor, is dying at sea, O'Neill uses fog imagery to express the difficulty of "seeing" life clearly. The repeated mention of fog implies the obscurity of purpose in mortal life, as when Yank believes it is foggy in the compartment in which he is dying, and as in Cocky's unintentionally ironic statement at the moment of Yank's death that, " 'the fog's lifted,' " or in his ironic expletive, " 'Gawd Blimey!' " (blind me) immediately following, at the final curtain.

What makes O'Neill's symbolic usages unique is his infusion of Christian spirituality blended with a naturalistic view of existence into his often-noted classical contexts. These radically diverse elements create a tension between a modernist, empirical perspective of reality and a traditional, nonempirical reality that O'Neill fashions for his characters through their attitudes and actions. The nature of life at sea was a ready-made subject matter through which these perspectives on existence could be presented.

All of O'Neill's one-act sea plays and most of his longer works signal in their titles some referential context at the same time that they suggest their literal content. An example of such contexts occurs in the title, *Bound East for Cardiff,* where "East" suggests the literal direction and destination of the ship, and the classical direction taken on the *nostos,* the long and arduous journey home for the ancient Greek sailors as celebrated by Homer. It also hints at the spiritual direction in a number of religious cosmographies.

The title of *The Long Voyage Home* (1917) is an ironic echo of Odysseus's epic journey. O'Neillian characters are not epic giants, but they do reflect the successes and failures of their cultures, as their outsized classical forebears did for their time and place. In the one-act sea plays as well as in the 1920 *Beyond the Horizon* and the 1921 *Anna Christie,* O'Neill depicts the distinctly modernist plight of his heroes and heroines through their sense of alienation from the land and its implication of social responsibilities. As sailors who must perform their daily *agons* at sea, O'Neill's seamen are expressions in themselves of the ailments of contemporary life. If we apply George Steiner's views from *After Babel* (1975) to O'Neill's ship and sea imagery, the diverse backgrounds and mixed nationalities aboard these vessels duplicate the problem of communication in a "fallen" modern world (ch. 1). Yank, in *"The Hairy Ape"* (1922), even more emphatically represents this failure in the world beyond the microcosm of the ship and his shipmates. Indeed, these sailors on board their ships form a rudimentary community, and in their attempts to establish a "home" on the sea, they express the universal need for the security of social intercourse.

Here, as in the world the ship represents, the desire to create a larger social unit based on a bond of shared values is subverted from within, and often as not the immediate pleasures and passions of individuals take precedence over communal needs. At times, the jumble of motives as well as of tongues results in some chaotic and violent act, such as the fight and stabbing of Paddy in *The Moon of the Caribbees* (1918). As O'Neill's symbolic settings urge, of course, there is no security in a fallen world where all things are transient and meaningful communication is not possible. The power and danger inherent in the physical sea provide O'Neill with a metaphor through which to characterize the radical insufficiency of the temporal order of things to fulfill the hopes and desires of a weak and impermanent mankind.

The journey "home" implied in the title of *Bound East for Cardiff* is revealed by O'Neill as the ironic ending of life when Yank dies, but the pathos of Yank's final concern with God's judgment, after he reveals that he "ain't never had religion," reflects O'Neill's presentation of the spiritual consequences of dwelling in a contingent universe. The absurdity of the attempt to grasp the spiritual by means of temporal and empirical perceptions emerges through an exchange between Yank and Driscoll, prompted by Yank's worry over the fact that he has stabbed a man in a "fair" fight. Fearing that God, as the controller of justice or fate, might or might not be just, Yank asks if Driscoll thinks that "He will hold it up agin me—God, I mean." Driscoll's response is a mere equivocation: "If there's justice in hiven, no!" The question remains unresolved, and Yank's belated concern with spirituality reinforces O'Neill's theme of the spiritual poverty and ultimate absurdity of modern man's secular orientation, a theme presented even more directly in *Lazarus Laughed* (1927).

Of all the members of the *Glencairn*'s crew, only Driscoll exhibits a constant religious sensibility; and by his frequent use of religious expressions and the signing of the cross, he is the originator of a verbal iconic statement that nontemporal fulfillments are possible, even though he can only pray for Yank "in some half-remembered prayer." He is thus an instance of the vital loss to mankind of what George Steiner refers to as "the prelapsarian vulgate of Eden" (ch. 1), through which all men had the ability to understand and speak "God's language."

Driscoll's "half-remembered" prayer in this context is O'Neill's reminder that there must have been a time before secular preoccupations when Driscoll (and mankind) knew the whole prayer and thus spoke intelligibly to God. Such instances evidence O'Neill's making his sea settings into symbolic locales, in much the same way as Crane had done in "The Open Boat," where the trials of a few individuals in a small boat on an immense sea image the condition of mankind in the world. For O'Neill, though, the solution is not so much a banding together for protection against the assaults of nature as it is a renewal of community through recognition of the spiritual impoverishment of the modern age, and a rediscovery of that forgotten spiritual "language" to which Steiner alludes.

The loss of spirituality in the modern world is a major theme that emerges from the characters in *Anna Christie*. In action taking place mostly in a waterfront saloon or on a coal barge, O'Neill brings together three major characters who act out in vague allegorical terms the universal quandary of love and its conflicting loyalties. Chris Christopherson, the old Swedish captain of the coal barge *Simeon Winthrop,* persists in blaming all error and suffering on "dat ole davil [devil] sea." Anna Christie, his daughter, who had recently been in hospital, in jail, and worked as a prostitute, is at the center of the action. The plot contains two main stories. The love between Anna and Mat Burke, a young Irish stoker on steam ships, is the complicating action that upsets what Chris had thought to be the status quo. The other story, that of the relationship between Anna and her father, is more profound.

Although Chris abandoned Anna as a child to relatives who live inland on a farm, he defends his past actions as necessities, and along with all problems in his life, the result of the occult malice of "dat ole davil sea." He repeats the phrase many times over in various contexts as O'Neill gives it the force of a poetic refrain with dark insinuations. Chris's justification for sending Anna inland when she was a child was to save her from what he insists are the diabolical forces of the sea, but his failure to realize the dangers on land for a young girl resulted in her almost complete moral debilitation when subsequently Anna was seduced at sixteen by her cousin. This act ultimately sent her on the streets as a common prostitute.

Sick and weak when she comes back to Chris after many years, Anna consents to live with him in his new role as solicitous father on the barge he commands. There, under the influence of the sea and the fog she begins to feel "clean" and normal until Mat Burke, a shipwrecked sailor, falls in love with her. Since Anna has kept her past a secret, a moral crisis is precipitated. Chris mightily resists the union of the pair until Anna, driven by the contention between her lover and her father, confesses her past to the dismay of both Burke, who leaves, and her father, who cannot believe it of his daughter. In the end, Mat returns, Chris grudgingly accepts a truce, and lover and father, unbeknown to each other, sign on to the same steamship. All go on living with the knowledge of the fundamental corruption of humanity revealed by Anna's story. Chris remains embittered by "dat ole davil sea," which, in the course of the dramatic action, has come to represent in his cosmology, if not in O'Neill's, all the forces of contingency, Luck, Fate, or God that rule over the uncertain lives of humanity.

Just a few years younger than O'Neill, and virtually unsuspected of having a sea consciousness, F. Scott Fitzgerald, in *The Great Gatsby* (1925), consciously employed the symbolic implications of his island-shore setting. *Gatsby* is the most significant American fiction to portray the hopelessness of the attempt to maintain the integrity of the self after Eliot's poem had condemned it to its arid prison. Fitzgerald's well-known concluding reference to how the early Dutch sailors saw the island as "a fresh, green breast of the new world" elucidates in retrospect, through its emphasis on the sailors' wonder at a natural world uncontaminated by European grasping for power and material advantage, the thematic oppositions of America-Europe, New World–Old World, and Sea-Land that underlie the more obvious critique of the American dream throughout the book.

In Auden's terms, of course, the Long Island setting images another doomed "paradisiacal garden-island," and the term is appropriate for the larger concerns in this novel, where the East and West Eggers establish false communities on their island, with money as their sole protection against the "sea" of existence. Gatsby's death symbolizes the victory of cultural sterility and amorality over the Romantic sensibility in America. For Fitzgerald, that meant the death of idealism and the passing of an American self that was capable of a vision of values beyond time and beyond personal history. Gatsby's belief that through an act of will he could change the past is one reflection of that theme. Another appears in the ironic parallel Fitzgerald establishes between Gatsby and the fabled King Arthur.

Gatsby, with his chivalric idealism, represents for America and Fitzgerald what Arthur represents for England and Sir Thomas Malory in *Le Morte D'Arthur:* the hope of the past as it contrasts and clashes with the pathos of the present. In Malory's account, when Arthur is mortally wounded, he sails for the paradisiacal island of Avalon; but Gatsby has no true island-garden to which he can voyage in order to escape the annihilation of time. Indeed, his attempt to return to Long Island and Daisy after the war in order to realize an illusion of bliss and felicity is a futile gesture through which Fitzgerald makes the telling irony that there are no island paradises in the modern world, a view later expressed in *Tender Is the Night* (1934), when Dick Diver's self-created "paradise" by the sea is invaded and eventually destroyed by forces he cannot control.

Gatsby's final "voyage" in a swimming pool aboard an air mattress is a mockery of Arthur's ceremonial sailing to Avalon over transcendental waters. Gatsby rests on a vessel and floats on water that flows toward a drain and sails on an accidental course. These details and Fitzgerald's emphasis on "accidental" point to human insignificance and the futility of attempts to realize dreams in a contingent universe. Like Eliot before him, Fitzgerald adapted the mythic realm of the waste land to suit a fictive expression of the moral aridity of his age, and that aridity is magnified by the presence of the island and water symbolisms.

Many other writers of the twentieth century followed Fitzgerald's example in adapting the wasteland myth as a means of presenting in fiction the failures of the modern age, but that imagery could not continue merely to corroborate Eliot's post-Victorian indictments. Writers of both popular and serious fiction began to explore other symbolisms more appropriate to changing values in a changing age, particularly in the politicized atmosphere of the first three decades of the twentieth century, where the sea proved to be one of the most responsive metaphors for the articulation of twentieth-century interests. These interests were served by a variety of popular adventure fictions, but not all were simple entertainments. Some of these merit citation in more serious lists.

In the early 1930s, for example, Charles Nordhoff and James Norman Hall published their immensely popular *Bounty* trilogy. These works contain many romance elements, but the authors advance historical parallels between Captain Bligh's cruel exploitation of the *Bounty*'s crew and the kinds of exploitations that prompted those events of the 1905 Russian Revolution centering on revolts in the fleet and at naval bases. With scrupulous attention to historical accuracy, Nordhoff and Hall created a version of real-life events to serve the ends of a thirties' vision of social justice. Nordhoff also edited some of his grandfather's nautical work in a 1940 collection, *In Yankee Windjammers*, where, again, the difficult life of the common sailor reflects the life of the common man. Similarly, such popular modernists as James Gould Cozzens in *S.S. San Pedro* (1931) and *Castaway* (1934), Gore Vidal, *Williwaw* (1946), and Herman Wouk, *The Caine Mutiny* (1951) employed historical realism, along with ship-sea and sea-island imagery, to explore the manners of their contemporary society and the effect of modern living on the individual.

Of all the modernists, though, the most distinguished writer of prose fiction in the twentieth century fully to explore the sea as both subject matter and metaphor was Ernest Hemingway. From the outset, Hemingway had experimented with a number of subjects—war and bullfighting the best known and most successful—until about the middle of his career when he settled on the sea as appropriate for his art. Two of his late novels, *The Old Man and the Sea* (1952) and *Islands in the Stream* (1970, posthumously), are his most prominent sea fictions, but he revealed an ocean consciousness as early as his high school juvenilia; and later, in a 1923 poem, "Oily Weather," published in *Poetry: A Magazine of Verse*, he used sexual slang to exploit the sea as a metaphor of the female principle on which the ship's "screw turns a throb."

In his first novel, *The Sun Also Rises* (1926), Hemingway turned to Eliot's sources in *The Waste Land* for much of his metaphorical material, but his use of the myth differs from Eliot's when he turns from an emphasis on the aridity and desiccation of a culture to the myth of the redeeming waters. For example, an emotionally exhausted Jake Barnes leaves behind the chaotic events of the fiesta in Pamplona and its pagan-Christian contrasts to immerse himself in the purifying sea at San Sebastian: "I was through with fiestas for a while," Jake says in a radically ironic understatement. Jake's descriptions of his immersions suggest that Hemingway intends a change or transfer of the meanings of the sea from the literal to symbolic: "I dove deep once, swimming down to the bottom." Or, again, "Then in the quiet water I turned and floated. Floating I saw only the sky, and felt the drop and lift of the swells" (ch. 19). The "bottom" and "sky" opposition not only reflect Jake's emotional traumas, but they also suggest his spiritual rejuvenation, as further imaged by his feeling of buoyancy: "It felt as though you could never sink," he says. The figurative usage here anticipates its appearance in the late novels as a means of asserting the resemblance between the character's emotional reality and the physical qualities of an objective fact.

After *The Sun Also Rises*, sea imagery appears with some frequency in the works of the 1930s. In "The Sea Change," a 1931 short story, its function is allusive and ironic. Borrowing from "Ariel's Song" in Shakespeare's sea-island drama *The Tempest*, the title suggests impermanence and mutability, while Hemingway adapts the sentiment to the moral universe and elicits a heavy irony from the part of Ariel's speech that alludes to transformations: " 'Nothing of him that doth fade / But doth suffer a sea change / Into something rich and strange' " (I.ii). Hemingway's truncation and ironic adaptation involves a man whose lover is leaving him because of a lesbian attachment. Initially embittered, the man in "Sea Change" seemingly accepts the situation. His is not an acceptance of a transvaluation of values, though; instead, he accepts in a bitterly realistic way Shakespeare's implication that existence resembles the sea in that all things are mutable and are destroyed eventually. The sea imagery here depends on recognition of its referent for its principal meaning, and the transference from the physical to the moral sphere in a context of sexual liberalism in the thirties gives it an ironic immediacy.

In still another guise, in the 1933 story "After the Storm," the sea demonstrates the immensity of the forces of nature during a hurricane off the Florida Keys. Through the land and sea opposition, Hemingway exploits his imagery as metaphor and his meanings again point to the moral sphere. When the protagonist, who is in trouble with the law over a brutal fight, leaves the land to search for an imagined five-million-dollar treasure in the hull of a sunken liner, he is unable to plunder the ship because he cannot break into it. All he sees through a porthole is a woman floating in the water with rings on one of her hands. Such details suggest that there are no treasures of the kind imagined by the protagonist, and his cold, calculating approach to tragedy suggests that he has already lost his "treasure" when he lost his humanity in the metaphorical sea of existence.

In *Green Hills of Africa* (1935), a work of personal history with big-game hunting as its ostensible subject, Hemingway overtly formulated the importance of the metaphorical implications of sea imagery. In the person of himself as hero, he examines questions concerning the artist, the materials of art, and the relationship of the artist to his milieu. In one meditation that is virtually a philosophical critique of the sea metaphor and of its value to art, Hemingway's persona reflects: "when, on the sea, you are alone with it and know that this Gulf Stream you are living with, knowing, learning about, and loving has moved as it moves, since before man, . . . and those that have always lived in it are permanent and of value because that stream will flow, as it has flowed. . . . The single, lasting thing [is] the stream" (ch. 8).

Hemingway's definition, in *Green Hills of Africa*, of the Gulf Stream as a metaphor of temporal existence through which the artist asserts the worth of that existence, its knowable reality, and its emotional accessibility is reinforced in a 1936 *Esquire* article, "On the Blue Water," where he asserts that "the Gulf Stream and the other great ocean currents are the last wild country there is left." Such metaphorical meanings prefigure fictional manifestations in *The Old Man and the Sea* and *Islands in the Stream.*

In a "middle" novel, *To Have and Have Not* (1937), one segment of which was written and published as a short story a year before the Gulf Stream metaphor appeared in *Green Hills,* the sea-stream imagery is the controlling figure in the narrative. All of the major action takes place on the sea or on islands, at Key West or Cuba, with the intervening Caribbean and Gulf Stream always in focus and contributing to meaning as a metaphorical context. In fact, Hemingway attempts closure by way of an image in his final sentence that is a metaphor of the human struggle in the sea of existence. Here, it is a tanker "hugging the reef as she made to the westward to keep from wasting fuel against the stream." Survival in this middle novel, it seems from this metaphor, is best served when one stays safely outside of, rather than going against the stream of being, as Harry Morgan, the protagonist, had attempted in his struggles.

The sea is a major source of metaphor in Hemingway's late novels, including the posthumous *Garden of Eden* (1986), with its seashore imagery. In three other late novels the sea points to Hemingway's shift toward an affirmative vision of existence. *Across the River and into the Trees* (1950) marks that turning point, although it is the only one of the three in which the sea does not serve as a fully controlling metaphor. Yet here, too, the sea represents existence, mainly the past and the present. From the outset, it is apparent that, because of a heart condition, the protagonist, Colonel Cantwell, has no future. The novel is set in Venice, and like James, Hemingway exploits the symbolic meanings attached to this "city in the sea." Cantwell has retreated to these islands in the hope of finding some value in his past, since the city represents a cause for which he fought during World War II.

The turn toward affirmation is apparent when Hemingway grants the possibility of secure "islands" in the colonel's psychic sea of past and present troubles by giving him a love in the present, a love that the colonel describes as "a great miracle" (ch. 6). Love is miraculous here because it transforms what had been a meaningless past into a coherent, meaningful time. The fable does not deteriorate into sentiment, however, and biological processes finally overcome the protagonist, as they do Santiago in Hemingway's crowning achievement in sea fiction, *The Old Man and the Sea*.

Santiago and Cantwell inhabit the same universe, but the fisherman's existence is closer to nature and his confrontation with the sea more clearly suggests the confrontation with existence faced by all humankind. Hemingway accomplishes this kind of suggestiveness through imagery that reveals the relationship between men like Santiago and the forces that work on them: "Everything about him was old except his eyes and they were the same color as the sea and were cheerful and undefeated" (10).

The sea is the controlling symbol throughout the work and reflects the generative power of the universe, which manifests itself in an ever-changing reality. Linking Santiago's eyes with the sea points to the extraordinary capacity with which he is endowed. As Hemingway finishes the physical description of Santiago, he completes the symbolization and effectually reconciles the seemingly dichotomous forces within reality. The gaunt, blotched Santiago has scars "as old as erosions in a fishless desert" (10). The meaning Hemingway urges in this desert-sea simile is that the plight of the flesh is of the same order as that of external nature in geological time. Neither form can resist change, but physical change is an effect of process, not process itself. As mirrored by Santiago's eyes, the dynamic of universal process is associated with the impulses of the generative sea. It is the power within him that Hemingway describes as "undefeated"; and as Santiago's actions define for the reader, this impulse in mankind is the will. In Santiago, the will is indomitable. The other fishermen do not choose to go "far out" into the Gulf Stream, remaining content to fish within sight of the island; but their way is the way of Auden's "communal surrender" that suppresses individual volition and exposes human beings to the contingencies of the flux in natural processes. As always for Hemingway, those processes are brute and irrational; but Santiago is a self-realized man: by willing and consciously choosing, he avoids the folly of surrender and fashions his own reality.

The Christ motif throughout the work supports Hemingway's celebration of the realized will, for the pattern of Christ gives dramatic emphasis to the theme of humankind's spiritual capacity. Like Christ's Passion, Santiago's voyage into the Gulf Stream affirms the ability of the individual to contend with the inchoate stuff of ongoing reality, to dominate that stuff, and give it shape and direction that have meaning and worth. The individual cannot avoid pain and suffering in this view but can exert his will to the extent of achieving a God-like dignity.

Both *The Old Man and the Sea* and *Islands in the Stream* are organically related in conception, for they grew out of what Hemingway called his "Seabook." Both were parts of a work that went through several mutations, becoming at one point a trilogy-in-progress entitled *The Sea When Young, The Sea When Absent,* and *The Sea in Being,* with the last title revealing a signification that may be applied to all of the versions. These evolutions and the fact that the editors did not publish all of the segments may suggest some confusion in the imaginative process, but the common denominator is the sea as image and metaphor of existence, a fact that remains constant in both the published and unpublished segments.

Because of his strength of will and affirmative view of existence, Santiago of *The Old Man and the Sea* is a hero different in kind from any others in Hemingway's works. By contrast, Thomas Hudson, the protagonist of *Islands in the Stream,* is a hero whose motive and conscience do not coincide closely with the reader's, although we fully approve of his last actions at the conclusion of the final, "At Sea," portion of the published narrative. What accounts for the immense differences between the conception of the hero as Santiago and the hero as Thomas Hudson is that these two are situated in worlds radically separated from each other.

In the "fable" of the Cuban fisherman, Hemingway characterizes the possibility of man's return to harmony with nature in an ahistorical situation. Biology and environment are forces in Santiago's world, and his perception of them coincides with their essence. He can appreciate the beauty of the marlin and understand its meaning; at the same time, he understands the meaning of the ugliness and brutality of the sharks that attack and destroy all but the skeleton of the giant marlin the old man has finally caught. Harmony is achieved by recognizing the dualism of the forces at work; what he knows and what he values in the world are not at odds. What human beings judge as good or bad in nature are actually of equal value.

The sea images these natural forces. The true note of discord in Santiago's world is not the sharks but the incursion of a crass and unfeeling society, seen at the conclusion in the tourists who penetrate the timeless world of the hero and tarnish his heroic actions by their ignorance. Santiago's struggle with the marlin is a symbol of the struggle of all men to achieve wholeness and reintegration with all being, and his return with the skeleton validates his attainment of that wholeness. Knowing his own capacities and limitations, he nevertheless chose to challenge these forces of biology and environment.

In both the printed and the unpublished manuscript versions of *Islands in the Stream,* Thomas Hudson does not know his own capacities or limitations as he embarks on his ordeal. Because of that ignorance, he does not understand his status as it relates to all being. As hero, Hudson's characterization rests on his role as an artist whose medium is visual. Hemingway exploits this facet of the characterization by attributing to the artist an ability to respond with a heightened sensitivity to the elements of modern life, and Hudson's responses in the action of the work effect such a large and near-tragic sweep that Hudson achieves considerable stature as a fictional hero. But as Hemingway has fashioned him, he is too much a modern man, living in a modernist universe, to be able to do anything more than reach tentatively toward tragic heights. In part, this is just the kind of statement that Hemingway wished to make through his character, but as usual, Hemingway's strong concern with the historical moment in which he sets his fictions functions here as well, and as the imagery of the titles of the various segments and of the early versions remind us, these events are but moments in the vast scheme of time and history.

Islands in the Stream moves beyond the limited sphere of the immediate social environment by presenting the artist as a reflector of the *zeitgeist* of his world. The time of the opening, "Bimini" section is 1936, and the second and third sections, "Cuba" and "At Sea," take place in 1943. If we are mindful of the major events that took place in the world during those years and consider Hemingway's concern with such larger historical contexts as World War I and the Spanish civil war in his earlier novels, the setting of "Bimini" in 1936 has ominous implications, for by the end of that year, the threat to democratic governance in Europe posed by the Franco rebellion in Spain and the consolidation of power by Hitler and Mussolini, as well as the Japanese military threat in Manchuria and China, were evident to all observers. These events provide the context for what amounts to the enveloping action of "Bimini." More narrowly, "Cuba" and "At Sea" are strongly influenced by the actual conflict of World War II.

The larger historical frame affects the meanings of the immediate action of the various parts, and clarifies both the characterization of the protagonist and the functional metaphor of the Sea in Being. Hudson's response to existence is suggested at the outset, in an image of Hudson's house, built on high ground between the harbor and the open sea. The house establishes an ironic metaphorical context that persists throughout the work. As it applies to Hudson, it figures his illusive feeling of security in having escaped from life and its exigencies. But in Auden's terms, it is a false "paradisiacal" island.

Like Adam after the Fall, Hudson abides in a sham Eden, for it is built on high ground just above the Gulf Stream, suggesting his belief that it is possible to divorce himself from the reality of the present. This attempt to avoid the flux of existence extends to Hudson's desire to escape from the pain of his personal past when he engages in a delusive battle to screen out the painful memories of that personal past and to focus solely on the pleasant ones. This conflict between the knowing in the present and the reality of the past implies a bleak future for him and signifies Hemingway's intention to create a hero who reflects the plight of modern consciousness. Throughout the work, the task of the hero and the goal of his quest are evident. Avoidance of reality is impossible, as the chaos of the queen's birthday celebration reveals, when Hudson's self-created sanctuary is penetrated by the corrupt rich on their yachts, and the ensuing fight between Rodger, his friend and alter ego, and one of the rich, reveals how delusive is the attempt to retreat to innocence in a fallen world. Escape and avoidance in Hemingway's universe is impossible, as the metaphorical formulation in *Green Hills of Africa* suggests, for "that stream will flow, as it has flowed" (ch. 8), and Hudson is eventually compelled to enter that stream by forces outside of his individual will.

As artist-hero, Hudson is doubly culpable for attempting to detach himself from the processes of life represented by the sea. His house has survived three hurricanes, but in 1936, evil poses a threat to the security of this smug world, as we are told through the imagery of the water near the house, which is safe to swim in during the day—but at night the sharks come in. The larger forces in nature are suggested when we are told that Hudson had studied tropical storms, and that he is willing to "go" with the house (ch. 1).

Hudson's willingness to die if his world dies too marks the extent of his vanity and pride. Hemingway portrays here the delusion of a self-centered modern consciousness when it presumes to understand the nature of larger

forces and pridefully assumes that it knows the parameters of the unknow-
able, that closure in experience is somehow possible; in this case, that death
can come in some grand cataclysm that sweeps everything away before it,
including consciousness. As Hudson discovers later, it does not come for
modern man so neatly and so finally as it comes in the King James account of
Saul, to whom Hudson and his sons bear a resemblance: "So Saul was dead,
and his three sons, and all of his house died together."

After Hudson receives the news of the death of his two younger sons in
an automobile accident, Hemingway reveals Hudson's sensibility as he discovers that what he thought he knew about existence was naive and delusive,
for the end of the world, as he phrases it, does not come as a cataclysmic
event, as in a great painting; instead, it comes in a radio message with the
mechanical request by the delivery boy to sign the "detachable" part of the
envelope.

Both the Cuba interlude and the final "At Sea" segment reveal just how
deeply Hudson's protective armor of isolation and detachment has been
pierced. In a pathetic epiphany, he sees his plight in the egocentric terms of
his violated insularity, when he decides that he should not have loved his sons
so much. His abortive attempts to come to terms with the nature of existence
are made to seem all the more specious with the addition of the war, when
Hudson learns of the death of his third son. As Hudson prepares to enter
that war, in a final and direct way, he sums up his despair in ironic terms
when he rationalizes that all he has left to "do" is duty. Seemingly, he has
experienced another epiphany, but this one is a further mockery and evasion.
As the last segment of the novel demonstrates, "duty" becomes an obsession
and another attempt to escape the pain of consciousness.

Throughout the final "At Sea" section, Hudson's obsessive search for a
German submarine crew that has murdered nine villagers leads him to for-
mulate a theory of a "death-house brotherhood," which in its inversion of
love as the Christian spiritual basis for brotherhood, reflects Hudson's spiritual dessication and, perforce, Hemingway's thematic purpose in creating
this character. Because he is a protagonist who reflects the conflict between
consciousness and reality in the modern world, Hudson's attitude indicts the
failure of modern man to accept spirituality and faith as mediating factors
that can reconcile the opposition in modern life between knowing and being.

The final portions of the work dramatize Hemingway's meaning, when
a captured German sailor, dying of gangrene, delivers the prophetic message of unfaith to Hudson: "nothing is important" (bk. 3, ch. 8). The nihil-
ism of this view reflects Hudson's own after the deaths of his sons, but this
extreme is counterbalanced by Hemingway when Willie, one of Hudson's

crew, tells the mortally wounded Hudson, "I love you, you son of a bitch, and don't die" (ch. 21). Willie's revelation provides an affirmative counter to Hudson's "death-house brotherhood"—and Hemingway's larger meaning is clear: love, not death, is the appropriate agency to form the common bond of humanity in the "sea in being." Hudson's response is a pathetic, halfway recognition, and with it he fulfills his role as a reflector of the plight of consciousness in the modern world.

Hemingway's contemporaneousness often reached peaks of insight and expression that are sometimes masked by subjects that hardly seem topical. Always, however, he fastened on the appropriate metaphors to represent the dynamics of his time. He had signaled just that interest in the affairs of the present (and unique ways of expressing it) when, in 1924, he published a collection of stories in Paris and titled it *in our time,* all in lowercase letters. After 1936, whenever he fastened on deep sea fishing as subject and the sea as metaphor in the later sea fictions, the marine metaphor provided his art with a vehicle for expression that transcended even those celebrated elements of style and description for which he had become universally known.

In that epoch in America that existed mainly between the two world wars, and among the so-called "high" moderns of American fiction, Hemingway best realized the potency of the sea as a source of symbols for his own time. He was obviously not alone in such a sea consciousness, but his was perhaps the strongest voice. Having learned from their own experience and from their realist and naturalist predecessors the futility of traditional and romantic sea-symbolizing, Hemingway, O'Neill, Fitzgerald, and other modernists found, nevertheless, that the sea, transformed by their own disillusioned perspective, served them well in expressing the aesthetic, emotional, and spiritual joys and despairs felt by many in the first half of this century.

JOSEPH DEFALCO

Chapter Fourteen

PROSE SINCE 1960

The contemporary American novel usually takes place in cities or suburbs and focuses on the individual's angst-filled isolation from meaningful personal relationships, traditional values, and social reality. Moreover, many writers explore these typical settings and themes in a style that emphasizes verbal play and linguistic fantasy. Such tendencies would seem to militate against nautical fiction, which stresses exotic locales, adventurous characters and plots, a strong narrative line, and the very values modern antiheroes eschew: courage, boldness, self-confidence, and a firm grip on reality.

It is remarkable, then, that so many novels of the last thirty years have used the sea as a fictional resource. Popular forms of fiction—detective stories, science fiction and fantasy, spy novels, and modern gothic thrillers such as Peter Benchley's *Jaws* (1974)—demonstrate that the traditional appeal of the sea as a place for high adventure and intense emotion is still with us. At the same time, such highly regarded authors as Peter Matthiessen, John Barth, and Charles Johnson have written books using the sea as their primary setting, while other serious writers have set important episodes at sea or sounded the psychological and symbolic thalassian depths. And a few lesser-known writers continue the tradition of nautical autobiography in narratives recounting personal experiences with the marine environment.

Although the aims and audiences of these authors differ widely, a new attitude toward the sea emerges in much recent writing. The old fear of the sea, the sense that it is an alien place hostile to humanity, often gives way to a sense of its being a positive and sustaining force for human life. If any trend marks the development of sea-conscious writing since 1960, it is this contemporary scientific notion that life on land depends on life in the sea. On this spaceship Earth, no creature or environment exists independent of the fate of another. Wetlands, estuaries, tides, fishing banks, and the purgative powers of the deepest ocean currents may in fact be more important to life on this planet than the plains and river valleys of dry land. And the mammals of the sea—dolphins, whales, seals—may be nearer kin to humankind than ever realized, perhaps even sharing with humanity the rare power of language. The old myths of sentient sea creatures, be they gods, fish, or monsters, gain new meaning with every advance in marine biology; the age-old dream of living among and communicating with the fish of the sea, blending with the water world as a mermaid, a Proteus, or a Captain Nemo, is now a credible fantasy.

It would take a separate essay to treat all the reasons for this development. Certainly, the ground was prepared by Anne Morrow Lindbergh's perennially popular *Gift from the Sea* (1955) and Rachel Carson's *The Sea Around Us* (1951). One cannot discount the pervasive influence of Jacques Cousteau's television specials, programs that encourage viewers to regard the sea as a home for humanity. This chapter can only trace the beginnings of a new consciousness toward the oceans in recent prose, and even then cannot claim to be exhaustive. For nautical settings, images, symbols, and themes pervade current literature and suggest that the United States has not forsaken its maritime heritage so much as transformed it to fit new values and concerns.

No writer has done more to popularize the sea and its creatures than Peter Benchley. After ten apprentice years as a journalist, Benchley turned to fiction (and exploited the old theme of the treacherous ocean) with *Jaws* (1974), a latter-day version of *Moby-Dick*, whose vengeful great white shark substitutes for the whale, and whose brutal shark fisherman, Quint, replaces Ahab. The book's Ishmael-figure is Martin Brody, the conscientious police chief of a village on Long Island. When the bloody, gruesome, and inevitably fatal shark attacks begin, corrupt village officials dismiss them as bad for business; but Brody persists in his efforts to face the truth and eventually, with the aid of Quint and an ichthyologist, destroys the monstrous two-and-a-half-ton fish. Like *Moby-Dick*, the book ends with a stove boat and a single survivor. Benchley's most thrilling and tightly written novel, *Jaws* combines a traditional fear of the sea with modern knowledge of shark behavior to lend credibility to the narrative.

Benchley's second and third novels move from the familiar shore to Atlantic islands and cloak the adventures in a sometimes tenuous web of maritime history. In *The Deep* (1976), a honeymoon couple scuba diving off Bermuda battles local drug runners over a sunken hoard of morphine ampules and Spanish gold. Benchley credits books by Cousteau and Kip Wagner, a treasure diver, for supplying many of the nautical details. In *The Island* (1979), a jaded New York writer discovers that descendants of the pirates of the Spanish Main still live on a remote island in the Turks-Caicos and ply their bloody ancestral trade among modern pleasure craft. Although both of these novels have considerable shock value, they fail to integrate historical and technical detail with plot and character.

A Benchley novel aimed at adolescents, *The Girl from the Sea of Cortez* (1982) tells the story of Paloma, a sixteen-year-old Mexican girl who learns from her fisher-father how to live in communion with the animals of the sea. Simply told, touching as a Steinbeck parable, and sensitive to the beauty and fragility of the ocean ecology, this book presents contemporary environmentalist attitudes charmingly yet forcefully. In contrast, *Beast* (1991) returns to the clichéd theme of sea monsters. Essentially a rewrite of *Jaws*, the novel describes a giant squid's repeated attacks on human beings in the waters off Bermuda.

The career of Hank Searls, a lesser-known popular author of nautical novels, moves like Benchley's (pre-*Beast*) from a hostile to a sympathetic view of the sea. A former navy officer, Searls resigned in 1954 when he was thirty-two to take up free-lance writing. After trying his hand at novels about space flight and other current topics, he drew on his experiences during World War II and wrote *The Hero Ship* (1969), a conventional story of his hero's rise from navy aviator to captain of an aircraft carrier. Ambitiously structured around a series of flashbacks, and loaded with accurate technical details, the novel suffers from shallow characterization and motivation. Searls's next attempt at sea fiction, *Overboard* (1977), uses as its narrative thread Mitch Gordon's frantic thirty-hour search for his wife, Lindy, who falls off their ketch at night somewhere between Bora Bora and Tahiti. Despite the impressive display of nautical detail and the eerie descriptions of Lindy's lonely vigil in the open sea, an elaboration on the fate of Melville's Pip, *Overboard* draws out a simple plot to untenable thinness and finally seems pointless.

After *Overboard*, Searls took up the chore of "novelizing" the screenplay of *Jaws 2*, the sequel to Steven Spielberg's film of Benchley's *Jaws*. Employing the same setting, characters, and plot as Benchley, Searls nevertheless stakes out his own claim to the material with his detailed descriptions of marine animal behavior. He tells much of the story from the point of view of the shark, a pregnant female. Less a monster than a mother trying to maintain her species, she is strangely sympathetic. The shark, rather than the Benchley characters Searls must adopt, becomes the true hero of the novel. Despite the predictable plot and the stale characterizations, Searls manages to add a new dimension to sea fiction in his sensitive and accurate depictions of marine fauna.

Searls's interesting novel *Sounding* (1982) extends the device of animal narration to over half the book, as two complementary plots alternate and eventually blend. The first follows the thoughts of an aging bull sperm whale. The second records the mounting tension aboard a Soviet nuclear submarine stranded six hundred feet deep. Bringing those two plots together is the novel's major theme: interspecies communication. Citing *Moby-Dick* as an early episode in the short history of human-whale interaction, Searls constructs an elaborate culture, history, and mythology for cetaceans that makes them more credible than his stereotyped human characters. The result is an original and well-informed fictive extrapolation from marine biology.

As the novels of Benchley and Searls demonstrate, stories of the sea continue to cross the boundary between the realistic and the fantastic as they have since the time of Homer. The high technology of well-built ships and accurate navigation competes with primal fears of sea monsters and unknown shores as science combines with myth to produce works of considerable power. Perhaps for this reason, science fiction has offered a hospitable harbor for nautical themes. The vast unknown distances of space bear comparison with ocean voyages of old, and many of the givens in sea fiction—isolation, alien environments, strange creatures, long journeys—can be easily adapted to science fiction. As in other forms of contemporary writing on the sea, however, the best science fiction recognizes the bonds between humanity and the sea.

Such a bond lies at the heart of Ursula K. LeGuin's monumental fantasy *The Earthsea Trilogy*. LeGuin's wizard-hero, Ged, is as adept at sailing small boats as he is at weaving spells. He feels safer at sea than on land, and at the conclusion of the first novel in the trilogy, *A Wizard of Earthsea* (1968), he even walks on water to confront his dark "shadow," an echo that fuses Jesus with "The Secret Sharer." During his "sea-quest" for this evil double, Ged duplicates the experiences of Odysseus, Beowulf, and Jonah, just to mention the most obvious sea myths LeGuin appropriates. In the second part of the trilogy, *The Tombs of Atuan* (1971), the sea is not mentioned until the final chapter, when it offers Ged escape and freedom from the evil he destroys on the island of Atuan. In contrast to Atuan's dark, arid, death-loving stillness and rigidity, the ocean is sustenance, movement, energy, and life. The third volume, *The Farthest Shore* (1972), is set almost entirely at sea. Ged and young prince Arren sail throughout the archipelago of Earthsea seeking a lapsed wizard who claims to have achieved immortality. Again, the sea represents life, the dry land death. Moreover, the sea symbolizes the proper relationship between the self and the world, the mortal and the immortal. The only society in the book with inner peace is the Children of the Open Sea, a primitive people who live on great rafts in the ocean and beyond the islands of Earthsea itself and, like Paloma in *The Girl from the Sea of Cortez*, live at one with the sea. LeGuin collected fugitive stories about Earthsea in *The Wind's*

Twelve Quarters (1975) and added a fourth volume to the series, *Tehanu: The Last Book of Earthsea*, in 1990. The latter reflects LeGuin's concern for gender equality more than the sea. One venture in realistic fiction, *Searoad: Chronicles of Klatsand* (1991), examines the intersecting lives of women in a small town on the Oregon coast, and is notable for its descriptions of coastal scenery.

Frank Herbert, author of *Dune* and many sequels, represents well the attraction that sea imagery has for the author of popular science fiction. In 1956, Herbert published *The Dragon in the Sea*, a novel about a twenty-first-century submarine; he later teamed with Bill Ransom to write *The Jesus Incident* (1979), a book that introduced the distant watery planet of Pandora. In *The Lazarus Effect* (1983), Herbert and Ransom focused exclusively on Pandora and its now-forgotten colony of human beings who, over the centuries, have evolved into two distinct varieties: mutant islanders who live on dry land, and relatively normal "mermen" who live underwater. As in LeGuin's trilogy, reader sympathies clearly lie with the more humane and peaceful people of the sea. So too in Joan Slonczewski's *A Door into Ocean* (1986), whose complicated plot pits a water planet of women "sharers" against a commercial, patriarchal, violent empire. This book's imagining of a water world may be its best achievement.

Kurt Vonnegut also speculates on the biological and evolutionary effects of the nautical environment in two novels that explore the relationship between human and marine ecology. This theme is adumbrated in *Cat's Cradle* (1963), when a crazed character drops into the sea a small piece of "Ice-9," a crystallized form of water. Immediately oceans, rivers, and lakes solidify, and the earth, no longer a "water planet," is doomed to slow and painful death by dehydration. In *Galápagos* (1985), Vonnegut views the sea as the eventual home for humanity. When an ovum-eating bacterium invades the human species, everyone on earth dies except a small band who had planned to take the "Nature Cruise of the Century" to the Galápagos islands. A sudden war disrupts the embarkation, and a few passengers manage to escape to the Galápagos on the luxurious cruise ship the *Bahia de Darwin*. The bizarre group of survivors (including one named after Joseph Conrad's James Wait) evolves over the course of one million years into a species that resembles seals more than human beings. Most important, their brains have shrunk so that they are no longer bothered with the "big brains" that gave earlier human beings so much trouble. An alternate title for the book, the narrator suggests, is a second "Noah's Ark" (ch. 1). As with LeGuin, Vonnegut combines biblical myths and nautical conventions to fabricate a tale that suggests human survival depends on living in harmony with the sea.

At the edge of science fiction lie a fine story of interspecies love and two ambitious dystopian novels set in the near future: Rachel Ingalls's *Mrs. Caliban* (1983), Ted Mooney's *Easy Travel to Other Planets* (1981), and John Calvin Batchelor's *Birth of the People's Republic of Antarctica* (1983). Ingalls's protagonist is a despairing housewife who has lost two children, and who is now emotionally dying in a sexless, loveless marriage. Into her kitchen steps a six-foot-seven-inch, froglike humanoid who has violently escaped from the scientists who had captured him in the Gulf of Mexico. Dorothy hides "Larry" in her guest room and builds a secret life with him that gives her a new, if disconcerting, perspective on her marriage, friendships, old griefs, and new desires. Their story is one of eros, tenderness, terror, wonder, and finally, of loss, as Larry unwillingly returns to the sea and leaves Dorothy bereft of hope, stranded.

Mooney extrapolates the dolphin communication studies of John Lilly and the anthropological theories of Claude Lévi-Strauss into a love affair between Melissa, a key assistant in a dolphin research project on St. Thomas, Virgin Islands, and Peter, a dolphin who understands and even speaks a few words of English. The electronically information-saturated world of the near future in which Melissa lives has disrupted the natural rhythms of life, love, and true communication. Television, radio, telephone answering machines, photography, and other forms of technological information-sharing have produced insurmountable alienation, even among Melissa's close friends and family members. In contrast, Peter's natural communication system blends all five senses and, like Searls's whale in *Sounding*, makes him more fully aware of other creatures than is any human being. His messages transcend language and gradually pull Melissa away from the self-destructive world of New York City back to direct contact between human being and dolphin in seawater. In a tragic and shocking conclusion, Melissa's pride, embarrassment, and fear—traits unknown to dolphins—drive her to kill Peter, the one creature offering the hope of honest, mutual communication.

Similar tragedy pervades Batchelor's philosophical morality tale. In 1995, in a world plagued by famine, war, disease, and tyranny, a young man named Grim Fiddle leads a small band of refugees from Sweden to the Falkland Islands in the ominously christened schooner *Angel of Death*. Beset by fierce storms, enemy vessels, and his own quenchless thirst for power and moral autonomy, Grim Fiddle grows ever more ruthless and tyrannical until his own people revolt and drive him to Antarctica. Although he founds his own "black kingdom" at the ends of the earth, he lives to regret his fall into tribalism and savagery and confesses his misdeeds at the end of his autobiographical story. Allusions to Norse legends, Christianity, and *Beowulf* undermine the modern romantic theme of escape from civilization to nature and lend the novel the considerable force of myth and allegory. For Batchelor, almost alone among contemporary writers, voyaging leads away from personal and social renewal and toward a violent, brutal future that confirms the old truths of human evil.

Nearer the factual end of the spectrum lie stories based on real characters and events, and some of the best yarns still come from writers of historical fiction. Descriptions of maritime commerce and naval battles play important roles in James Clavell's epics *Tai-Pan* (1966) and *Shōgun* (1975), immense novels that dramatize in great detail the opening of the Western shipping trade with the Orient. James Michener chronicles several billion years of history as he describes both the geological and cultural development of the Hawaiian Islands in *Hawaii* (1959). His imaginative reconstruction of settlers' voyages from the South Seas, New England, and China hold special interest for readers of nautical history. In *Chesapeake* (1978), Michener offers a four-hundred-year profile of the unique bay's cultural evolution. His well-crafted portraits of Quaker boatbuilders and "watermen," sailors who brave the bay's unpredictable weather to catch fish, crabs, and oysters, are particularly poignant reminders of America's maritime heritage. Three of Bermuda-born Brian Burland's historical fictions occur at sea: *Stephen Decatur* (1975), a fictionalized biography of the American naval hero; *Surprise* (1975), the story of a slave who escapes from Bermuda in the 1840s to found his own colony; and *A Fall from Aloft* (1969), the narrative of a young boy's daring voyage from Bermuda to England aboard a Liberty ship early in World War II. Sterling Hayden's one venture into fiction, *Voyage: A Novel of 1896* (1976), borrows from Joseph Conrad by describing a brutal trip 'round the Horn in an iron-hulled barque loaded with smoldering coal. Subplots involving politics, labor unrest, scientific exploration, and the sybaritic life of captains of industry overburden an otherwise appealing and technically credible tale of fin-de-siècle seamanship. A more truly Conradian tale, one that ironically analyzes the psychology of command, is included in Mark Helprin's *Ellis Island & Other Stories* (1981). "Letters from the *Samantha*," narrated as letters from a proper British captain to the owner, records the tensions aboard a merchant ship in 1909 when the captain "incorrectly" takes aboard a disconcertingly human Madagascar ape and can only rid himself of the nightmarish creature by throttling it with his own hands. And F. W. Belland's *True Sea* (1984), a personal novel about coming of age on a tiny sea island on Florida's west coast from 1909–25, sensitively records the changes in coastal life that come with a new railroad, a world war, and prohibition, yet still portrays the sea's changeless appeal as a final place of escape.

Two historical novels by Henry Carlisle hold special interest for readers of Melville, for they retell incidents crucial to *Billy Budd, Sailor* and *Moby-Dick* from the perspective of a minor player in each tragedy. In *Voyage to the First of December* (1972), Carlisle uses Robert Leacock, the ship's actual surgeon, to describe the infamous *Somers* affair, an instance of suspected mutiny and summary execution aboard a United States brig in 1842. Through Leacock's fictional journal account of the matter, as well as newspaper clippings, the record of the court of inquiry, and occasional correspondence, Carlisle casts the controversial events in a lurid but plausible Freudian glow. *The Jonah Man* (1984) is a fictional autobiography of George Pollard, captain of the *Essex* when it was stove by a whale in 1820. This incident moved the mate, Owen Chase, to write the narrative that later inspired Melville's final apocalyptic chapter in *Moby-Dick*. By covering Pollard's whole life from childhood on Nantucket through two disastrous commands to an ignominious retirement as nightwatchman, Carlisle invests his protagonist with something of the fated solitude that appealed to Melville in *Clarel*, where Pollard is alluded to as "a Jonah" (I.37, l. 90). As a man who survived one disaster by cannibalizing his cousin only to endure a second shipwreck on his next voyage out, Pollard stands in quiet, stoic testimony to the "power of blackness" and the stubborn humanity that persists in spite of it.

Of less quality and interest than either nautical fantasies or histories are the durable and often extremely popular tales of adventure and mystery on the high seas, a subgenre that has persisted since Frederick Marryat. Journeymen writers Paul Gallico, John Hersey, and Morris West have added sea adventure to their considerable stock of fictional themes in four well-paced narratives. Gallico's *Poseidon Adventure* (1969), a tale of escape from a capsized ocean liner in mid-Atlantic, is one of the most popular sea novels of the period. It incorporates and updates all the most exciting conventions of maritime adventure: an improperly ballasted ship, an incompetent captain, a sudden accident, mendacious versus courageous survivors, a gruesome array of drowned corpses, and a gradual winnowing of the unfit before rescue. A sequel, *Beyond the Poseidon Adventure* (1978), strains the original idea to the breaking point. But both novels struck the popular imagination forcefully enough to become successful motion pictures. More realistically, Hersey's *Under the Eye of the Storm* (1967) takes two modern young couples on a pleasure cruise into a hurricane to expose the shallowness of contemporary values and the impossibility of certainty in a world dominated by primitive force. Although the storm scene bears comparison with Conrad's "Typhoon" (as do the narrative tricks), Hersey's irony toward his fussy protagonist lacks control and purpose and thwarts his attempts at greater psychological depth. In *The Navigator* (1976), West leaves realism behind in the story of Gunnar Thorkild, a young professor who follows Polynesian legend and finds a large island in the South Pacific. A more positive version of Batchelor's Grim Fiddle, Thor-kild discovers peace, love, and spiritual fulfillment on his island home. Two extremely popular authors, John D. MacDonald and Clive Cussler, round off this group. MacDonald's rowdy detective, Travis McGee, lives on a house-boat in a Fort Lauderdale marina and moves through twenty-one formulaic mysteries solving crimes involving everything from treasure hunts and sal-vage operations to floating prostitution. Former advertising agent Cussler blends sex, violence, and international intrigue in numerous thrillers about Dirk Pitt, a daredevil salvage expert whose adventures are as improbable as his name.

A final subgenre of popular fiction, naval adventure, has diminished in popularity and frequency as World War II fades into the past. But a few older writers keep the tradition alive. In *The Winds of War* (1971), Herman Wouk recounts the events leading up to Pearl Harbor, and in *War and Remembrance* (1978), he returns to the waters he sailed so well in his first novel, *The Caine Mutiny* (1951), dramatizing accurately the Battle of Midway and other vast naval encounters in the Pacific. Sloane Wilson, best known for *The Man in the Gray Flannel Suit* (1955), has tried his hand at sea fiction with the partly auto-biographical *Ice Brothers* (1979), a vivid recreation of the U.S. Coast Guard's Greenland Patrol in the early days of World War II. A companion volume, *Pacific Interlude* (1982), moves his characters into the latter days of the war and places them aboard a huge tanker during Douglas MacArthur's drive to retake the Philippines. Commander Edward Beach, author of the novel that set the standard for modern submarine fiction, *Run Silent, Run Deep* (1955), has written two sequels: *Dust on the Sea* (1972) focuses on submarine warfare in the Yellow Sea during World War II, while *Cold Is the Sea* (1978) updates his characters to 1960 and describes a struggle between American and Russian nuclear submarines under the Arctic Sea. Two younger writers who have written bestsellers in this subgenre have followed Beach's lead and focused on the cold-war tensions and high technology of nuclear sub-marines. Mark Rascovich's *The Bedford Incident* (1963) concerns the silent war between Soviet subs and NATO destroyers in the North Atlantic, while Tom Clancy's *The Hunt for Red October* (1984) tells the exciting but improb-able story of a Soviet submarine captain who defects with his ship and crew to the United States. Less exciting, perhaps, but far truer to human nature and to probability is William Brinkley's *The Last Ship* (1988), in which an American destroyer and a Russian submarine are the only apparent survivors of a nuclear war. Sometimes echoing or alluding to works by Conrad and Melville, this intelligent story, narrated by the American captain, also com-plicates the situation intriguingly: among the destroyer's crew are a number of women. Brinkley really did not know how to create credible women in his comic *Don't Go Near the Water* (1956), but he has learned a lot since then—dramatizing, for example, the legitimate and decisive power of the female minority that determines the relations among the surviving men and women.

With such a diverse appropriation of the sea by popular writers, one might not expect to find the more academic novelists of this period (who have often tended toward urban settings, suburban mores, or linguistic fantasies) pursuing nautical themes. Their nautical tendencies seem captured in the title John Updike chose for one of his miscellanies: *Hugging the Shore* (1983). Nevertheless, a surprising number of academic writers weave sea imagery and settings into complex narratives with contemporary themes. Some go further than a Benchley or Searls in focusing their tales on the sea, while others outstrip even Ted Mooney and Kurt Vonnegut in recognizing humanity's need for sea values. In the hands of the best current writers, the sea becomes an important resource for symbolizing the timeless themes of freedom, change, and opportunity, even while its darker, cannibal undercurrent flows on.

Peter Matthiessen's *Far Tortuga* (1975) may well be (as has been claimed) the best sea novel of the period, perhaps the best American sea novel since *Moby-Dick*. A pointed dramatization of the decline of the Cayman Island turtle fishery during the 1960s, the novel resonates with larger implications of cultural, moral, and environmental decline. With its present-tense narration, fragmented structure, and pidgin English conversations, the novel demands almost as much of its readers as anything by Joyce or Faulkner; but the presence of nautical conventions keeps the narrative accessible and clearly structured. A classic "last voyage" story, the novel travels waters familiar to all readers of sea fiction: the gathering of the crew, the preparation for the voyage, ominous portents, a proud captain, mutinous rumblings, dangerous islands, capricious weather, a final grand shipwreck, and one lonely survivor. Matthiessen knows the Caribbean and its impoverished turtlers who strive to maintain the pride and dignity of their trade in the face of overpopulation, political chaos, diminishing natural resources, and increasing violence and amorality. Crew members blame the selfishness and aimlessness that surround them in this "modern time, mon" (46), but the best of them struggle, like Hemingway's Santiago, to maintain their humanity in the face of moral anarchy. Captain Raib Avers, the last of the old-time turtlers, epitomizes the knowledge, skill, dedication, and integrity traditionally associated with the sea captain; he has unusual psychological depth, however, because he is also arrogant, contemptuous, aloof, and harsh. Speedy, the single survivor, blends traditional and contemporary values. He is hardworking and loyal, but fully capable of a sudden, self-protective violence that isolates him from the rest of the crew. Nautical conventions interweave with modern narrative techniques in a sea story as compelling and profound as Joseph Conrad's *The Nigger of the "Narcissus."*

As a threnody to a dying trade, *Far Tortuga* prefigures *Men's Lives: The Surfmen and Baymen of the South Fork* (1986), Matthiessen's history of the declining fishing communities on eastern Long Island. The harsh lives of American fishermen and their families reify the fictional experience of Raib Avers in a bleak and poignant blend of text and photographs reminiscent of James Agee and Walker Evans's Depression-era classic, *Let Us Now Praise Famous Men* (1941). Always sensitive to the human sacrifice the sea demands, Matthiessen is a supreme psychological realist with a genuine tragic sense—the profound sense of human limitation.

As if extrapolating from Matthiessen, John Casey and Paul Watkins explore the emotional and physical hardships of Rhode Island fishermen, another battered but hardy breed of misfits, isolatoes, and mere survivors. Casey's *Spartina* (1989), winner of a National Book Award, offers a compelling and rich psychological portrait of a man caught in the same tides of change that destroy Matthiessen's characters. But Casey's middle-aged hero triumphs over natural and psychological storms by learning to be like the deep-rooted marsh grass of the title—to bend with the waves and filter the good from the bad. Beset by financial woes and midlife marital ennui that leads to an affair, a brush with cocaine smuggling, and other desperate liaisons, Dick Pierce struggles to finish building his own boat—a fifty-foot wooden fishing vessel—and steers it alone through a powerful hurricane to save it from destruction in dock. The experience, akin to saving one's child, gives Dick a better understanding of his personal resources, the ebb and flow of life, and his need for others. He ends feeling "like the salt marsh, the salt pond at high water, brimming" (375), chastened and made more self-aware by the unremitting power and redemptiveness of the sea. Watkins's *Calm at Sunset, Calm at Dawn* (1989) offers a tawdrier version of the Rhode Island fishery, an unprincipled and amoral world similar to Jack London's in *The Sea-Wolf*. Shipping out against his fisher-father's wishes on a Newport scallop trawler, the *Gray Ghost* (an echo of Wolf Larsen's *Ghost*), twenty-year-old novice James Pfeiffer is quickly initiated into an occupation dominated by suspicion, fear, and violence. Watkins's sea is unremittingly harsh and cold, a place of dangerous work and sudden death where machinery breaks and injures careless men while the captain navigates his vessel along the fringes of the law. The personal cost is enormous. Some fishermen suffer from recurrent fantasies of shark attack or drowning, others turn to alcohol, and some—like James's own father—reluctantly take up drug-running to wrest a living from their profitless trade. Narrated as a fictional autobiography, the novel lacks the broader sympathies of Matthiessen's and Casey's work, yet its gritty detail and steely characterizations challenge idealistic notions of nautical redemption and remind us of the sea's brutal power over men.

The stark realism of Matthiessen, Casey, and Watkins finds corroboration in John McPhee's account of the contemporary United States Merchant Marine, *Looking for a Ship* (1990). McPhee's razor-sharp prose offers detailed portraits of the crew and captain of the *Stella Lykes*, a huge cargo ship on a six-week voyage from Charleston, South Carolina, to the west coast of South America. Echoing the novelists' theme of decline, McPhee shows how a web of government, business, and union regulations has strangled American shipping and made American sailors an endangered species. Additional problems with unregulated foreign competition, illegal immigrants as stowaways, incompetent port authorities, and even pirates (McPhee reports on two encounters) render the Merchant Marine "dead in the water," like the *Stella Lykes* at the end. Yet some nobility remains in the sailors, aging men who maintain their pride and skill in the face of their profession's slow demise.

In those academic writers who deemphasize realism in favor of symbolism and postmodernist play, a more optimistic view of the sea often resurfaces. The leading practitioner is John Barth, who has increasingly exploited nautical themes and conventions in his fictions. Drawing on his native Chesapeake Bay region and his personal interest in sailing, Barth uses the sea as setting, comic resource, and metaphor for art. In his first novel, *The Floating Opera* (1956), the narrator, Todd Andrews, compares the novel itself to a showboat idly drifting in the ebb and flow of a tidal river. Readers, like spectators on the banks, must exercise their imaginations to find a coherent plot in the wandering and digressive "floating opera" of Andrews's story. Barth takes the comic approach in *The Sot-Weed Factor* (1960), part of which satirically describes a six-month voyage from England to Maryland in the early 1700s and parodies the wretched "sea-poetry" such trips sometimes inspired. After surviving the usual storm, piracy, and near-mutiny, the hero, Ebenezer Cooke, actually walks the plank in this entertaining mockery of the nautical epic. *Lost in the Funhouse* (1968) uses the hyperboles of long-distance swimming in the chapter "Night-Sea Journey," a first-person narrative of a spermatazoan's upstream swim toward union with the ovum; and in "Water-Message" and the Homeric parody "Anonymiad," Barth elaborates the message-found-in-a-bottle motif into a complex metaphor for the paradoxically intimate and detached relationship between author and reader.

Barth's fascination with the similarities between storytelling and seamanship reaches its peak in *Sabbatical: A Romance* (1982). "If life is like a voyage, reader, a voyage may be like life" we are told; and as the story of this voyage of self-discovery develops, the reader realizes a voyage may also be like a book. Fenwick Turner, former CIA agent, and his wife, Susan Seckler, professor of American literature, are returning from a nine-month sabbatical cruise on their punningly named sailboat *Pokey, Wye I*. During a storm they stumble onto a secret CIA base and begin to fear The Company is seeking revenge on Fenwick for his exposé, *KUDOVE*. As they tack their way up Chesapeake Bay and back to their professional routines, they reflect on their past lives and their uncharted future. Simultaneously they navigate the waters of the Chesapeake and the dangerous shoals of conflicting individual desires that threaten their devoted marriage. Witty, literate, and sophisticated self-conscious narrators, they also debate the nature of good storytelling as Barth crafts what must be the first piece of nautical metafiction. Since sailing literalizes metaphors for both life and art—taking a different tack, keeping things shipshape—it refreshes our jaded sensibilities and provides a new awareness of the possibility for freedom and independence. "Realism is your keel and ballast of your effing Ship of Story, and a good plot is your mast and sails," Fenn tells Susan. "But magic is your wind, Suse. Your literally marvelous is your mother-effing wind" (137). By the end of the novel, Fenn and Susan link the historical vision of Francis Scott Key with the fantasies of Edgar Allan Poe's hero Arthur Gordon Pym, and see that "It is not that the end of the voyage interrupts the writing, but that the interruption of the writing ends the voyage" (365). Art directs life even as charts direct ships, and language, as the multiple puns on the name of their sturdy sailboat *Pokey* imply, directs everything. *Sabbatical* legitimizes sea fiction as a serious academic subgenre and ensures that the most sophisticated contemporary literary critics will have to know a jib from a stay.

Two massive, even more aggressively postmodern fictions continue Barth's appropriation of nautical metaphors but lack the "keel and ballast" that keep *Sabbatical* on course. *The Tidewater Tales: A Novel* (1987) uses the familiar Chesapeake Bay setting to follow Katherine and Peter Sagamore on a two-week sail in the sloop *Story* as Katherine, eight-and-a-half-months pregnant, tries to coax her husband out of a prolonged case of writer's block. The novel elaborates the basic themes of *Sabbatical* with refashionings of the *Odyssey*, the *Arabian Nights, Don Quixote,* and other Barth favorites, and reintroduces Wye Island, CIA exposés, and some character types from the earlier novel. Although nautical motifs contribute little to the novel, the larger structure of sailing provides the isolation, freedom, and responsibility necessary to confront artistic and spiritual solipsism. Simply by maintaining a log, Peter writes. And by navigating and reading maps, he negotiates otherness. When Katherine delivers twins, named Adam and Eve, artistic and biological reproduction unite to imply renewed opportunities for storytelling. In *The Last Voyage of Somebody the Sailor* (1991), Barth extrapolates from the 1,001 nights to tell the story of Simon William Behler, a psychically distressed journalist seeking to reestablish his emotional and professional life. In multiple incarnations, time periods, and locales, Behler becomes Baylor, Bey-el-Loor, B-Bibi-Bill, and Simmon-Simon-Somebody as he travels between his time and Sinbad's, the exotic world of Scheherazade's medieval Arabia. In these multiple personas, Behler takes seven voyages, each intercut by (usually erotic) interludes descending in number from seven to one. A skilled sailor himself, Behler navigates this dual and recursive narrative with some confusion but immense gusto as he fights pirates, builds a replica of Joshua Slocum's *Spray*, battles storms, survives shipwreck, nearly drowns, and recounts the intense loves of his lives. By integrating these incidents more fully into the structure and characterizations, *Somebody* is a much better example of Barth's imaginative application of nautical materials than *The Tidewater Tales*. Nevertheless, because these works consciously replicate a replica—Sir Richard Burton's *Arabian Nights*—both "ships of story" tack uneasily between realism and fantasy in a frothy and choppy metafictional sea.

In *Sailing* (1988), Susan Kenney also exploits the vitalistic metaphor of small-boat sailing, but balances implications of renewal with a clinically minute account of the ravages of inoperable cancer. When he discovers he has a terminal case, Phil Boyd, a college professor, takes up sailing and finds solace in it. His wife, Sara, initially shares in the new pastime. Soon, however, Phil becomes obsessive about sailing, grows more distant from Sara, and after battling the disease for five years, sails off alone to meet his death. Sara, driven to confront her childhood fears of death and rejection, learns to deal with the inevitable and sympathize more deeply with her husband's plight. Although Phil turns back at the very end, he returns to death and suffering. Sailing becomes a complex metaphor for balancing life and death, the quality versus the quantity of living, much as a sailor manages his boat by balancing the conflicting forces of wind and sail, friction and gravity, to decide whether to maximize speed or minimize distance.

A more tightly conceived symbolic appropriation of nautical themes occurs in Lois Gould's *A Sea-Change* (1976), which allegorizes the behavior of the wrasse or "sea-wife" fish in an unusual tale of emergent female assertiveness and power. Jessie Waterman, a former model, and her old school chum Kate weather a hurricane in a summer house off the coast of Maine. During the storm Jessie mysteriously begins to assert a dark and violent male side of herself, a side she names "BG" after a black gunman who raped her with a pistol. Despite a threat to her new personality from a coast guardsman who seeks refuge during the storm, Jessie completes her unusual "sea-change" and moves to an island farther offshore where she finally assumes a male identity. When the dominant male of a wrasse harem dies, the dominant female switches gender and becomes the new dominant male. Just so, Jessie asserts her freedom from sexual stereotypes and male victimization in a sex change inspired by the sea's transformative ecology.

Another writer who, like Barth and Gould, has mined the sea for its metaphorical possibilities, is John Hawkes. Although he has not written any strictly nautical fiction, two of his many novels, almost companion pieces, use the sea to symbolize the need for comic acceptance rather than tragic struggle. *Second Skin* (1964) is narrated by Skipper, a former World War II naval officer. From his present peaceful life on a wandering tropical island, he recalls two sea experiences that threatened his identity. The first was a wartime mutiny aboard his vessel during which he was beaten and brutally sodomized by the young ringleader, Tremlow—"a perfect Triton." The second occurred aboard the *Peter Poor*, an unkempt New England fishing boat, where he got seasick and fantasized that his daughter was willingly seduced by Captain Red, the boat's unsavory owner. Hawkes treats both incidents comically, using them to point out Skipper's repressed homosexuality and foolish mock-heroism. Lyrical descriptions of the sea counterpoint Tremlow's brutality and Skipper's opacity, and imply the tranquility one can achieve by viewing nature directly. The "second skin" of the title is the tight-fitting yellow oil-skin Skipper dons on the *Peter Poor*. It symbolizes the social and emotional constriction that occurs when people deny their natural selves, and contrasts sharply with Skipper's freedom on the magical sea island where, half-naked and deeply tanned, he is at peace with both his environment and himself. In Hawkes's *Adventures in the Alaskan Skin Trade* (1985), the Skipper-like hero Uncle Jake Deauville abandons life in the overcivilized, death-directed lower forty-eight for the rugged Alaskan frontier in the 1930s. He buys and refits a huge old fishing boat and cruises up and down the south coast of Alaska seeking adventure and fortune. Although the sea functions more as a backdrop than a symbolic locale, many of the best comic scenes involve Jake's frustrations with his bulky and recalcitrant boat and his clumsy attempts at seamanship and navigation. By showing his mock-epic hero repeatedly failing traditional nautical tests of manhood, Hawkes satirizes the outdated view of heroism and adventure Uncle Jake represents.

Two recent novels about the Middle Passage—the infamous return voyage of slave cargo ships from West Africa to the Caribbean and American South—blend historical and postmodernist fictional techniques in stories set almost entirely at sea. Joanna Scott's *Closest Possible Union* (1988), an allegory of American guilt for the slave trade, combines dream, fantasy, and superstition in the narration of a naive adolescent voyaging on his father's covert Guineaman at some vague time in the early nineteenth century. Anxious and guarded, Tom journeys into a nightmare of increasing ambiguity and disorder. After the "blackfish oil" comes aboard, the captain gradually becomes incapacitated by melancholy and alcoholism, most of the crew abandons ship, and the slaves assume control of the vessel and cannibalize the captain, simultaneously ingesting and exorcising their oppressor. The book ends as inconclusively as Poe's *Pym* (with which it has numerous affinities) as the shattered ship drifts aimlessly between Africa and America, irreconcilable cultural opposites. Charles Johnson's *Middle Passage* (1990), a much more successful reconstruction of a similar voyage located in 1830, tells its tale through the cynical eyes of a freed black confidence man, Rutherford Calhoun. In this metafictional novel that not only draws upon many antecedents from the *Odyssey* to *Moby-Dick* but is also heavily influenced by "The King's Indian," a nautical metafiction by Johnson's collegiate mentor John Gardner, incongruities abound, from Calhoun's position as crew member on a slaver and Captain Falcon's uncanny erudition, to Johnson's inaccurate nautical terminology and anachronistic style. Yet, a powerful plot maintains interest. Once the slaves are brought aboard—the "Allmuseri," a mysterious tribe with supernatural powers—Calhoun finds himself caught between divided loyalties: his Americanness and his captain versus his Africanness and the Allmuseri. Since Captain Falcon has also captured the Allmuseri god, a dark force crated and boiling in the hold, the Africans lay claim to spiritual powers stronger than the chains of slavery. Aided by two characters straight out of Melville's "Benito Cereno," the Allmuseri seize control of the ship and try to steer it back to Africa. After Captain Falcon commits suicide, Calhoun, a classic *picaro*, sides with the slaves, but then escapes a shipwreck by plotting with the white cook. Even more fantastic than Scott's novel, Johnson's uses the slave trade as a metaphor to explore contemporary problems of racial guilt and black identity. Neither novel seeks historical or nautical verisimilitude, but each dramatizes the dehumanizing effects of the Middle Passage on whites and blacks alike. Consequently, each suffers somewhat as sea fiction, yet both extend the range of the genre and reinforce the prevailing idea of the sea's regenerative powers. A National Book Award winner, as was Casey's *Spartina*, *Middle Passage* demonstrates well the adaptability, symbolic potential, and enduring appeal of America's checkered maritime experience.

A final clutch of novels by some of the most respected writers of the period shows how the sea's symbolic connection with freedom and personal liberation pervades contemporary fiction. In Joseph Heller's contemporary classic

Catch-22 (1961), the sea offers the only sure escape from the madness of war. After months of quiet planning, the aptly named Orr successfully paddles from the Mediterranean to Sweden in a yellow life raft, surviving on codfish and rations. Orr's open-boat journey, perhaps the most positive event in the novel, inspires the protagonist Yossarian to make his own break for freedom at the end. The mental patients in Ken Kesey's *One Flew Over the Cuckoo's Nest* (1962) find release from the antiseptic imprisonment of the asylum in a rowdy and sensuous one-day fishing trip led by Randle P. McMurphy, the confidence man turned savior. McMurphy and his crew of twelve, like Jesus and the disciples on the Sea of Galilee, experience the heady exhilaration of sea air and salt water, and come back to land changed men, unwilling to accept the rigid and dehumanizing restrictions of Big Nurse Ratched's machinelike hospital. For Katherine Anne Porter in *Ship of Fools* (1962), a transatlantic voyage in 1932 forces a mixed bag of aristocratic, bourgeois, and lower-class Germans, Mexicans, and Americans to confront the tedium and sterility of their lives. Although Porter barely mentions the sea, this very omission suggests her characters' distance from natural freedom, their dependence on the captain as well as on the decadent rituals of the passenger ship, and the spiritual emptiness of the German soul on the eve of Hitler's rise to power. In a comic version of the same theme, Wright Morris chronicles Professor Arnold Soby's sabbatical voyage to and around the Mediterranean in *What a Way to Go* (1962). Soby's halting, clumsy pursuit of youth and love among his German companions almost seems to parody Porter's gloomy *Narrenschiff* and, like Mark Twain in *Innocents Abroad*, suggests that voyaging has a humorous as well as a serious side. Finally, for William Styron and Bernard Malamud, the sea offers salvation from slavery and nuclear destruction, two of the greatest threats to American life. In Styron's *Confessions of Nat Turner* (1967), the enslaved hero's quest for freedom is framed by his recurrent dream of drifting down a river toward the oceanic roar of the sea, a symbol for him, as for Langston Hughes and Charles Johnson, of a path back to Africa where the black American can recover his lost identity and racial pride. In Malamud's post–World War III fantasy, *God's Grace* (1982), the sea is a literal savior for young paleologist Calvin Cohn. Submerged in a small research submarine during the brief war, Cohn survives to replicate Robinson Crusoe's experience by establishing a new life on a tropical island. In another fantasy of interspecies communication, he teaches chimpanzees to talk and even successfully mates with one. Although his dream of refounding the human race dies violently, he has experienced "grace" by renouncing human pride and living more harmoniously with nature.

The popular and the academic novelists depend on a third kind of writer for the view of the sea's benevolence that often characterizes contemporary fiction. This final category embraces those adventurers and scientists who have recorded their personal experiences with the sea in autobiographies and reflective narratives. Numerous books and essays could be included in this group, for many yachtsmen, sailors, marine biologists, and oceanographers have written of their pursuits. A few works stand out both for their popularity and their quality.

Two autobiographies, Sterling Hayden's *Wanderer* (1963) and Robin Lee Graham's *Dove* (1972), narrate voyages of escape from the stifling socialization of land to the freedom and openness of the sea. Known best as a movie star, Hayden actually began adulthood as a sailor. On his first command in 1938 he sailed a brig from Boston to Tahiti, attracted the notice of movie producers, and entered the emotional maelstrom of Hollywood. Over the years he found increasing solace on his ninety-eight-foot schooner *Wanderer*, a vessel built in 1893 and carefully refitted by Hayden himself. In 1959, when a California court threatened to deprive him of his children in a wrenching custody battle, he sailed off with them and a makeshift crew to Tahiti, making headlines and proving that the romantic allure of the sea is not dead. Hayden self-consciously and somewhat too literarily cites both *Walden* and *Mutiny on the Bounty* as sources of inspiration; but his thorough knowledge of the ocean still offers a moral clarity and physical purity unavailable on dry land. *Dove* records Graham's long solo voyage around the world in 1965, when he was only sixteen. At sea, Graham finds a clearer perspective on civilization and, like many of his generation, comes to disavow the materialism and violence of American life during the Vietnam War. He reads Joshua Slocum, Charles Darwin, and the Bible, and when he returns to the United States after thirty-one thousand miles of lonely cruising, he moves to Montana and dedicates his life to primitive Christian idealism. At times a naive innocence undermines the integrity of Graham's adventure. Philosophically, however, Graham remains a real-life counterpart to Malamud's Calvin Cohn, a young man who discovers spiritual values in solo voyaging.

In grim contrast to Hayden's and Graham's inspiriting relationships with the sea stands Steven Callahan's terrifying autobiographical tale of modern survival, *Adrift: Seventy-Six Days Lost at Sea* (1986). Determined to follow in the wake of such romantic single-handed voyagers as Robert Manry, Callahan builds his own boat and in 1980 sets off on a solo Atlantic crossing and return. On the second leg a huge wave swamps his twenty-one-foot cruiser, forcing him to abandon ship in a violent storm 450 miles northwest of the Cape Verde Islands. For the next two-and-a-half months he survives in a rubber life raft through determination, luck, and knowledge. Following the principles of survival in Dougal Robertson's *Survive the Savage Sea* (1973), Callahan operates a solar still, catches fish, and even snags a bird and eats it raw, as did William Bligh and the remnant of his crew from HMS *Bounty*. Despite repeated threats from sharks and one from a whale, Callahan survives to tell of the relativity of good and evil, his own insignificance in the larger scheme of creation, and the blind indifference of the sea.

A completely different tone controls William Buckley's three glib and witty accounts of blue-water yachting, *Airborne: A Sentimental Journey* (1976), *Atlantic High: A Celebration* (1982), and *Racing Through Paradise* (1987), which are less books about the sea than opportunities for Buckley to parade his wide knowledge of politics, art, literature, and current celebrities. By the third volume in this series, Buckley's narrative loses vitality, suggesting that nautical autobiographies require brave deeds as well as bold words.

It remains the province of marine naturalists to provide the most moving accounts of our emerging reorientation toward the sea as a place of life and beauty that we must preserve, not exploit. Lee Maril, Victor Scheffer, and Jack Rudloe hover along that vague generic boundary between scientific and humanistic writing as they grapple with the most difficult environmental and biological problems surrounding marine ecology. Maril's *Cannibals and Condos* (1986) takes a sociologist's perspective on the Texas Gulf Coast and concludes that the seemingly infinite ocean is indeed threatened by the greed and unthinking development of coastal populations. Scheffer, biologist for the U.S. Fish and Wildlife Service since 1937, has written many books about his vocation, but two are especially noteworthy: *The Year of the Whale* (1969) and *The Year of the Seal* (1970), both "fictions based on facts." One describes a sperm whale's first twelve months of life, while the other recounts a golden seal mother's bearing and nurturing of her pup. Each story is told objectively and unsentimentally from the perspective of the mammal involved, allowing Scheffer to individualize his account without distorting the biological and ecological realities the animals confront.

Both for the quantity and literary value of his prose, Jack Rudloe, a marine biologist who provides specimens to university laboratories all over the United States, comes closer to being a nautical Thoreau than anyone else writing today. Two of Rudloe's books, *The Erotic Ocean: A Handbook for Beachcombers* (1971) and *The Time of the Turtle* (1979), are fairly technical discussions of their subjects. But three others, *The Sea Brings Forth* (1968), *The Living Dock at Panacea* (1977), and *The Wilderness Coast* (1988), personalize his job and offer insights about the relationship between humanity and the sea far surpassing John Steinbeck's attempt in this genre, *Sea of Cortez* (1941).

The best of these is *The Living Dock at Panacea*, Rudloe's account of one year in the life of his dock. Like Thoreau, he structures his book around the seasons, beginning in summer and moving through winter storms, spring floods, a midsummer hurricane that destroys his dock, to final renewal when he rebuilds and continues the endless natural cycle. He praises the amazing fertility and resilience of the sea and coastline, their ability to withstand disasters natural and human, and explores his ambivalent feelings toward his trade. In an especially powerful chapter about collecting a gallon of blood from a live shark, an experience that bears comparison with James Dickey's poem "The Shark's Parlor," he acknowledges his atavistic blood-lust even as he feels remorse for animals that die uselessly. He realizes that he too is a predator, and that he is justified only if his predation serves a higher purpose. Properly used, nothing is wasted or truly destroyed; only human beings, with their greed and shortsightedness, violate this natural cycle. Rudloe's book offers poignant firsthand testimony to the contemporary view of the sea as necessary to human existence. He uses his own intense but ambiguous relationship with the sea to articulate a marine ethic as probing and challenging as Thoreau's speculations in "Higher Laws."

American prose since 1960 continues many of the traditional themes of nautical literature: Benchley and Searls give us sea monsters, Matthiessen and Casey realistic voyages of self-discovery, and Hayden and Rudloe personal accounts of one man's love for the sea. And of course the sea remains an exciting backdrop for stories of action, mystery, adventure, and social analysis. The ship, it seems, will always be a microcosm of society, and the sea a place for escape and solitary reflection. But the great new theme of the sea as a home for humanity, a comparatively unthreatening place for human fulfillment, is an important development. Writers such as Ingalls, Gould, Mooney, Benchley, and Searls look to the future of maritime fiction in their fantasies of intimate human involvement with the sea and its creatures. And a further development, the self-conscious appropriation and mutation of nautical literary conventions by such postmodern writers as John Barth and Charles Johnson, shows that nautical themes not only transcend potboiling romances and war stories but also survive literary sea-changes. Just as the sea is increasingly viewed by environmentalists as an essential part of human ecology, so the long tradition of sea literature is employed by many serious writers as a valuable resource for vitalizing contemporary American fiction.

DENNIS BERTHOLD

Bibliography

This bibliography, first of all, documents the works and authors specifically referred to in the text. Secondly, it presents a record of the scholarship and criticism to which the chapters are indebted. Beyond these basic and obligatory functions, the first list contains many primary works on which the text is silent, but which, for one reason or another, are judged of interest to readers of this book, and worth mention as part of a practical bibliographical record assembled from many sources and easily accessible here.

The scope of the book, and therefore of its bibliography, does have some limitations. We have made no attempt to cover juvenile literature, modern romance novels, magazine and newspaper fiction, and verse, pamphlets, reform tracts, broadsides, other ephemera, and unpublished materials—though an occasional title from such categories does appear. Furthermore, the primary bibliography does not list all sea-related works by popular authors such as Joseph Holt Ingraham, Justin Jones, "Ned Buntline," and several others, each of whom turned out many such. Instead, representative titles are cited, enough to show that the particular author made sea literature a specialty.

The citations after 1800 include publisher as well as place of publication when known. Exceptionally long titles are usually abbreviated at the first full stop; ellipsis is used for the reader's benefit, indicating that the omitted part of the title contributes to explaining the special character of that item. In each citation, the first American edition is generally noted first. If another edition (perhaps a revised or expanded one), or a British one, is to the point, however, it is cited instead. Entries include, whenever possible, definitive or standard editions, and frequently cite recent, accessible reprints too.

The secondary bibliography focuses on substantive twentieth-century commentary, in English, on American literature and the sea. Works in other languages are minimally represented. Though it tries to cover all pertinent scholarship and criticism, like the primary bibliography, and for similar reasons, it is not comprehensive or definitive. In addition, some works on the subject that, in the judgment of the contributors, are seriously flawed or of negligible value have been omitted. In deciding which of the many thousands of articles and books on American literature to cite (a problem particularly acute when it comes to authors of high critical stature), the discriminator used has been the centrality of the sea or of maritime matters to the discussion.

Primary Bibliography

Abbey, Charles A. *See* Gosnell, Harpur Allen, ed.

Adams, Bill. *Fenceless Meadows.* New York: Stokes, 1923.

———. "Way for a Sailor." *Great Sea Stories of All Nations and All Times.* Ed. H. M. Tomlinson. Garden City: Garden City Publishing, 1937. 585–605.

———. *Wind in the Topsails.* London: Harrap, 1931.

Adams, Henry. *The Education of Henry Adams.* Boston: Houghton, 1918.

Aiken, Conrad. *Ushant.* Boston: Little, 1952.

Albee, Edward. *Seascape: A Play.* New York: Athenaeum, 1975.

Alcott, Louisa May. *Work.* Boston: Roberts Brothers, 1873.

Aldrich, Thomas Bailey. *The Ballad of Babie Bell and Other Poems.* New York: Rudd & Carleton, 1859.

Allen, Harriet. MS diary. Kendall Whaling Museum, Sharon, MA.

Allyn, Gurdon L. *The Old Sailor's Story, ...* Norwich, CT: Wilcox, 1879.

Ames, Nathaniel. *A Mariner's Sketches.* Providence, RI: Cory, Marshall and Hammond, 1830.

———. *Nautical Reminiscences.* Providence, RI: Marshall, 1832.

———. *An Old Sailor's Yarns.* New York: Dearborn, 1835.

Arnold, Birch. *See* Bartlett, Alice Eloise.

Auden, W. H. *The Enchafèd Flood; or, The Romantic Iconography of the Sea.* New York: Random House, 1950. New York: Vintage, 1967.

Averill, Charles E. *Blackbeard; or, The Bloodhound of the Bermudas: A Tale of the Ocean's Exciting Incidents.* New York: De Witt, 1868.

———. *The Female Fishers; or, The Beautiful Girl of Marblehead. A Thrilling Story of the Sea Coast.* Boston: Wiley, 1848.

———. *The Secrets of the High Seas; or, The Mysterious Wreck in the Gulf Stream.* Boston: Williams, 1849.

———. *The Wreckers; or, The Ship Plunderers of Barnegat.* Boston: Gleason, 1848.

Baars, Fred D. *The Story of the Seas: A Romance in Reality of a Sailor's Life.* Arkadelphia, AR: Baars & Neeley, 1896.

Bailey, Joseph. *God's Wonders in the Great Deep.* New York, 1750.

Baker, Louis A. *Harry Martingale; or, Adventures of a Whaleman in the Pacific Ocean.* Boston, 1848.

Balano, Dorothea. *The Log of the Skipper's Wife.* Ed. James W. Balano. Camden, ME: Down East Books, 1979.

Baldwin, Bates [John E. Jennings]. *The Tide of Empire.* New York: Holt, 1952.

Ball, Charles. *Slavery in the United States: A Narrative of the Life and Adventures of Charles Ball, a Black Man.* Lewistown, PA: Shugert, 1836. New York: Negro UP, 1969.

Ballou, Maturin Murray. *See* Murray, Lieut. [pseud.]; Forester, Frank [pseud.].

Barbeau, Marius. "Voyageur Songs." *Beaver* June 1942: 15–19.

Barber, John W. *A History of the Amistad Captives.* New Haven: Barber, 1840.

Barker, Benjamin. *The Bandit of the Ocean; or, The Female Privateer.* New York: Dewitt, c. 1855.

——— . *Blackbeard; or, The Pirate of Roanoke.* Boston: Gleason, 1847.

——— . *Corilla; or, The Indian Enchantress. A Romance of the Pacific and Its Islands.* Boston: Flag of Our Union Office, 1847.

——— . *The Nymph of the Ocean; or, The Pirate's Betrothal.* Boston: United States Pub. Co., 1846.

Barnard, Charles H. *A Narrative of the Sufferings and Adventures of Capt. Charles H. Barnard in a Voyage Round the World.* New York: J. Lindon, 1829. Rpt. in *The Sea, the Ship and the Sailor.* Ed. Elliot Snow. Salem: Marine Research Society, 1925.

Barnard, John. *Ashton's Memorial.* Boston, 1725.

Barth, John. *The Floating Opera.* New York: Appleton, 1956.

——— . *The Last Voyage of Somebody the Sailor.* Boston: Little, 1991.

——— . *Lost in the Funhouse.* New York: Doubleday, 1968.

——— . *Sabbatical: A Romance.* New York: Putnam's, 1982.

——— . *The Sot-Weed Factor.* Garden City, NY: Doubleday, 1960.

——— . *The Tidewater Tales: A Novel.* New York: Putnam's, 1987.

Bartholomew, Benjamin. "A Relation of the Wonderful Mercies of God . . . the 19 of October 1660 in the Ship Exchange." *In the Trough of the Sea.* Ed. Donald P. Wharton. Westport, CT: Greenwood, 1978. 120–26.

Bartlett, Alice Eloise [Birch Arnold]. *Mystery of the Monogram.* Detroit: Detroit and Cleveland Navigation Company, 1904.

——— . *Spirit of The Inland Seas.* Detroit: Detroit and Cleveland Navigation Company, 1910.

Bassett, Fletcher S. *Legends and Superstitions of the Sea and of Sailors.* London, 1885.

[Batchelder, Eugene]. *A Romance of the Sea Serpent; or, The Ichthyosaurus.* Cambridge: Bartlett, 1849.

Batchelor, John Calvin. *The Birth of the People's Republic of Antarctica.* New York: Dial, 1983.

Bates, Mrs. D. B. *Incidents on Land and Water.* Boston: French, 1857.

Beach, Edward. *Cold Is the Sea.* New York: Holt, 1978.

——— . *Dust on the Sea.* New York: Holt, 1972.

——— . *Run Silent, Run Deep.* New York: Holt, 1955.

Beale, Thomas. *The Natural History of the Sperm Whale.* London, 1839.

Belland, F. W. *The True Sea.* New York: Holt, 1984.

Benchley, Nathaniel. *Sail a Crooked Ship.* New York: McGraw-Hill, 1960.

Benchley, Peter. *Beast.* New York: Random House, 1991.

——— . *The Deep.* Garden City, NY: Doubleday, 1976.

——— . *The Girl from the Sea of Cortez.* Garden City, NY: Doubleday, 1982.

——— . *The Island.* Garden City, NY: Doubleday, 1979.

————. *Jaws*. Garden City, NY: Doubleday, 1973.

Benham, Howard Curtis. *Shipwrecks: A Romance of 1899*. Buffalo, NY: Hausauer, 1899.

Bennet, George, and Daniel Tyerman. *Journal of Voyages and Travels*. 2 vols. Ed. James Montgomery. Boston: Crocker & Brewster, 1832.

Bennett, Frederick D. *Narrative of a Whaling Voyage Round the Globe, from the Year 1833 to 1836*. London, 1840. New York and Amsterdam: N. Israel and Da Capo Press, 1970.

Bibb, Henry. *Narrative of the Life and Adventures of Henry Bibb, an American Slave, Written by Himself*. New York: Bibb, 1850.

Biddle, Richard. *A Memoir of Sebastian Cabot; With a Review of the History of Maritime Discovery*. Philadelphia: Carey and Lea, 1831.

Binns, Archie. *Lightship*. New York: Reynal and Hitchcock, 1934.

The Biography of a Bottle. By a friend of temperance. Boston: Perkins, Marvin, 1835.

Bird, Robert Montgomery. *The Adventures of Robin Day*. Philadelphia: Lea & Blanchard, 1839.

————. "The Ice Island." *Philadelphia Monthly Magazine* 1 (1827): 109–14.

Bishop, Elizabeth. *The Complete Poems*. New York: Farrar, 1969.

Bishop, John Peale. *The Collected Poems*. Ed. Allen Tate. New York: Scribners, 1948.

Bishop, William Henry. "Deodand." *Choy Susan and Other Stories*. Boston: Houghton, 1885.

Blackbird, Andrew. *History of the Ottawa and Chippewa Indians of Michigan*. Ypsilanti, MI: The Ypsilantian Job Printing House, 1887.

[Blake, John Lavris]. *Ramon, the Rover of Cuba, and Other Tales*. New York: Nafis & Cornish, 1843.

Block, Jack [pseud.]. "The Cruise of the Mohawk." *American Monthly Magazine* 5 (1835): 417–25.

Bloomfield, Leonard. *Menomini Texts*. New York: Stechert, 1928.

Bok, Gordon. *Time and the Flying Snow*. Sharon, CT: Folk Legacy Records, 1977.

Bolton, John. *An Account of the Loss of the American Clipper Ship Tontine*. Pittsburgh: Callow, 1861.

————. *Account of the Loss of the (American) Ship Omartel (on the Banks) by . . . One of Her Crew. . . .* N.p., [1859].

The Book of Negro Folklore. Ed. Langston Hughes and Arna Bontemps. New York: Dodd, Mead, 1958.

Booth, Philip. *Relations: Selected Poems, 1950–1985*. New York: Viking, 1986.

Botsford, Edmund. *The Spiritual Voyage, Performed in the Ship Covert*. Charleston: Hoff, 1814.

Bowker, Francis E. *Blue Water Coaster*. Camden, ME: International Marine, 1972.

————. *Hull Down*. New Bedford: Reynolds, 1963.

Boyer, Samuel Pellman. *Naval Surgeon: Blockading the South, 1862–1866*. Ed. Elinor Barnes and James A. Barnes. Bloomington: Indiana UP, 1963.

Brackenridge, Henry Marie. *Voyage to South America. . . .* Baltimore, 1818.

Bradford, William. *Of Plimouth Plantation: 1620–1647*. Ed. Samuel Eliot Morison. New York: Knopf, 1952.

Brady, Cyrus Townsend. *For the Freedom of the Sea: A Romance of the War of 1812.* New York: Scribner's, 1898.

———. *The Grip of Honor: A Story of Paul Jones and the American Revolution.* New York: Scribner's, 1900.

"Breakers! A Scene at Sea." *Knickerbocker* Dec. 1835: 495–500.

Brehm, Victoria, ed. *Sweetwater, Storms, and Spirits: Stories of the Great Lakes.* Ann Arbor: U of Michigan P, 1990.

Brewer, Lucy. *The Female Marine; or, Adventures of Miss Lucy Brewer.* Boston, 1816.

Brewster, Mary. *"She Was a Sister Sailor": The Whaling Journals of Mary Brewster, 1845–1851.* Ed. Joan Druett. Mystic: Mystic Seaport Museum, 1993.

Bridge, Horatio. *See* Hawthorne, Nathaniel, ed.

The Brigantine; or, Admiral Lowe. By an American. New York: Crowen & Decker, 1839.

[Briggs, Charles F.] *The Adventures of Harry Franco.* New York: Saunders, 1839.

Briggs, Charles F. *The Trippings of Tom Pepper; or, The Results of Romancing.* New York: Burgess & Stringer, 1847.

———. "A Veritable Sea Story." *Knickerbocker* Feb. 1844: 151–52.

———. *Working a Passage; or, Life in a Liner.* New York: Allen, 1844.

Brinkley, William. *Don't Go Near the Water.* New York: Random, 1956.

———. *The Last Ship.* New York: Viking, 1988.

———. *The Ninety and Nine.* Garden City, NY: Doubleday, 1966.

Brown, Alexander Crosby, ed. *Longboat to Hawaii: An Account of the Voyage of the Clipper Ship* HORNET *of New York Bound for San Francisco in 1866.* Cambridge, MD: Cornell Maritime P, 1974.

Brown, Alice. "A Sea Change." *The Country Road.* Boston: Houghton, 1906. 147–89.

Brown, William Wells. *Clotel; or, The President's Daughter.* London: Partridge, 1853. New York: Macmillan, 1970.

———. *Narrative of William W. Brown, A Fugitive Slave, Written by Himself.* [Boston: Anti-Slavery Office, 1847.] Rpt. in *Five Slave Narratives: A Compendium.* New York: Arno, 1968.

Browne, Benjamin F. *See* Hawthorne, Nathaniel, ed.

Browne, J. Ross. *Etchings of a Whaling Cruise.* New York, 1846.

Bryant, William Cullen. *The Poetical Works of William Cullen Bryant.* Ed. Henry C. Sturges. New York: Appleton, 1903.

———, ed. *Picturesque America.* New York: Appleton, 1872.

Bryant, William M. *The Old Sailor: The Thrilling Narrative of the Life and Adventures of Elias Hutchins.* Biddeford, ME, 1853.

"The Bucaneer." *Ladies Companion* Sept. 1840: 247–52.

Buckley, William F. *Airborne: A Sentimental Journey.* New York: Macmillan, 1976.

———. *Atlantic High: A Celebration,* Garden City, NY: Doubleday, 1982.

———. *Racing Through Paradise.* New York: Random, 1987.

Bunnell, David C. *The Travels and Adventures of David C. Bunnell, During Twenty-Three Years of a Sea-faring Life.* Palmyra, NY: Grandin, 1831.

Buntline, Ned [Edward Zane Carroll Judson]. *The Black Avenger of the Spanish Main; or, The Fiend of Blood.* Boston: Gleason, 1847.

———. "A Chronicle of Our Navy." *Knickerbocker* Dec. 1846: 527–31.

———. "A Dream That Was Not All a Dream." *Knickerbocker* Sept. 1846: 244–47.

———. *The Red Revenger; or, Pirate King of the Floridas.* Boston: Gleason, 1848.

———. "Running a Blockade in the Last War." *Knickerbocker* Apr. 1847: 306–9.

———. "A Visit to Lafitte." *Knickerbocker* Mar. 1847: 254–61.

Burdick, Austin C. [Sylvanus Cobb]. *The Storm Children . . . .* Boston: Gleason, 1853.

Burland, Brian. *A Fall from Aloft.* New York: Random, 1969.

———. *Stephen Decatur.* London: Allen and Unwin, 1975.

———. *Surprise.* New York: Harper, 1975.

Burney, James. *A Chronological History of the Discoveries in the South Seas.* London, 1803.

Burroughs, William S. *Cities of the Red Night.* New York: Holt, 1981.

Burton, William E. "A Cape Codder Among the Mermaids." *Burton's Gentleman's Magazine* Dec. 1839: 287–92.

Burts, Robert. "The Escape: A Tale of the Sea." *Knickerbocker* Sept. 1836: 270–75.

———. "The Flying Dutchman." *Knickerbocker* Nov. 1836: 545–47.

———. "Jack Marlinspike's Yarn." *Knickerbocker* Aug. 1836: 202–9.

———. "The Man Overboard." *Military and Naval Magazine* Oct. 1835: 92–94.

———. "The Privateer." *Knickerbocker* Dec. 1836: 650–55.

———. *The Scourge of the Ocean.* Philadelphia: Carey and Hart, 1837.

———. *The Sea-King.* Philadelphia: Hart, Carey, and Hart, 1851.

Bushnell, William H. *Snow Drift. An Indian Tale.* Boston: Elliot, Thomas & Talbot, 1865.

Butnam, Captain. *Narrative of the Capture, Arrival, and Examination of the Sixteen Pirates, Taken on the Coast of Africa.* N.p., 1833.

Byles, Mather. *Poems on Several Occasions.* Boston, 1744.

Byrne, Frank. "The Cruise of the Gentile." *Graham's Magazine* 32 (1848): 133–47, 205–17.

Byron, George Gordon, Lord. *Poetical Works.* New York: Oxford UP, 1967.

Byron, J. *The Narrative of the Honourable John Byron. . . .* London, 1768.

Callahan, Steven. *Adrift: Seventy-Six Days Lost at Sea.* Boston: Houghton, 1986.

Carlisle, Henry. *The Jonah Man.* New York: Knopf, 1984.

———. *Voyage to the First of December.* New York: Putnam, 1972.

Carnes, J. A. *Journal of a Voyage from Boston to the West Coast of Africa.* Boston: Jewett, 1852.

Carse, Robert. *The Beckoning Waters.* New York: Scribners, 1953.

Carson, Rachel. *The Sea Around Us.* New York: Oxford UP, 1951.

Carter, Isabel Hopestill. *Shipmates: A Tale of the Seafaring Women of New England.* New York: Scott, 1934.

Carter, John Henton. *The Log of Commodore Rollingpin: His Adventures Afloat and Ashore.* New York: Carleton, 1874.

Casey, John. *Spartina.* New York: Knopf, 1989.

Cather, Willa. *My Antonia.* Boston: Houghton, 1918.

Catherwood, Mary Hartwell. *Mackinac and Lakes Stories.* New York: Harper, 1899.

Chaplin, Gordon. *The Fever Coast Log: At Sea in Central America.* New York: Simon & Schuster, 1992.

"A Chapter of Sea Life." *New-England Magazine* Jan. 1833: 47–51.

"A Chapter on Sharking." *Knickerbocker* Jan. 1836: 14–24.

Chase, Owen. *Narrative of the Most Extraordinary and Distressing Shipwreck of the Whale-Ship Essex, of Nantucket.* New York: Gilley, 1821.

Chase-Riboud, Barbara. *Echo of Lions.* New York: Morrow, 1989.

Cheever, Rev. Henry T. *The Whale and His Captors; or The Whaleman's Adventures, the Whale's Biography.* New York: Harper, 1849.

———, ed. *Journal and Memorials of Captain Obadiah Congar, for Fifty Years Mariner and Shipmaster from the Port of New York.* New York: Harper, 1851.

———, ed. *Ship and Shore in Madeira, Lisbon and the Mediterranean. By Rev. Walter Colton (late) of the United States Navy.* New York: Barnes, 1851.

Chiles, Webb. *Open Boat Across the Pacific.* New York: Norton, 1982.

Chopin, Kate. *The Awakening.* Chicago: Stone, 1899.

Clancy, Tom. *The Hunt for Red October.* Annapolis: Naval Institute P, 1984.

Clark, Henry A. "The Cruise of the Raker." *Graham's Magazine* 33 (1848): 69–74, 129–36, 188–96, 257–66.

Clark, Joseph G. *Lights and Shadows of Sailor Life, Adventures of United States Exploring Expeditions, and Reminiscences of an Eventful Life on the "Mountain Wave."* Boston: Mussey, 1848.

Clavell, James. *Shōgun.* New York: Atheneum, 1975.

———. *Tai-Pan.* New York: Atheneum, 1966.

"The Clerk's Yarn." *Knickerbocker* Mar. 1837: 268–75.

Cleveland, Richard J. *In the Forecastle; or, Twenty-Five Years a Sailor.* New York: Manhattan Pub. Co., [184?].

———. *A Narrative of Voyages and Commercial Enterprises.* 2 vols. Cambridge: Owen, 1842.

Clew Garnett [J. W. Gould]. *Forecastle Yarns.* Ed. Edward Gould. New York: New World, 1843.

Clinch, J. H. "The Pirate." *Ladies Companion* 7 (1837): 244–48.

Cobb, Josiah. *See* A Younker [pseud.].

Cobb, Sylvanus. *See* also Burdick, Austin C. [pseud.].

———. *The Golden Eagle; or, The Privateer of '76.* Boston: Gleason, 1850.

———. *Marco; or, The Female Smuggler.* Boston, 1857.

———. *The Sea Lion; or, The Privateer of the Penobscot.* New York: French, 1853.

Codman, John. *See* Ringbolt, Captain [pseud.].

Coggeshall, George. *Voyages to Various Parts of the World Between 1799 and 1844.* New York: Appleton, 1851.

Coker, Daniel. *Journal of Daniel Coker, a Descendant of Africa, from the Time of His Leaving New York in the Ship Elizabeth, Captain Sebor, on a Voyage for Sherbro, in Africa. . . .* Baltimore: Coale, 1820.

Colcord, Lincoln. *The Drifting Diamond.* New York: Macmillan, 1912.

———. *The Game of Life and Death.* New York: Macmillan, 1914.

———. *An Instrument of the Gods and Other Stories of the Sea.* New York: Macmillan, 1922.

———. *Sea Stories from Searsport to Singapore, Selected Works of Lincoln Colcord.* Ed. Donald Mortland. Thorndike, ME: North Country P, 1987.

Colnett, James. *A Voyage to the South Atlantic and Round Cape Horn into the Pacific Ocean.* London, 1798.

Colton, Rev. Walter. *Deck and Port; or, Incidents of a Cruise in the United States Frigate Congress to California.* New York: Barnes, 1850.

Columbus, Christopher. *Four Voyages to the New World.* Gloucester: Smith, 1961.

———. *The Log of Christopher Columbus.* Trans. Robert H. Fuson. Camden, ME: International Marine, 1987.

Colvocoresses, Lieut. George M. *Four Years in a Government Exploring Expedition. . . .* New York: Cornish, Lamport, 1852.

Comstock, William. *The Life of Samuel Comstock, the Terrible Whaleman.* N.p., 1829.

Connolly, James Brendan. *Out of Gloucester.* New York: Scribner's, 1902.

Conrad, Joseph. *The Nigger of the "Narcissus."* London: Heinemann, 1897.

Cook, Frederick A. *Through the First Antarctic Night, 1898–1899.* New York: Doubleday and McClure, 1900. Montreal: McGill-Queen's UP, 1980.

Cooper, James Fenimore. *Afloat and Ashore; or, The Adventures of Miles Wallingford.* Philadelphia, 1844.

———. *The Bravo.* Philadelphia: Carey & Lea, 1831.

———. *Early Critical Essays (1820–22).* Gainesville, FL: Scholars' Facsimiles & Reprints, 1955.

———. "Elaborate Review." *Proceedings of the Naval Court Martial in the Case of Alexander Slidell Mackenzie. . . .* New York: Langley, 1844. Delmar, NY: Scholars' Facsimiles & Reprints, 1992.

———. *The History of the Navy of the United States of America.* 2 vols. Philadelphia: Lea and Blanchard, 1839.

———. *Homeward Bound; or, The Chase: A Tale of the Sea.* Philadelphia: Carey, Lea, and Blanchard, 1838.

———. *Jack Tier; or, The Florida Reef.* New York: Burgess, Stringer, 1848. Printed in *Graham's Magazine* (Nov. 1846–Mar. 1848) as "The Islets of the Gulf; or Rose Budd."

———. *The Letters and Journals of James Fenimore Cooper.* Ed. James F. Beard. 6 vols. Cambridge: Harvard UP, 1960–68.

———. *Lives of Distinguished American Naval Officers.* Philadelphia: Carey and Hart, 1846.

———. *Mercedes of Castile; or, The Voyage to Cathay.* Philadelphia: Lea and Blanchard, 1840.

———. *Ned Myers; or, A Life Before the Mast.* Philadelphia: Lea and Blanchard, 1843. Ed. William S. Dudley. Annapolis: Naval Institute P, 1989.

———. *Novels.* 32 vols. New York: Townsend, 1859–61.

———. *The Pathfinder; or, The Inland Sea.* New York: Gregory, 1840. Ed. Richard D. Rust. Albany: State U of New York P, 1981.

———. *The Pilot: A Tale of the Sea.* New York: Wiley, 1824. Ed. Kay Seymour House. Albany: State U of New York P, 1986.

———. *The Red Rover.* Philadelphia: Carey, Lea & Carey, 1828. Ed. Warren S. Walker. Lincoln: U of Nebraska P, 1963.

———. *The Sea-Lions; or, The Lost Sealers.* New York: Stringer and Townsend, 1849. Ed. Warren S. Walker. Lincoln: U of Nebraska P, 1965.

———. *The Spy.* New York: Wiley & Halsted, 1821.

———. *The Two Admirals: A Tale.* Philadelphia: Lea and Blanchard, 1842. Ed. Donald A. Ringe, James A. Sappenfield, and E. N. Feltskog. Albany: State U of New York P, 1990.

———. *The Water-Witch; or, The Skimmer of the Seas: A Tale.* Philadelphia: Carey and Lea, 1830.

———. *The Wing-and-Wing; or, Le Feu-Follet: A Tale.* Philadelphia: Lea and Blanchard, 1842.

Cooper, Rev. W. H. *Incidents of Shipwreck; or, The Loss of the San Francisco.* Philadelphia, 1855.

Costello, Frederick Hankerson. *Master Ardick, Buccaneer.* New York: Appleton, 1896.

Coulter, John. *Adventures in the Pacific.* Dublin: Curry, 1845.

Coyote Was Going There. Ed. Jarold Ramsey. Seattle: U of Washington P, 1977.

Cozzens, James Gould. *Castaway.* New York: Longmans, 1934.

———. *S.S. San Pedro.* New York: Harcourt, 1931.

Crafts, William. *A Selection in Prose and Poetry from the Miscellaneous Writings of the Late William Crafts.* Charleston: Sebring & Burges, 1828.

Crane, Hart. *The Collected Poems of Hart Crane.* Ed. Waldo Frank. New York: Liveright, 1933.

Crane, Stephen. "Flanagan and His Short Filibustering Adventure." *Illustrated London News* 28 Aug. 1897: 279–82. Charlottesville: UP of Virginia, 1970. Vol. 5 of *The Works of Stephen Crane.* Ed. Fredson Bowers. 10 vols. 1969–1976.

———. *Maggie: A Girl of the Streets.* New York: Privately printed, 1893. Charlottesville: UP of Virginia, 1969. Vol. 1 of *The Works of Stephen Crane.* Ed. Fredson Bowers. 10 vols. 1969–1976.

———. "The Open Boat." *Scribner's Magazine* Jan. 1897: 728–40. Charlottesville: UP of Virginia, 1969. Vol. 5 of *The Works of Stephen Crane.* Ed. Fredson Bowers. 10 vols. 1969–1976.

———. "Stephen Crane's Own Story." *New York Press* 7 Jan. 1897: 1–2. Charlottesville: UP of Virginia, 1971. Vol. 9 of *The Works of Stephen Crane.* Ed. Fredson Bowers. 10 vols. 1969–1976.

Crapo, Thomas. *Strange But True: Life and Adventures of Captain Thomas Crapo and Wife.* Ed. William J. Cowan. New Bedford: Crapo, 1893.

Crevecoeur, J. Hector St. John. *Letters from an American Farmer.* Dublin: Exshaw, 1782. Gloucester, MA: Smith, 1968.

Crosby, Ralph Mitchell. *We Have Met the Enemy.* Indianapolis: Bobbs Merrill, 1940.

Crowninshield, Mary. *Latitude 19°: A Romance of the West Indies.* . . . New York: Appleton, 1898.

"The Cruise of the Enterprize." *United States Magazine and Democratic Review* July 1839: 33–42.

Cuffee, Paul. *Memoir of Captain Paul Cuffee, a Man of Colour.* Liverpool: Egerton Smith, 1811.

Cuffee, Paul, [Jr.]. *Narrative of the Life and Adventures of Paul Cuffee, a Pequot Indian: During Thirty Years Spent at Sea and in Travelling in Foreign Lands.* Vernon, NY: Bill, 1839.

Cullen, Countee. *On These I Stand.* New York: Harper, 1947.

Curtis, Stephen. *Brief Extracts from the Journal of a Voyage Performed by the Whale Ship M——y, of New Bedford, Massachusetts, Commencing May 25, 1841 and Terminating August 1, 1844.* Boston, 1844.

Curwood, James Oliver. *The Courage of Captain Plum.* New York: Hurst, 1908.

——. *Falkner of the Inland Seas.* New York: Grosset, 1931.

Cussler, Clive. *Cyclops: A Novel.* New York: Pocket, 1986.

——. *Deep Six.* New York: Viking, 1989.

——. *Iceberg.* New York: Dodd, 1975.

——. *The Mediterranean Caper.* New York: Pyramid Publications, 1973.

——. *Night Probe!* New York: Bantam, 1981.

——. *Pacific Vortex!* New York: Bantam, 1983.

——. *Raise the Titanic!* New York: Viking, 1976.

——. *Vixen 0–3.* New York: Viking, 1978.

Dana, Richard Henry, Jr. *The Seaman's Friend.* Boston: Little, Brown, Loring, 1841.

——. *To Cuba and Back.* Boston: Ticknor and Fields, 1859.

——. *Two Years Before the Mast.* New York: Harper, 1840.

Dana, Richard Henry [Sr.]. *Poems and Prose Writings.* 2d ed. 2 vols. New York: Baker & Scribner, 1850.

Danforth, F. T. *The Startling, Thrilling and Interesting Narrative of the Life, Sufferings, Singular and Surprising Adventures of Fanny Templeton Danforth.* . . . Philadelphia, 1849.

Danforth, Harry [Charles Jacobs Peterson]. *The Algerine, and Other Tales.* Boston: Gleason, 1846.

——. "The Black Rover." *Sartain's Magazine* Jan.–Mar. 1849: 37–44, 119–26, 202–10.

——. "Getting to Sea." *Graham's Magazine* 26 (1844): 105–9.

——. "Off Calais." *Graham's Magazine* 28 (1845): 214–18.

——. "The Union Jack." *Graham's Magazine* 23 (1843): 105–9.

Dann, John C., ed. *The Nagle Journal: A Diary of the Life of Jacob Nagle, Sailor, from the Year 1775 to 1841.* New York: Weidenfeld & Nicolson, 1988.

——. *The Revolution Remembered: Eyewitness Accounts of the War for Independence.* Chicago: U of Chicago P, 1980. Chapter 9 reprints narratives of "Maritime Combat" by Abel Woodworth, Elnathan Jennings, Thomas Marble, John Ingersoll, and Joseph Saunders.

Darwin, Charles. *The Descent of Man, and Selection in Relation to Sex.* London: Murray, 1871.

———. *On the Origin of Species.* . . . London: Murray, 1859.

Davidson, Louis B., and Edward J. Doherty. *Captain Marooner.* New York: Crowell, 1952.

Davis, Lemuel Clarke. *A Stranded Ship: A Story of Sea and Shore.* New York: Putnam, 1869.

Davis, Richard Harding. *The Lion and Unicorn.* New York: Scribner's, 1904.

Dawes, Rufus. *Nix's Mate.* New York: Colman, 1839.

"The Dead Man's Sermon." *Knickerbocker* Sept. 1845: 203–12.

Dean, Harry. *The Pedro Gorino: The Adventures of a Negro Sea-Captain in Africa.* . . . Boston: Houghton, 1929.

Dean, John. *A Narrative of the Sufferings, Preservation, and Deliverance . . . in the Nottingham-Galley.* London, 1711.

De Blois, Henrietta. "Off to the Pacific." *Newport, RI, Mercury,* 21 Feb.–15 Aug. 1885.

De Felitta, Frank. *Sea Trial.* New York: Avon, 1980.

Delano, Amasa. *A Narrative of Voyages and Travels in the Northern and Southern Hemispheres.* Boston: Howe, 1817.

Delano, Reuben. *Wanderings and Adventures of Reuben Delano, Being a Narrative of Twelve Years in a Whaleship.* Worcester, MA: Drew, 1846.

Delany, Martin. *Blake; or, The Huts of America. Weekly Anglo-African* (1859, 1861–62). Boston: Beacon, 1970.

A Description of a Great Sea-Storm. . . . London, 1671.

Desrosiers, Leo-Paul. *Les Engagés du Grand Portage.* Paris: Gallimard, 1938.

Detzer, Karl. *The Marked Man: A Romance of the Great Lakes.* Indianapolis: Bobbs, 1927.

Dewhurst, Henry William. *The Natural History of the Order Cetacea and the Oceanic Inhabitants of the Arctic Regions.* London, 1834.

Dexter, Elisha. *Narrative of the Loss of the Whaling Brig William and Joseph, of Martha's Vineyard.* Boston, 1848.

Dibdin, Charles. *The Professional Life of Mr. Dibdin, Written by Himself. Together with the Words of Six Hundred Songs.* London, 1803.

Dickey, James. *Buckdancer's Choice.* Middletown, CT: Wesleyan UP, 1965.

Dickinson, Emily. *The Letters of Emily Dickinson.* Ed. Thomas H. Johnson and Theodora Ward. 3 vols. Cambridge: Harvard-Belknap, 1958.

———. *The Poems of Emily Dickinson.* Ed. Thomas H. Johnson. 3 vols. Cambridge: Harvard-Belknap, 1955.

Dietz, Steven. *Ten November.* New York: Theatre Communications Group, 1987.

Dillard, Annie. "Ship in a Bottle." *Harper's Magazine* Sept. 1989: 68–71.

Dilworth, Sharon. "The Long White." *The Long White.* Iowa City: U of Iowa P, 1988.

Doane, Benjamin. *Following the Sea.* Halifax: Nimbus Publishing and the Nova Scotia Museum, 1987.

Dodson, Kenneth. *Away All Boats.* Boston: Little, 1954.

Doner, Mary Frances. *Blue River.* Garden City, NY: Doubleday, 1946.

———. *Glass Mountain.* Garden City, NY: Doubleday, 1942.

———. *Not by Bread Alone.* New York: Doubleday-Doran, 1941.

———. *The Wind and the Fog.* New York: Avalon, 1963.

Donne, John. "A Sermon Preached to the Honorable Company of the Virginian Plantation, 1622." *Seventeenth-Century Prose and Poetry.* Ed. Robert P. T. Coffin and Alexander M. Witherspoon. 2d ed. New York: Harcourt, 1957. 114–15.

———. "Sermon LXXII." *Seventeenth-Century Prose and Poetry.* Ed. Robert P. T. Coffin and Alexander M. Witherspoon. 2d ed. New York: Harcourt, 1957. 89–102.

Donnelly, Ignatius. *Atlantis: The Antediluvian World.* New York: Harper, 1882.

D[oolittle], H[ilda]. *Collected Poems of HD, 1912–1944.* New York: Boni and Liveright, 1925. *H.D.: Collected Poems, 1912–1944.* Ed. Louis L. Martz. New York: New Directions, 1983.

Douglass, Frederick. "The Heroic Slave." *Frederick Douglass' Paper,* Mar. 1853. *Three Classic African-American Novels.* Ed. William L. Andrews. New York: Mentor, 1990. 25–69.

———. *Life and Times of Frederick Douglass.* Hartford, CT: Park, 1881. New York: Collier, 1962.

———. *Narrative of the Life of Frederick Douglass, an American Slave.* Boston: 1845. New York: Viking Penguin, 1982.

Downey, Joseph T. *The Cruise of the Portsmouth, 1845–1847.* Ed. Howard Lamar. New Haven: Yale UP, 1958.

Downs, Barnabas, Jr. *A Brief and Remarkable Narrative . . . .* Boston: Russell, 1786.

Drago, Henry Sinclair. *Where the Loon Calls.* New York: Macaulay, 1928.

Drake, Joseph Rodman. *The Life and Works of Joseph Rodman Drake.* Ed. Frank Lester Pleadwell. Boston: Merrymount, 1935.

Drake, Richard. *Revelations of a Slave Smuggler.* New York: DeWitt, 1860.

Drake, Richard, and G. F. Dow. *Slave Ships and Slaving.* Salem: Marine Research Society, 1927.

Dring, Thomas. *See* Green, Albert, ed.

Druett, Joan. *See* Brewster, Mary.

Duer, John K. *The Matricide.* New York: Graham, 1846.

Duer, John K., ed. *The Nautilus, a Collection of Select Nautical Tales and Sea Sketches, with a Full Narrative of the Mutiny of the Somers.* New York: Winchester, 1843.

Dunbar, Paul Laurence. *The Complete Poems of Paul Laurence Dunbar.* New York: Dodd, 1913.

Duncan, Archibald. *The Mariner's Chronicle, Being a Collection of the Most Interesting Narratives of Shipwrecks, Fires, Famines, and Other Calamities Incident to the Life of Maritime Enterprise.* 4 vols. London: Cundee, 1804.

Dunham, Jacob. *Journal of Voyages.* New York, 1850.

Dunlap, William. *Yankee Chronology.* New York, 1812.

Durand, James. *The Life and Adventures of James R. Durand . . . 1801 to 1816.* Rochester, NY: E. Peck, 1820. Rpt. as *James Durand, an Able Seaman of 1812.* Ed. George S. Brooks. New Haven: Yale UP, 1926.

Dutton, Fred W. *Life on the Great Lakes. Inland Seas* Fall 1981–Spring 1984.

Eberhart, Richard. *Collected Poems, 1930–1976.* New York: Oxford, 1976.

Eliot, T. S. *Collected Poems, 1909–1935.* New York: Harcourt, 1936.

———. *The Four Quartets.* New York: Harcourt, 1943.

Elivas, Knarf [Frank Mackenzie Savile]. *John Ship, Mariner; or, By Dint of Valor.* New York: Stokes, 1898.

Ellis, William. *Polynesian Researches.* 4 vols. New York: Harper, 1833.

Ellison, James. *The American Captive.* Boston: Belcher, 1812.

Ellms, Charles. *Shipwrecks and Disasters at Sea.* Boston: Dickinson, 1836.

Ely, Ben-Ezra Stiles. *"There She Blows": A Narrative of a Whaling Voyage in the Indian and South Atlantic Oceans.* Philadelphia: Simon, 1849. Ed. Curtis Dahl. Middletown, CT: Wesleyan UP, 1971.

Emberg, Ralph. "Out of the Trough." *Phantom Caravel.* Boston: Humphries, 1948.

Emerson, Ralph Waldo. *Poems by Ralph Waldo Emerson.* Boston: Houghton, 1885. New York: Houghton, 1904. Vol. 9 of *The Complete Works of Ralph Waldo Emerson.* Ed. E. W. Emerson. 12 vols. 1903–4.

Engelmann, George J. *The American Girl of Today: Modern Education and Functional Health.* Washington, D.C., 1900.

Equiano, Olaudah. *The Interesting Narrative of the Life of Olaudah Equiano; or, Gustavus Vassa, the African, Written by Himself.* London, 1789. Boston: Knapp, 1837. Rpt. in *Great Slave Narratives.* Ed. Arna Bontemps. Boston: Beacon, 1969.

Erskine, Charles. *Twenty Years Before the Mast [incl.] . . . Under the Command of the Late Admiral Charles Wilkes.* Boston: Erskine, 1890. Washington: Smithsonian, 1985.

Evans, Robley D. *A Sailor's Log: Recollections of Forty Years of Naval Life.* New York: Appleton, 1901.

"An Execution at Sea." *Knickerbocker* Mar. 1836: 285–88.

Fanning, Edmund. *Voyages Round the World.* New York: Collins and Hanna, 1833.

———. *Voyages to the South Seas, . . . Between the Years 1830–1837.* New York, 1838. Rpt. as *Voyages and Discoveries in the South Seas.* Salem: Maine Research Society, 1924. New York: Dover, 1989.

Fanning, Nathaniel. *Narrative of the Adventures of an American Navy Officer. . . .* New York, 1806.

"The Farewell at Sea." *Boston Pearl and Literary Gazette* 1 Aug. 1835: 373–75.

Fillmore, John. *A Narrative of the Singular Sufferings of John Fillmore and Others on Board the Noted Pirate Ship. . . .* Aurora, NY: Clapp, 1837. Buffalo: Buffalo Historical Society, 1907. Publications, vol. 10.

Fitzgerald, F. Scott. *The Great Gatsby.* New York: Scribner's, 1925.

———. *Tender is the Night.* New York: Scribner's, 1934.

Flavel, John. *Navigation Spiritualized; or, A New Compass for Sea-Men.* Boston: Boone, 1726.

Forester, Frank [Maturin Murray Ballou]. *Albert Simmons; or, The Midshipman's Revenge.* Boston: Gleason, 1845.

Fraser, Eliza Anne. *Narrative of the Capture, Sufferings, and Miraculous Escape of Mrs. Eliza Fraser.* New York: Webb, 1837.

Fraser, Margaret. "Journal on Board of Ship *Sea Witch,* 1852–53." Peabody Museum, Salem, MA.

"The Freebooter." *Parlour Journal* 2 (1834): 113–15.

Freneau, Philip. *Poems of Freneau.* Ed. Harry Hayden Clark. New York: Hafner, 1960.

———. *The Poems of Philip Freneau.* Ed. F. L. Pattee. 3 vols. Princeton, 1902–7.

Frost, John. *Pictorial History of the American Navy.* New York: Leavitt & Allen, 1845.

Frost, Robert. *Complete Poems of Robert Frost.* New York: Holt, 1949.

Fuller, Iola. *The Gilded Torch.* New York: Putnam, 1957.

Gallico, Paul. *Beyond the Poseidon Adventure.* New York: Delacorte, 1978.

———. *The Poseidon Adventure.* New York: Coward-McCann, 1969.

Gann, Ernest K. *Twilight for the Gods.* New York: Sloane, 1956.

Gardner, John. *The King's Indian.* New York: Knopf, 1974.

Garnett, Captain Mayn Clew [Thornton Jenkins Hains]. *The White Ghost of Disaster: The Chief Mate's Yarn.* New York: Dillingham, 1912.

Garth, David. *Fire on the Wind.* New York: Putnam, 1957.

Gillette, Virginia M., and Josephine Wunsch [J. Sloan McLean]. *The Aerie.* Los Angeles: Nash, 1974.

Godfrey, William C. *Narrative of the Last Grinnel Arctic Exploring Expedition in Search of Sir John Franklin, 1853–55.* Philadelphia: Lloyd, 1857.

Godman, Stuart Adair. *The Ocean-Born: A Tale of the Southern Seas.* New York: Bunce, [1852].

Goff, George Paul. *Johnny Quickstep's Whaling Voyage.* San Francisco, 1894.

Goodrich, Frank B. *Man upon the Sea; or, A History of Maritime Adventure, Exploration and Discovery, from Earliest Ages to the Present Time. . . .* Philadelphia: Lippincott, 1858.

Goodrich, Marcus. *Delilah.* New York: Holt, 1941.

Gosnell, Harpur Allen, ed. *Before the Mast in the Clippers: The Diaries of Charles A. Abbey, 1856 to 1860.* New York: Derrydale, 1937. Dover, 1989.

Gould, John W. *See also* Clew Garnett [pseud.].

———. *Forecastle Yarns.* Baltimore: Taylor, 1845.

———. *John W. Gould's Private Journal, of a Voyage from New York to Rio de Janeiro.* New York, 1838.

Gould, Lois. *A Sea-Change.* New York: Simon, 1976.

Graham, Robin Lee, with Derek L. T. Gill. *Dove.* New York: Harper, 1972.

Grandy, Moses. *Narrative of the Life of Moses Grandy; Late a Slave in the United States of America.* Boston: Johnson, 1844. Rpt. in *Five Slave Narratives: A Compendium.* New York: Arno, 1968.

Greely, Adolphus W. *Three Years of Arctic Service: An Account of the Lady Franklin Bay Expedition of 1881–84 and the Attainment of the Farthest North.* 2 vols. New York: Scribner's, 1886.

Green, Albert, ed. *Recollections of the Jersey Prison Ship from the Manuscript of Captain Thomas Dring.* Providence: Brown, 1829. New York: Corinth, 1961.

Gregory, Hugh McCulloch. *The Sea Serpent Journal: Hugh McCulloch Gregory's Voyage Around the World in a Clipper Ship, 1854–55.* Ed. Robert H. Burgess. Charlottesville: UP of Virginia, 1975.

Griswold, Rufus, ed. *The Female Poets of America.* 2d ed. Philadelphia: Baird, 1853.

Gunter, Archibald Clavering. *See* Warneford, Lieutenant [pseud.].

Habersham, A. W. *The North Pacific Surveying and Exploring Expedition; or, My Last Cruise.* Philadelphia: Lippincott, 1857.

Hains, Thornton Jenkins. *See also* Garnett, Captain Mayn Clew [pseud.].

——— . *The Black Barque.* Boston: Page, 1905.

——— . *The Strife of the Sea.* New York: Baker and Taylor, 1903.

——— . *Tales of the South Seas.* Portland: Brown Thurston, 1894.

——— . *The Voyage of the Arrow.* Boston: Page, 1906.

——— . *The Wreck of the Conemaugh.* Philadelphia: Lippincott, 1900.

Hale, Sarah Josepha. *Harry Guy, the Widow's Son. A Story of the Sea.* Boston: Mussey, 1848.

Haley, Alex. *Roots: The Saga of an American Family.* Garden City, NY: Doubleday, 1976.

Haley, Nelson. *Whale Hunt: The Narrative of a Voyage by Nelson Cole Haley, Harpooner in the Ship Charles W. Morgan, 1849–1853.* New York: Ives Washburn, 1948. Mystic: Mystic Seaport Museum, 1990.

Hall, F. H. *In the Lamb-White Days.* New York: Bobbs-Merrill, 1975.

Hall, James Norman. *The Tale of a Shipwreck.* Boston: Houghton, 1934.

Halleck, Fitz-Greene. *The Poetical Works of Fitz-Greene Halleck.* New York: Appleton, 1847.

Hallet, Richard Matthews. *The Lady Aft.* Boston: Small, Maynard, 1915.

——— . *The Rolling World.* Boston: Houghton, 1938.

——— . *Trial by Fire: A Tale of the Great Lakes.* Boston: Small, Maynard, 1916.

Halliard, Jack [pseud.]. *Voyages and Adventures of Jack Halliard with Captain Morrell.* Boston: Russell, Odion, 1833.

Halyard, Harry [pseud.]. *The Doom of the Dolphin; or, The Sorceress of the Sea.* Boston: Gleason, 1848.

——— . *The Ocean Monarch; or, The Ranger of the Gulf.* Boston: Gleason, 1848.

——— . *Wharton the Whale Killer! or, The Pride of the Pacific.* Boston: Gleason, 1848.

Hammon, Briton. *A Narrative of the Uncommon Sufferings, and Surprizing Deliverance of Briton Hammon, a Negro Man.* Boston, 1760. Vol. 8 of *The Garland Library of Narratives of North American Indian Captivities.* New York: Garland, 1978.

Harlow, Frederick Pease. *The Making of a Sailor; or, Sea Life Aboard a Yankee Square-Rigger.* Salem, MA: Marine Research Society, 1928. New York: Dover, 1988.

Harper, Michael. *Images of Kin: New and Selected Poems.* Urbana: U of Illinois P, 1977.

Harper, Robert S. *Trumpet in the Wilderness.* New York: Mill, 1940.

Harrison, David. *The Melancholy Narrative of the Distressfull Voyage . . . of the Sloop Peggy.* London, 1766.

Harry Harpoon; or, The Whaleman's Yarn. New York, n.d.

Hart, Joseph C. *Miriam Coffin; or, The Whale-fishermen.* New York: Carvill, 1834.

Hass, Robert. *Field Guide.* New Haven: Yale UP, 1973.

Haven, Gilbert, and Thomas Russell. *Father Taylor, the Sailor Preacher.* . . . Boston: Hunt & Eaton, 1871.

Havighurst, Walter. *Signature of Time.* New York: Macmillan, 1949.

Hawes, William P. *Sporting Scenes and Sundry Sketches.* New York: Gould, Banks, 1842.

Hawkes, John. *Adventures in the Alaskan Skin Trade.* New York: Simon, 1985.

———. *Second Skin.* New York: New Directions, 1964.

Hawkins, Christopher. *The Adventures of Christopher Hawkins.* New York: Bushnell, 1864. New York: New York Times, 1968.

Hawks, Francis L. *Narrative of the Expedition of an American Squadron to the China Seas and Japan, Performed in the Years 1852, 1853 and 1854 Under the Command of Commodore Matthew Calbraith Perry, United States Navy.* 3 vols. Washington: Nicholson, 1856.

Hawthorne, Nathaniel. *The Scarlet Letter.* Boston: Ticknor, Reed & Fields, 1850. Columbus: Ohio State UP, 1963. Vol. 1 of *The Centenary Edition of the Works of Nathaniel Hawthorne, 1804–1864.* Ed. Fredson Bowers. 20 vols.

———. *Twice-Told Tales.* Boston: American Stationers Co., 1837. Columbus: Ohio State UP, 1974. Vol. 9 of *The Centenary Edition of the Works of Nathaniel Hawthorne, 1804–1864.* Ed. Fredson Bowers. 20 vols.

Hawthorne, Nathaniel, ed. *Journal of an African Cruiser,* by Horatio Bridge. New York: Wiley and Putnam, 1845.

———. "Papers of an Old Dartmoor Prisoner, by Benjamin F. Browne." *The United States Magazine and Democratic Review* Jan.–Sept. 1846: 31–39, 97–111, 200–212, 360–68, 457–65. Rpt. as *The Yarn of a Yankee Privateer.* New York: Funk and Wagnalls, 1926.

Hayden, Robert. *Angle of Ascent: New and Selected Poems.* New York: Liveright, 1975.

Hayden, Sterling. *Voyage: A Novel of 1896.* New York: Putnam, 1976.

———. *Wanderer.* New York: Knopf, 1963.

Hazard, Sarah Congdon. "Around the Horn: Journal of the Captain's Wife." *Newport History* 38 (1965): 131–49.

Hazel, Harry [Justin Jones]. *The Doomed Ship; or, The Wreck of the Arctic Regions.* Philadelphia: Peterson, 1864.

———. *The Flying Yankee; or, The Cruise of the Clippers.* New York: Long, 1853.

———. *Fourpe Tap; or, The Middy of the Macedonian.* Boston: Jones, 1847.

———. *Red King, the Corsair Chieftain.* New York: Long, [1850].

———. *The Smuggler King! or, The Rovers of the Antilles.* Boston, 1855.

Hazen, Jacob. *Five Years Before the Mast; or, Life in the Forecastle Aboard of a Whaler and Man-of-War.* Philadelphia: Hazard, 1854.

Hazzard, Samuel. "Extracts from a Sea Book." *The Legendary.* Ed. Nathaniel Parker Willis. Boston: Goodrich, 1828. 146–81.

———. "A Mystery at Sea." *American Monthly Magazine* 1 (1829): 303–12.

H.D. *See* D[oolittle], H[ilda].

Hearn, Lafcadio. *Chita: A Memory of Last Island.* New York: Harper, 1889.

——— . " 'Ti Canotié." *Two Years in the French West Indies.* Vol. 1. Boston: Houghton, 1890.

Heggen, Thomas. *Mr. Roberts.* Boston: Houghton, 1946.

Heller, Joseph. *Catch-22.* New York: Simon, 1961.

Helprin, Mark. *Ellis Island & Other Stories.* New York: Delacorte, 1981.

Hemingway, Ernest. *Across the River and into the Trees.* New York: Scribner's, 1950.

——— . *By-Line Ernest Hemingway: Selected Articles and Dispatches of Four Decades.* Ed. William White. New York: Scribner's, 1967.

——— . *Ernest Hemingway: Selected Letters, 1917–1961.* New York: Granada, 1981.

——— . *The Garden of Eden.* New York: Scribner's, 1986.

——— . "The Great Blue River." *Holiday* July 1949: 60+. Rpt. in *By-Line Ernest Hemingway: Selected Articles and Dispatches of Four Decades.* Ed. William White. New York: Scribner's, 1967. 403–16.

——— . *Green Hills of Africa.* New York: Scribner's, 1935.

——— . *Islands in the Stream.* New York: Scribner's, 1970.

——— . "Oily Weather." *Poetry: A Magazine of Verse* 21 (1923): 193.

——— . *The Old Man and the Sea.* New York: Scribner's, 1952.

——— . "On the Blue Water: A Gulf Stream Letter." *Esquire* Apr. 1936: 31+. Rpt. in *By-Line Ernest Hemingway: Selected Articles and Dispatches of Four Decades.* Ed. William White. New York: Scribner's, 1967. 236–44.

——— . "Out in the Stream: A Cuban Letter." *Esquire* Aug. 1934: 19+. Rpt. in *By-Line Ernest Hemingway: Selected Articles and Dispatches of Four Decades.* Ed. William White. New York: Scribner's, 1967. 172–78.

——— . *The Sun Also Rises.* New York: Scribner's, 1926.

——— . *To Have and Have Not.* New York: Scribner's, 1937.

Henry, George. *Life of George Henry: Together with a Brief History of the Colored People in America.* Henry, 1894. Freeport, NY: Books for Libraries P, 1971.

Herbert, Charles. *A Relic of the Revolution. . . .* Boston: Peirce, 1847.

Herbert, Frank. *The Dragon in the Sea.* Garden City, NY: Doubleday, 1956.

Herbert, Frank, and Bill Ransom. *The Jesus Incident.* New York: Berkeley, 1979.

——— . *The Lazarus Effect.* New York: Putnam, 1983.

Herbert, Henry William. *Ringwood the Rover: A Tale of Florida.* Philadelphia: Graham, 1843.

——— . *Tales of the Spanish Seas.* New York: Burgess, Stringer, 1847.

Hernton, Calvin. *Scarecrow.* Garden City, NY: Doubleday, 1974.

Hersey, John. *Under the Eye of the Storm.* New York: Knopf, 1967.

Hewitt, Mary E. *The Songs of Our Land and Other Poems.* Boston: Ticknor, 1846.

Higginson, Thomas Wentworth. See *Thalatta.*

Hine, E. Curtis. *Orlando Melville; or, The Victims of the Press-Gang, a Tale of the Sea.* Boston: Gleason, 1848.

The History of Constantius and Pulchera. 3d ed. Norwich, CT, 1794.

Hoffman, Daniel G. *An Armada of Thirty Whales*. New Haven: Yale UP, 1954.

———. *The City of Satisfactions*. New York: Oxford UP, 1963.

———. *The Little Geste and Other Poems*. New York: Oxford UP, 1960.

Holbrook, Samuel F. *Threescore Years: An Autobiography, Containing Incidents of Voyages and Travels, Including Six Years in a Man-of-War*. Boston: French, 1857.

Holcomb, Eliah. *A Wonderful Providence in Many Incidents at Sea*. Buffalo, 1849.

Holden, Horace. *A Narrative of the Shipwreck, Captivity and Sufferings of Horace Holden and Benj. H. Nute*. . . . Boston: Russell, Shattuck, 1836. Fairfield, WA: Galleon, 1975.

Holmes, Oliver Wendell. *The Poetical Works of Oliver Wendell Holmes*. London: Routledge, 1852. Boston: Houghton, 1975.

A Home on the Deep; or, The Mariner's Trials on the Dark Blue Sea. By a Son of the Ocean. Boston, 1857.

Hough, Henry Beetle. *Long Anchorage*. New York: Appleton, 1947.

Hovey, Richard, and Bliss Carmen. *More Songs from Vagabondia*. Boston: Copeland & Day, 1896.

Howe, Henry, ed. *Life and Death on the Ocean: A Collection of Extraordinary Adventures*. . . . Cincinnati: Howe, 1855.

Howells, William Dean. *The Rise of Silas Lapham*. Boston: Ticknor, 1885. Bloomington: Indiana UP, 1971. Vol. 12 of *A Selected Edition of William Dean Howells*. Ed. Walter J. Meserve and David N. Nordloh. 32 vols. 1968–83.

———. *A Sea Change; or, Love's Stowaway*. Boston: Ticknor, 1888. *The Complete Plays of W. D. Howells*. Ed. Walter J. Meserve. New York: New York UP, 1960. 269–99.

———. *Venetian Life*. New York: Hurd & Houghton, 1866.

Howes, Ebenezer. *Some Account of the Proceedings on Board Ship Niagara of Boston, Commanded by One Ebenezer Howes Jr. During Her Passage From Liverpool to Boston in April and May, 1836, with Remarks upon the Rights of Seamen*. Boston: Hart, 1836.

Howland, S. A. *Steamboat Disasters and Railroad Accidents in the United States*. . . . Worcester: Dorr, Howland, 1840.

Hoxse, John. *The Yankee Tar*. Northampton: Metcalf, 1848.

Hubbard, John. *A Monumental Gratitude Attempted, in a Poetical Relation of the Danger and Deliverance of Several of the Members of Yale-College in Passing the Sound*. New London, 1727.

[Huet, M. M.]. *Davis the Pirate; or, The True History of the Freebooters of the Pacific*. New York: Long, [1853].

———. *Morgan the Buccaneer; or, The True History of the Freebooters of the Antilles*. New York: Long, [1853].

Hughes, Langston. *The Big Sea*. New York: Knopf, 1940. New York: Hill, 1963.

———. *I Wonder as I Wander: An Autobiographical Journey*. New York: Rinehart, 1956. New York: Hill, 1964.

———. *Selected Poems*. New York: Knopf, 1959. New York: Vintage, 1974.

Hunt, Levi. *A Voice from the Forecastle of a Whale Ship*. Buffalo: Reese, 1848.

Huntress, Keith. *Narratives of Shipwrecks and Disasters*. Ames: Iowa State UP, 1974.

Hurston, Zora Neale. *Their Eyes Were Watching God*. Philadelphia: Lippincott, 1937.

Ichabod [pseud.]. "The Ice Ship." *Bower of Taste* 1 (1828): 257–60.

Ingalls, Rachel. *Mrs. Caliban.* London: Faber, 1983.

Ingraham, Joseph Holt. *The Cruiser of the Mist.* New York: Burgess, Stringer, 1845.

———. *Freemantle; or, The Privateersman: A Nautical Romance of the Last War.* Boston: Redding, 1845.

———. *Lafitte: The Pirate of the Gulf.* New York: Harper, 1836.

———. *The Spectre Steamer and Other Tales.* Boston: United States Publishing Co., 1846.

———. *Winwood; or, The Fugitive of the Seas.* New York: Williams, 1846.

Ireland, Howard. *A Green Mariner: A Landsman's Account of a Deep-Sea Voyage.* Philadelphia: Lippincott, 1900.

Iron, N. C. *The Double Hero: A Tale of Sea and Land During the War of 1812.* New York: Beadle and Company, [1861].

Irving, Washington. *Astoria; or, Anecdotes of an Enterprise Beyond the Rocky Mountains.* Philadelphia: Carey, Lea, & Blanchard, 1836. Ed. Richard Rust. Boston: Twayne, 1976.

———. *The Life and Voyages of Christopher Columbus.* London and New York, 1828. Ed. John Harmon McElroy. Boston: Twayne, 1981.

———. *Miscellaneous Writings II.* Ed. Wayne Kime. Boston: Twayne, 1981.

———. *The Sketch Book of Geoffrey Crayon, Gent.* New York and London, 1819–20. Ed. Haskell Springer. Boston: Twayne, 1978.

———. *Voyages and Discoveries of the Companions of Columbus.* Philadelphia: Carey and Lea, 1831. Ed. James W. Tuttleton. Boston: Twayne, 1986.

———. *Wolfert's Roost.* New York: Putnam, 1855. Ed. Roberta Rosenberg. Boston: Twayne, 1979.

Isaacs, Nicholas. *Twenty Years Before the Mast.* New York: Beckwith, 1845.

Jacobs, Harriet. *Incidents in the Life of a Slave Girl. Written by Herself.* Boston, 1861. Cambridge: Harvard UP, 1987.

Jacobs, Thomas Morrell. *Scenes, Incidents, and Adventures in the Pacific Ocean . . . During the Cruise of the Clipper Margaret Oakley Under Captain Benjamin Morrell.* New York: Harper, 1844.

James, Henry. "The Aspern Papers." *The Complete Tales of Henry James.* Ed. Leon Edel. Philadelphia: Lippincott, 1962.

———. *The Complete Tales of Henry James.* Ed. and intro. Leon Edel. 12 vols. Philadelphia: Lippincott, 1961–64.

———. *Hawthorne.* London: Macmillan, 1879.

———. "The Patagonia." *The Complete Tales of Henry James.* Ed. Leon Edel. Philadelphia: Lippincott, 1962.

———. *The Portrait of a Lady.* Boston: Houghton, 1881.

———. *What Maisie Knew.* New York: Stone, 1897.

———. *The Wings of the Dove.* New York: Scribners, 1902.

Janeway, James. *Mr. James Janeway's Legacy to His Friends.* London, 1674.

Janvier, Thomas Allibone. *In Great Waters.* New York: Harper, 1901.

————. *In the Sargasso Sea, a Novel.* New York: Harper, 1898.

Jarman, Robert. *Journal of a Voyage to the South Seas, in the "Japan," Employed in the Sperm Whale Fishery.* London, 1848.

Jeffers, Robinson. *The Double Ax.* New York: Random, 1948.

————. *The Selected Poetry of Robinson Jeffers.* New York: Random, 1938.

Jenkins, Thomas H. *Bark Kathleen Sunk by a Whale, as Related by the Captain.* New Bedford, MA: Hutchinson, 1902.

Jennings, John E. *See also* Baldwin, Bates [pseud.].

————. *River to the West.* Garden City, N.Y.: Doubleday, 1948.

————. *The Salem Frigate, A Novel.* Garden City, N.Y.: Doubleday, 1946.

————. *The Sea Eagles: A Story of the American Navy During the Revolution.* Garden City, NY: Doubleday, 1950.

————. *The Tall Ships.* New York: McGraw-Hill, 1959.

Jewett, Sarah Orne. *A Country Doctor.* Boston: Houghton, 1884.

————. *The Country of the Pointed Firs.* Boston: Houghton, 1896.

————. *Deephaven.* Boston: Osgood, 1877.

————. *A Marsh Island.* Boston: Houghton, 1885.

————. *The Tory Lover.* Boston: Houghton, 1901.

Joffe, Natalie F. *Fox Texts.* Leiden, Germany: Brill, 1907.

Johnson, Charles. *Middle Passage.* New York: Atheneum, 1990.

Johnson, Edward. *Wonder-Working Providence of Sion's Savior in New England.* London, 1654.

Johnson, James Weldon, and J. Rosamond Johnson. *The Books of American Negro Spirituals.* New York: Viking, 1964.

Jones, John Paul. *See* Sands, Robert C., ed.

Jones, Justin. *See* Hazel, Harry [pseud.]; Merry, Captain [pseud.].

Jones, Randolph. *The Buccaneers: A Historical Novel of the Times of William III and Louis XIV.* New York: Authors' Publishing Co., 1878.

Jones, Thomas H. *The Experience of Thomas H. Jones, Who Was a Slave for Forty-Three Years, Written by a Friend as Given to Him by Brother Jones.* Worcester: Howland, [1849].

Jones, William. *Ojibway Texts.* New York: Stechert, 1917–19.

Jordan, Tristram. *Voyages.* Ed. Alfred T. Hill. New York: McKay, 1977.

Josselyn, John. *An Account of Two Voyages.* London, 1674.

Jourdain, Silvester. *A Plaine Description of the Barmudas.* London, 1610.

Judah, Samuel B. *The Buccaneers; A Romance of Our Country, in Its Ancient Day.* New York, 1827.

Judson, Edward Zane Carroll. See Buntline, Ned.

Kane, Elisha Kent. *Arctic Explorations: The Second Grinnell Expedition in Search of Sir John Franklin, 1853, '54, '55.* 2 vols. Philadelphia: Childs, 1856.

————. *The United States Expedition in Search of Sir John Franklin.* New York: Harper, 1854.

Keeler, John. *The South Sea Islanders, with a Short Sketch of Captain Morrell's Voyage . . . by John Keeler, One of the Crew.* New York: Snowden, 1831.

Kendall, Edmund Hale. *See* Sampson, Abel.

Kenney, Susan. *Sailing.* New York: Viking, 1988.

Kesey, Ken. *One Flew Over the Cuckoo's Nest.* New York: Viking, 1962.

King, Charles, ed. *By Land and Sea.* Philadelphia: Hamersly, 1891.

Kintziger, Louis J. *Bay Mild.* Milwaukee: Bruce, 1945.

Kipling, Rudyard. "Letters to the Family." *Letters of Travel.* Vol. 19 of *Collected Works of Rudyard Kipling.* New York: AMS, 1970.

Klemmer, Harvey. *Harbor Nights.* New York: Sun Dial P, 1937.

Knight, John B. *A Journal of a Voyage in the Brig Spy of Salem, 1832–34.* Salem: Marine Research Publications, 1925.

Knoblock, Curt G. *Above Below.* Hancock, MI: Book Concern, 1952.

Kotzebue, Otto von. *Voyage of Discovery in the South Seas.* London, 1821.

LaFarge, Oliver. *Long Pennant.* Boston: Houghton, 1933.

Lambkin, Amelia MacMillan. *Buckskin and Ermine.* Cedar Rapids, IA: Torch P, 1928.

Lane, Carl Daniel. *The Fleet in the Forest.* New York: Coward, 1943.

Lane, Lunsford. *The Narrative of Lunsford Lane, Formerly of Raleigh, N.C.* Boston: Lane, 1842. Rpt. in *Five Slave Narratives: A Compendium.* New York: Arno, 1968.

Langman, Christopher. *A True Account of the Voyage of the Nottingham-Galley.* London, 1711.

Lanier, Sidney. "The Marshes of Glynn." In *A Masque of Poets.* Ed. George Parsons Lathrop. Boston: Roberts, 1878. 88–94.

Larcom, Lucy. *A New England Girlhood: Outlined from Memory.* Boston: Houghton, 1889.

Lawrence, Mary Chipman. *The Captain's Best Mate: The Journal of Mary Chipman Lawrence on the Whaler Addison.* Ed. Stanton Garner. Providence: Brown UP, 1966.

Lay, William, and Cyrus W. Hussey. *A Narrative of the Mutiny, on Board the Ship Globe, of Nantucket in the Pacific Ocean, Jan. 1824.* New London: Lay, 1828.

Lazarus, Emma. "The New Colossus." In *Catalogue of the Pedestal Fund Art Loan Exhibition, at the National Academy of Design.* New York: DeVinne, Dec. 1883. 9.

Leader, Mary. *Triad.* New York: Coward, 1973.

Ledyard, John. *A Journal of Captain Cook's Last Voyage.* Hartford, CT, 1783. Chicago: Quadrangle, 1963.

Leech, Samuel. *Thirty Years from Home; or, A Voice from the Main Deck.* Boston: Tappan and Dennett, 1843.

Leggett, William. *Naval Stories.* New York: Carvill, 1834.

——— . *Tales and Sketches by a Country Schoolmaster.* New York: Harper, 1829.

Le Guin, Ursula K. *The Farthest Shore.* New York: Atheneum, 1972.

——— . *Searoad: Chronicles of Klatsand.* New York: Harper Collins, 1991.

——— . *Tehanu: The Last Book of Earthsea.* New York: Atheneum, 1990.

——— . *The Tombs of Atuan.* New York: Atheneum, 1971.

———. *The Wind's Twelve Quarters.* New York: Harper, 1975.

———. *A Wizard of Earthsea.* Emeryville, CA: Parnassus, 1968.

Leland, Charles G. *The Algonquin Legends of New England.* Boston: Houghton, 1884.

Levertov, Denise. *Collected Earlier Poems: 1940–1960.* New York: New Directions, 1979.

———. *The Jacob's Ladder.* New York: New Directions, 1961.

———. *Relearning the Alphabet.* New York: New Directions, 1970.

"Life at Sea: Hunting a Devil Fish." *Military and Naval Magazine* Jan. 1836: 364–66.

Life in a Whale Ship; or, The Sports and Adventures of a Leading Oarsman. Written by an American Author and Based upon the Cruise of an American Whale Ship in the South Atlantic and Indian Ocean, During the Years 1836–7–8. Boston: Bradley, 1841?

Life on Board a Man-of-War. . . . Glasgow: Blackie, 1829.

Lindbergh, Anne Morrow. *Gift from the Sea.* New York: Pantheon, 1955.

Lindsley, A. B. *Love and Friendship; or, Yankee Notions.* New York: Longworth, 1809.

Little, George. *Life on the Ocean; or, Twenty Years at Sea.* Baltimore: Armstrong and Berry, 1843.

Lodge, George Cabot. *Song of the Wave and Other Poems.* New York: Scribner, 1898.

The Log of the Bon Homme Richard. Intro. Louis Middlebrook. Mystic, CT: Marine Historical Association, 1936.

London, Jack. *The Cruise of the Dazzler.* New York: Century, 1902.

———. *Great Short Works of Jack London.* Ed. Earle Labor. New York: Harper, 1965.

———. *The Human Drift.* New York: Macmillan, 1917.

———. *John Barleycorn.* New York: Grosset and Dunlap, 1913. London: Bodley Head, 1964. Vol. 2 of *The Bodley Head Jack London.* Ed. Arthur Calder-Marshall. 4 vols. 1964–66.

———. *Martin Eden.* New York: Grosset and Dunlap, 1906. London: Bodley Head, 1965. Vol. 3 of *The Bodley Head Jack London.* Ed. Arthur Calder-Marshall. 4 vols. 1964–66.

———. *The Mutiny of the Elsinore.* New York: Macmillan, 1914.

———. *On the Makaloa Mat.* New York: Macmillan, 1919.

———. *The Sea-Wolf.* New York: Macmillan, 1904. Ed. Matthew J. Bruccoli. Boston: Houghton, 1964.

Longfellow, Henry Wadsworth. *The Complete Poetical Works of Longfellow.* Boston: Osgood, 1876. Boston: Houghton, 1922.

———. *Tales of a Wayside Inn.* Boston: Ticknor & Fields, 1863.

Longfellow, Samuel. See *Thalatta.*

"Loss of the Ship Tonquin, Near the Mouth of the Columbia River." *National Intelligencer,* Washington, DC, June 22, 1813.

"The Lost Fisherman." *Burton's Gentleman's Magazine* Nov. 1838: 340–47.

"The Lost Sailor." *Military and Naval Magazine* Jan. 1836: 366–67.

Low, Gorham P. *The Sea Made Men.* New York: Revell, 1937.

Lowell, James Russell. *The Complete Poetical Works of James Russell Lowell.* Ed. Horace E. Scudder. Boston: Houghton, 1896.

Lowell, Robert. *For Lizzie and Harriet.* New York: Farrar, 1973.

———. *Poems: 1938–1949.* London: Faber, 1950.

Lynch, W. F. *Narrative of the United States' Expedition to the River Jordan and the Dead Sea.* Philadelphia: Lea, 1849.

———. *Naval Life; or, Observations Afloat and on Shore. The Midshipman.* New York: Scribner, 1851.

Mabie, Louise. *The Lights Are Bright.* New York: Harper, 1914.

McCormick, Jay. *November Storm.* New York: Doubleday, 1943.

MacDonald, John D. *Condominium.* Philadelphia: Lippincott, 1977.

———. *The Deep Blue Goodbye.* Greenwich, CT: Fawcett, 1964.

———. *A Flash of Green.* New York: Simon, 1962.

McFee, William. *Aliens.* London: Arnold, 1914. Garden City, NY: Doubleday, Page, 1918.

———. *Derelicts.* New York: Book League of America, 1938.

MacHarg, William, and Edwin Balmer. *The Indian Drum.* New York: Grosset, 1917.

McKay, Claude. *Banjo.* New York: Harper, 1929.

———. *Home to Harlem.* New York: Harper, 1928.

———. *A Long Way from Home.* New York: Furman, 1937. New York: Harcourt, 1970.

———. *Selected Poems of Claude McKay.* New York: Harcourt, 1953.

McKenna, Richard. *The Left-Handed Monkey Wrench.* Ed. Robert Shenk. Annapolis: Naval Institute P, 1984.

———. *The Sand Pebbles.* New York: Harper, 1962.

Mackey, Mary. *The Dear Dance of Eros.* Seattle: Fjord, 1987.

McLean, J. Sloan. *See* Gillette, Virginia M.

McNally, William. *Evils and Abuses in the Naval and Merchant Services Exposed.* Boston: Cassady, 1839.

McPhee, John. *Looking for a Ship.* New York: Farrar Straus Giroux, 1990.

McRoberts, Walter. *Rounding Cape Horn, and Other Sea Stories.* Peoria, IL: Hill, 1895.

Macy, Capt. W. H. *There She Blows! or, The Log of the Arethusa.* Boston: Lee & Shepard, 1877.

Macy, Obed. *The History of Nantucket. . . .* Boston: Hilliard, Gray, 1835.

Maffitt, John Newland. *Nautilus; or, Cruising Under Canvas.* New York: United States Pub. Co., 1871.

Mahan, Alfred T. *The Influence of Sea Power upon History, 1660–1783.* Boston: Little, 1890.

———. *The Interest of America in Sea Power, Present and Future.* Boston: Little, 1898. Englewood Cliffs: Prentice, 1967.

[Maitland, James A.] *The Cabin Boy's Story: A Semi-Nautical Romance.* New York: Garrett, 1854.

———. *The Pirate Doctor; or, The Extraordinary Career of a New-York Physician.* New York: Garrett, [185–?].

Malamud, Bernard. *God's Grace.* New York: Farrar, 1982.

Manry, Robert. *Tinkerbelle.* New York: Harper, 1965.

Maril, Lee. *Cannibals and Condos.* College Station, TX: Texas A&M UP, 1986.

The Mariner's Chronicle: Containing Narratives of the Most Remarkable Disasters at Sea. . . . New Haven: Durrie and Peck, 1834.

The Mariner's Chronicle of Shipwrecks, Fires, Famines and Other Disasters at Sea. Philadelphia: Harding, 1849.

The Mariner's Library, or Voyager's Companion. . . . Boston, 1833.

Marshall, Paule. *Brown Girl, Brownstones.* New York: Random, 1959.

——— . *Praisesong for the Widow.* New York: Dutton, 1983.

——— . *Reena and Other Stories.* Old Westbury, NY: Feminist P, 1983.

Martingale, Hawser [John Sherburne Sleeper]. *Jack in the Forecastle; or, Incidents in the Early Life of Hawser Martingale.* Boston: Crosby, Nichols, Lee, 1860.

——— . *Mark Rowland, A Tale of the Sea.* Boston: Loring, 1867.

——— . *Salt Water Bubbles; or, Life on the Wave.* Boston: Reynolds, 1854.

——— . *Tales of the Ocean, and Essays for the Forecastle.* Boston: Dickinson, 1841.

Mason, Arthur. *The Cook and the Captain Bold.* Boston: Atlantic Monthly P, 1924.

——— . *The Flying Bo'sun.* New York: Holt, 1920.

——— . *Ocean Echoes.* New York: Holt, 1922.

Mather, Cotton. *Magnalia Christi Americana; or, The Ecclesiastical History of New England.* 7 vols. London, 1702. Ed. Thomas Robbins. 2 vols. Hartford: Andrus, 1853.

Mather, Increase. *An Essay for the Recording of Illustrious Providences.* Boston, 1684.

Mather, May. "A Story of the Sea." *Our Young Folks* July 1868: 387–89.

Mather, Richard. *Journal. Dorchester Antiquarian and Historical Society Collections,* no. 3 (Boston, 1850).

Mathews, Cornelius. *Behemoth, a Legend of the Mound Builders.* New York: Langley, 1839.

Matthiessen, F. O. *American Renaissance.* New York: Oxford UP, 1941.

Matthiessen, Peter. *Blue Meridian.* New York: Random, 1971.

——— . *Far Tortuga.* New York: Random, 1975.

——— . *Men's Lives: The Surfmen and Baymen of the South Fork.* New York: Random, 1986.

——— . *Race Rock.* New York: Harper, 1954.

——— . *Raditzer.* New York: Viking, 1961.

May, Caroline, ed. *The American Female Poets.* Philadelphia: Lindsay & Blakiston, 1848.

Mayer, Brantz. *Captain Canot; or, Twenty Years of an African Slaver.* New York: Appleton, 1854. New York: Boni, 1928.

Mayo, William S. "The Captain's Story." *United States Magazine and Democratic Review* Apr. 1846: 305–11.

——— . "The Escape of the Atlanta." *Ladies Companion* May 1843: 24–26.

——— . "A Reel Pirate." *United States Magazine and Democratic Review* 22 Mar. 1848: 263–69.

Melville, Herman. *Battle-Pieces and Aspects of the War.* New York: Harper, 1866.

——— . "Benito Cereno." *Putnam's Monthly Magazine* Oct. 1855: 353–67; Nov.: 459–73; Dec.: 633–44. *The Piazza Tales and Other Prose Pieces.* Ed. Harrison Hayford,

Alma A. MacDougall, G. Thomas Tanselle, and others. Evanston: Northwestern-Newberry, 1987. 46–117.

———. *Billy Budd, Sailor.* Ed. Harrison Hayford and Merton M. Sealts, Jr. Chicago: U of Chicago P, 1962.

———. *Clarel: A Poem and Pilgrimage in the Holy Land.* New York: Putnam's, 1876. Ed. Walter E. Bezanson. New York: Hendricks, 1960.

———. *Collected Poems of Herman Melville.* Ed. Howard P. Vincent. Chicago: Packard, 1947.

———. *The Confidence-Man: His Masquerade.* New York: Dix, Edwards, 1857. Ed. Harrison Hayford, Hershel Parker, and G. Thomas Tanselle. Evanston: Northwestern-Newberry, 1984.

———. "The Encantadas, or Enchanted Isles." *Putnam's Monthly Magazine* Mar. 1854: 311–19; Apr.: 345–55; May: 460–66. *The Piazza Tales and Other Prose Pieces.* Ed. Harrison Hayford, Alma A. MacDougall, G. Thomas Tanselle, and others. Evanston: Northwestern-Newberry, 1987. 125–73.

———. "Hawthorne and His Mosses." *New York Literary World* 17 Aug. 1850: 125–27. *The Piazza Tales and Other Prose Pieces.* Ed. Harrison Hayford, Alma A. MacDougall, G. Thomas Tanselle, and others. Evanston: Northwestern-Newberry, 1987.

———. *Israel Potter, His Fifty Years of Exile.* New York: Putnam's, 1855. Ed. Harrison Hayford, Hershel Parker, and G. Thomas Tanselle. Evanston: Northwestern-Newberry, 1982.

———. *John Marr and Other Sailors with Some Sea-Pieces.* New York: The De Vinne Press, 1888.

———. *Journals.* Ed. Howard Horsford and Lynn Horth. Evanston: Northwestern-Newberry, 1988.

———. *The Letters of Herman Melville.* Ed. Merrell R. Davis and William H. Gilman. New Haven: Yale UP, 1960.

———. *Mardi: And a Voyage Thither.* New York: Harper, 1849. Ed. Harrison Hayford, Hershel Parker, and G. Thomas Tanselle. Evanston: Northwestern-Newberry, 1970.

———. *Moby-Dick; or, The Whale.* New York: Harper, 1851. Ed. Harrison Hayford and Hershel Parker. Norton Critical Edition. New York: Norton, 1967.

———. *Omoo: A Narrative of Adventures in the South Seas.* London: Murray, 1847. Ed. Harrison Hayford, Hershel Parker, and G. Thomas Tanselle. Evanston: Northwestern UP, 1968.

———. *Pierre; or, The Ambiguities.* New York: Harper, 1852. Ed. Harrison Hayford, Hershel Parker, and G. Thomas Tanselle. Evanston: Northwestern-Newberry, 1971.

———. *Redburn: His First Voyage.* New York: Harper; London: Bentley, 1849. Ed. Harrison Hayford, Hershel Parker, and G. Thomas Tanselle. Evanston: Northwestern UP, 1969.

———. Rev. of Browne's *Etchings of a Whaling Cruise* and Codman's *Sailor's Life and Sailor's Yarns. New York Literary World* 6 Mar. 1847: 105–6. *The Piazza Tales and Other Prose Pieces*. Ed. Harrison Hayford, Alma A. MacDougall, G. Thomas Tanselle, and others. Evanston: Northwestern-Newberry, 1987. 205–11.

———. Rev. of Cooper's *The Sea-Lions. New York Literary World* 28 Apr. 1849: 370. *The Piazza Tales and Other Prose Pieces*. Ed. Harrison Hayford, Alma A. MacDougall, G. Thomas Tanselle, and others. Evanston: Northwestern-Newberry, 1987. 235–36.

———. *Timoleon*. New York, 1891.

———. "The Town-Ho's Story." *Harper's New Monthly Magazine* Oct. 1851: 658–65.

———. *Typee: A Peep at Polynesian Life*. New York: Wiley & Putnam, 1846. Ed. Harrison Hayford, Hershel Parker, and G. Thomas Tanselle. Evanston: Northwestern-Newberry, 1968.

———. *White-Jacket; or, The World in a Man-of-War*. New York: Harper, 1850. Ed. Harrison Hayford, Hershel Parker, and G. Thomas Tanselle. Evanston: Northwestern-Newberry, 1970.

———. *The Works of Herman Melville*. Standard Edition. 16 vols. London: Constable, 1922–24. New York: Russell and Russell, 1963.

[Mercier, Henry James, and William Gallop]. *Life in a Man-of-War; or, Scenes in "Old Ironsides" . . . By a Foretopman*. Philadelphia: Bailey, 1841. Boston: Houghton, 1927.

Meredith, William. *The Open Sea and Other Poems*. New York: Knopf, 1958.

Merry, Captain [Justin Jones]. *The Flying Dutchman; or, The Wedding Guest of Amsterdam*, New York: Long [c. 1852].

———. *Sweeney Todd; or, The Ruffian Barber, a Tale of the Terrors of the Seas and the Mysteries of the City*. New York: Long [c. 1853].

Merwin, Samuel. *His Little World: The Story of Hunch Badeau*. New York: Barnes, 1903.

———. *The Road to Frontenac*. New York: Collier, 1901.

Merwin, W. S. *The Drunk in the Furnace*. London: Hart-Davis, 1960.

———. *Green with Beasts*. New York: Knopf, 1956.

Michener, James. *Chesapeake*. New York: Random, 1978.

———. *Hawaii*. New York: Random, 1959.

———. *The Watermen*. New York: Random, 1979.

Miller, Joaquin. "Columbus." In *Christopher Columbus*. Ed. John Marcus Dickey. Chicago, 1892. 235–36.

Minturn, Robert B. *From New York to Delhi, by Way of Rio de Janeiro, Australia, and China*. New York: Appleton, 1858.

Mitchell, Donald G. "A Man Overboard." *Southern Literary Messenger* Jan. 1848: 10–11.

Mizzen, Mat [Henry Llewellyn Williams]. *The Black Cruiser; or, The Scourge of the Sea*. New York: DeWitt, [186–?].

———. *The Flying Arrow; or, The Pirate's Revenge*. New York: DeWitt, [185–?].

Montgomery, James Eglinton. *Our Admiral's Flag Abroad. The Cruise of Admiral D. G. Farragut*. New York: Putnam, 1869.

Mooney, Ted. *Easy Travel to Other Planets*. New York: Farrar, 1981.

Moore, Marianne. *The Complete Poems of Marianne Moore*. New York: Macmillan/Viking, 1967.

Morgan, Charles P. *The Phantom Cruiser; or, The Pilot of the Gulf.* Boston: Gleason, 1864.

Morgan, Dodge. *The Voyage of the "American Promise."* Boston: Houghton, 1989.

Morrell, Abby Jane. *Narrative of a Voyage to the Ethiopic and South Atlantic Ocean, Indian Ocean, Chinese Sea, North and South Pacific Oceans, in the Years 1829, 1830, 1831.* New York: Harper, 1833. Upper Saddle River, NJ: Gregg, 1970.

Morrell, Benjamin. *A Narrative of Four Voyages.* New York: Harper, 1832.

Morris, Wright. *What a Way to Go.* New York: Atheneum, 1962.

Morrison, Mary Gray. *The Sea-Farers: A Romance of a New England Coast Town.* New York: Doubleday, 1900.

Morrison, Toni. *Song of Solomon.* New York: Knopf, 1977.

———. *Tar Baby.* New York: Knopf, 1981.

Mortimer, Charles. *Captain Antle, the Sailor's Friend.* Boston: Damrell & Upham, 1898.

[Murphey, Charles]. *Thrilling Whaling Voyage Journal, Containing 220 Stanzas, in Poetry Composed by the 3D Mate on Board Ship Dauphin, of Nantucket.* Mattapoisett: Atlantic, 1877.

Murphy, Robert Cushman. *Logbook for Grace.* New York: Macmillan, 1947.

Murray, Lieut. [Maturin Murray Ballou]. *Fanny Campbell, The Female Pirate Capt. A Tale of the Revolution.* New York: French, 1844.

———. *The Naval Officer; or, The Pirate's Cave, A Tale of the Last War.* New York: Brady, 1845.

———. *Roderick the Rover; or, the Spirit of the Wave.* Boston: Gleason, 1848.

———. *The Sea Witch; or, The African Quadroon. A Story of the Slave Coast.* New York: French, 1855.

Murray, Captain Thomas. "Some Recollections." *Inland Seas* Jan. 1946: 28–32.

Murrell, William Meacham. *Cruise of the Frigate Columbia Around the World, Under the Command of Commodore George C. Read, in 1838, 1839, and 1840.* Boston: Mussey, 1840.

Narrative of Calamitous and Interesting Shipwrecks & c., ... Philadelphia: Carey, 1810.

A Narrative of the Capture of the United States Brig Vixen. New York: Office of "The War," 1813.

"Naval Life." *Military and Naval Magazine* Mar. 1834: 60–66; July: 364–68; Sept.: 5–14.

Nemerov, Howard. *The Collected Poems of Howard Nemerov.* Chicago: U of Chicago P, 1977.

Nevens, William. *Forty Years at Sea.* Portland: Thurston, Fenley, 1846.

Newhall, Charles L. *The Adventures of Jack; or, A Life on the Wave.* Southbridge, MA, 1859.

Nickerson, Thomas. *The Loss of the Ship "Essex" Sunk by a Whale, and the Ordeal of the Crew in Open Boats.* Ed. Helen W. Chase and Edouard A. Stackpole. Nantucket: Nantucket Historical Assoc., 1984.

"A Night at Sea." *Philadelphia Monthly Magazine* 15 Sept. 1828: 338–43.

Noble, Louis Legrand. *After Icebergs with a Painter: A Summer Voyage to Labrador and Around Newfoundland.* New York: Appleton, 1861.

"The Nobleman and the Fisherman." *New-England Magazine* Apr. 1834: 280–89.

[Nordhoff, Charles]. *Life in a Whale Ship; or, The Sport and Adventures of a Leading Oarsman.* Boston: Bradley, [1841?].

———. *Man-of-War Life: A Boy's Experiences in the United States Navy.* Cincinnati: Moore, Wilstach and Keys, 1855.

———. *The Merchant Vessel: A Sailor Boy's Voyages to See the World.* New York: Dodd, Mead, 1855.

———. *Nine Years a Sailor.* Cincinnati: Moore, Wilstach and Keys, 1857.

Nordhoff, Charles. *Life on the Ocean* . . . . Cincinnati: Wilstach, Baldwin, 1874. Rpt. as *In Yankee Windjammers.* New York: Dodd, 1940.

Nordhoff, Charles, and J. N. Hall. *Men Against the Sea.* Boston: Little, 1934.

———. *Mutiny on the Bounty.* Boston: Little, 1932.

———. *Pitcairn's Island.* Boston: Little, 1934.

Norris, Frank. *McTeague.* New York: Doubleday, 1899.

———. *Moran of the Lady Letty.* New York: Doubleday, 1898.

Northup, Solomon. *Twelve Years a Slave.* . . . Auburn, Buffalo, and London: Derby and Miller, 1853.

Norwood, Henry. "A Voyage to Virginia." *Collection of Voyages and Travels.* Ed. Awnsham Churchill. 2d ed. London, 1732. 161–86.

Oates, Joyce Carol. "Ladies and Gentlemen." *Harper's Magazine* Dec. 1990: 54–58.

The Ocean Queen; or, The Seaman's Bride. New York: Dick & Fitzgerald, [185–?].

"The Old Seaman, a Sketch from Nature." *Port Folio* June 1823: 456–59.

Olmsted, Francis Allyn. *Incidents of an Whaling Voyage.* New York: Appleton, 1841.

Olmsted, Gideon. *The Journal of Gideon Olmsted: Adventures of a Sea Captain During the American Revolution.* Ed. Gerald W. Gawalt and Charles W. Kreidler. Washington: Library of Congress, 1978.

Olson, Charles. *Archaeologist of Morning.* New York: Grossman, 1973.

———. *The Maximus Poems.* Ed. George F. Butterick. New York: Jargon/Corinth, 1960. Berkeley: U of California P, 1983.

O'Neill, Eugene. *The Emperor Jones, Anna Christie, "The Hairy Ape."* New York: Modern Library, 1937.

———. *Lazarus Laughed.* New York: Boni and Liveright, 1927. *Nine Plays.* New York: Random, 1959.

———. *Seven Plays of the Sea.* New York: Vintage, 1972.

Onion, Stephen B. *Narrative of the Mutiny on Board Schooner Plattsburgh.* Boston: Goss, 1819.

Orr, Myron. *White Gold.* Detroit: Capper, Harmon, Slocum, 1936.

Orson [pseud.]. "Life at Sea." *Knickerbocker* July 1836: 66–70.

Osborne, Lieut. E. *The Polar Regions; or, A Search After Sir John Franklin's Expedition.* New York: Barnes, 1854.

Osgood, Frances S. *Poems.* New York: Clark & Austin, 1846.

Oxenhorn, Harvey. *Tuning the Rig: A Journey to the Arctic.* New York: Harper, 1990.

Paddack, William C. *Life on the Ocean or Thirty-five Years at Sea.* . . . Cambridge: Riverside, 1893.

Palmer, James Croxall. *Thulia: A Tale of the Antarctic.* New York: Colman, 1843.

Palmer, John. *Awful Shipwreck . . . of the Ship Francis Spaight.* Boston: Perry, 1837.

Parker, Lucretia. *Piratical Barbarity; or, The Female Captive. . . .* New York: Parker, 1825.

Parker, William Harwar. *Recollections of a Naval Officer, 1841–1865.* New York: Scribner's, 1883. Annapolis: Naval Institute P, 1985.

The Patriotic Sailor; or, Sketches of the Humors, Cares, and Adventures of a Naval Life. By a thoroughbred seaman. Baltimore: Bool, 1829.

Patterson, Samuel. *Narrative of the Adventures and Sufferings of Samuel Patterson.* Palmer, MA, 1817. Fairfield, WA: Galleon, 1967.

Paulding, Hiram. *Journal of a Cruise in the United States Schooner Dolphin . . .* [incl.] *Pursuit of the Mutineers of the Whaleship Globe.* New York: Carvill, 1831.

Paulding, James Kirke. "The Ghost." *The Atlantic Souvenir for 1830.* Philadelphia: Carey, Lea and Carey, 1830.

[Payson, Edward]. *The New Age of Gold; or, The Life and Adventures of Robert Dexter Romaine. Written by Himself.* Boston: Sampson, 1856.

Pearce, Perce. *Seaman Si.* Chicago: Great Lakes Bulletin, 1918.

Peck, William Henry. *The Confederate Flag on the Ocean: A Tale of the Cruises of the Sumter and Alabama. . . .* New York: Van Evrie, Horton, 1868.

Peirce, Nathanael. *The Remarkable Deliverance of Capt. Nathanael Peirce.* Boston, 1756.

Pellowe, William C. S. *Tales from a Lighthouse Cafe.* Adrian, MI: Raison River Publishing, 1960.

Pennington, James J. W. C. *The Fugitive Blacksmith; or, Events in the History of James J. W. C. Pennington, Pastor of a Presbyterian Church, New York, Formerly a Slave in the State of Maryland.* In *Great Slave Narratives.* Ed. Arna Bontemps. Boston: Beacon P, 1969.

Penny, Joshua. *The Life and Adventures of Joshua Penny. . . .* New York, 1815.

Percival, James Gates. *The Poetical Works of James Gates Percival.* 2 vols. Boston: Ticknor & Fields, 1859.

Perils of the Sea; Being Authentic Narratives of Remarkable and Affecting Disasters Upon the Deep, with Illustrations of the Power and Goodness of God in Wonderful Preservations. New York: Harper, 1833.

Perry, Matthew C. *See* Hawks, Francis L.

Peterson, Charles J. *See also* Danforth, Harry, [pseud.].

———. *The American Navy. . . .* Philadelphia: Smith, 1856.

———. *Cruising in the Last War.* 2 vols. Philadelphia: Peterson, 1850.

"The Phantom Ship." *New-England Magazine* Aug. 1832: 122–27.

Phelps, Elizabeth Stuart. *The Madonna of the Tubs.* Boston: Houghton, 1886.

Phelps, Matthew. *Memoirs and Adventures of Captain Matthew Phelps.* Ed. Anthony Haswell. Bennington, VT: Haswell, 1802.

Pickthall, Marjorie Lowry Christie. *The Bridge: A Story of the Great Lakes.* New York: Century, 1922.

Picturesque America. See Bryant, William Cullen, ed.

Pidgeon, Harry. *Around the World Single Handed: The Cruise of the Islander.* New York: Appleton, 1932.

"A Piratical Sketch." *Ladies Companion* 9 (1838): 9–11.

Poe, Edgar Allan. *Collected Works of Edgar Allan Poe.* 3 vols. Ed. T. O. Mabbott. Cambridge: Harvard UP, 1978.

——— . *The Narrative of Arthur Gordon Pym. Of Nantucket.* New York: Harper, 1838.

Pomfret, John E., ed. *California Gold Rush Voyages, 1848–1849. Three Original Narratives.* San Marino: Huntington Library, 1954.

Porter, David. *Journal of a Cruise in the Pacific Ocean in the United States Frigate Essex, in the Years 1812, 13, 14.* 2 vols. Philadelphia: Bradford and Inskeep, 1815. Ed. R. D. Madison and Karen Hamon. Annapolis: Naval Institute P., 1986.

Porter, Katherine Anne. *Ship of Fools.* New York: Little, 1962.

Potter, Israel. *Life and Remarkable Adventures of Israel R. Potter.* Providence, 1824.

Pound, Ezra. *The Cantos.* London: Faber & Faber, Revised Collected Edition, 1975.

——— . *Personae: The Collected Poems of Ezra Pound.* New York: Boni & Liveright, 1926.

Prescott, R. E. *Historical Tales of the Huron Shore Region, and Rhymes.* Lincoln, MI: Alcona Country Herald, 1934.

Raine, Norman Reilly. "The Deep Water Mate." *McLean's Magazine.* 1 Oct. 1928: 6+.

Rascovich, Mark. *The Bedford Incident.* New York: Atheneum, 1963.

Ratigan, William. *The Adventures of Captain McCargo.* New York: Random, 1956.

——— . *Soo Canal!* Grand Rapids, MI; Eerdmans, 1954.

Read, Thomas Buchanan. *Poems.* 2 vols. Boston: Ticknor & Fields, 1847.

"Recollections of a Sailor." *Military and Naval Magazine* July 1835: 340–45; Sept.: 43–49.

Reed, Charles Bert. *Four Way Lodge.* Chicago: Covici, 1924.

"The Reefer's First Cruise." *American Monthly Magazine* 1 Mar. 1834: 25–32, 105–12, 185–92.

Reynolds, Jeremiah N. "Mocha Dick or the White Whale of the Pacific." *Knickerbocker* May 1839: 377–92.

——— . *Pacific and Indian Oceans; or, The South Sea Surveying and Exploring Expedition.* New York: Harper, 1841.

——— . *Voyage of the United States Frigate Potomac Under Commodore Downs. . . .* New York: Harper, 1834.

Rhodes, James A. *Cruise in a Whale Boat.* New York: New York Pub. Co., 1848.

Rich, Adrienne. *Poems Selected and New, 1950–1974.* New York: Norton, 1975.

Rich, R. *Newes from Virginia. The Lost Flocke Triumphant.* London, 1610.

Richard, Mark. *Fishboy.* New York: Doubleday, 1993.

Richter, Conrad. *The Sea of Grass.* New York: Knopf, 1937.

Ricketson, Anna. *The Journal of Annie Holmes Ricketson on the Whaleship A. R. Tucker, 1871–1874.* Ed. Philip F. Purrington. New Bedford: Old Dartmouth Historical Society, 1958.

Riesenberg, Felix. *Endless River.* New York: Harcourt, Brace, 1931.

——— . *Mother Sea.* New York: Kendall, 1933.

——— . *Under Sail.* New York: Macmillan, 1918.

Riesenberg, Felix, and Archie Binns. *The Maiden Voyage.* New York: Day, 1931.

Ringbolt, Captain [John Codman]. *Sailor's Life and Sailor's Yarns.* New York: Francis, 1847.

Roberts, Edmund. *Embassy to the Eastern Courts of Cochin-China, Siam, and Muscat.* New York: Harper, 1857.

Robertson, Dougal. *Survive the Savage Sea.* New York: Praeger, 1973.

Robertson, Morgan. *Down to the Sea.* New York: Harper, 1905.

———. *Futility; or, The Wreck of the Titan.* New York: Mansfield, [1898]. Rpt. as *The Wreck of the Titan; or, Futility.* New York: McClure's, [1912].

———. *Land Ho!* New York: Harper, 1905.

———. *Over the Border.* New York: McClure's, n.d.

———. *Shipmates.* New York: Appleton, 1901.

———. *Sinful Peck.* New York: McClure's, n.d. Harper, 1903.

———. *Three Laws and the Golden Rule.* New York: McKinlay, n.d.

———. *Where Angels Fear to Tread.* New York: Century, 1899.

———. *Where Angels Fear to Tread and Other Tales of the Sea.* [New York]: McClure's Magazine and Metropolitan Magazine, [1899].

Robinson, Edwin Arlington. *Collected Poems.* New York: Macmillan, 1921.

Roethke, Theodore. *The Collected Poems of Theodore Roethke.* Garden City, NY: Doubleday, 1966.

Romance of the Deep! or, The Cruise of the Aeronaut, . . . During a Three Years' Voyage in an American Whale Ship. Boston: Redding, 1846.

Ross, W. Gilles, ed. *Arctic Whalers, Icy Seas: Narratives of the Davis Strait Whale Fishery.* Annapolis: Naval Institute P, 1987.

Rowson, Susanna. *Charlotte Temple.* London, 1791. Ed. Clara M. Kirk and Rudolph Kirk. New York: Twayne, 1964.

———. *The Slaves of Algiers.* Philadelphia, 1794.

Rudloe, Jack. *The Erotic Ocean: A Handbook for Beachcombers.* New York: World, 1971.

———. *The Living Dock at Panacea.* New York: Knopf, 1977.

———. *The Sea Brings Forth.* New York: Knopf, 1968.

———. *The Time of the Turtle.* New York: Knopf, 1979.

———. *The Wilderness Coast: Adventures of a Gulf Coast Naturalist.* New York: Talley, Dutton, 1988.

Ruschenberger, William S. W. *A Voyage Round the World.* Philadelphia: Carey, Lea & Blanchard, 1838.

Ruskin, John. *Stones of Venice.* 3 vols. London: Smith, Elder, 1851–53.

Russell, David. *Autobiography of David Russell, a Boston Boy and True American.* Boston, 1857.

Ryther, John. *The Seaman's Preacher: Consisting of Nine Short and Plain Discourses on Jonah's Voyage. . . .* Cambridge: Hilliard, 1806.

St. Pierre, Bernadin de. *Paul et Virginie.* Paris, 1789.

Sampson, Abel. *The Wonderful Adventures of Abel Sampson, Related by Himself; Written by Edmund Hale Kendall, Esq. . . .* Lawrence, MA, 1847.

Samuels, Samuel. *From the Forecastle to the Cabin.* New York: Harper, 1887. New York: Library Editions, 1970.

Sands, Robert C., ed. *Life and Correspondence of John Paul Jones.* New York: Fanshaw, 1830.

Santayana, George. *Sonnets and Other Verses.* Chicago: Stone and Kimball, 1894.

Sargent, Charles Lennox. *The Life of Alexander Smith, Captain of the Island Pitcairn. . . .* Boston: Goss, 1819.

Sargent, Epes. *American Adventure by Land and Sea.* 2 vols. New York: Harper, 1842.

———. *Songs of the Sea with Other Poems.* Boston: Munroe, 1847.

Saunders, Ann. *Narrative of the Shipwreck and Sufferings of Miss Ann Saunders.* Providence: Crossman, 1827.

Saunders, Daniel. *A Journal of the Travels and Sufferings of Daniel Saunders, Jun.* Salem, 1794.

Savile, Frank Mackenzie. *See* Elivas, Knarf.

Scenes on Lake Huron; A Tale, Interspersed with Interesting Facts, in a Series of Letters. New York, 1836.

Scheffer, Victor B. *The Year of the Seal.* New York: Scribner's, 1970.

———. *The Year of the Whale.* New York: Scribner's, 1969.

Schley, Winfield Scott. *Forty-Five Years Under the Flag.* New York: Appleton, 1904.

Schoolcraft, Henry Rowe. *Algic Researches.* New York: Harper, 1839.

Schumann, Mary. *My Blood and My Treasure.* New York: Dial P, 1941.

Schwatka, Frederick. *The Long Arctic Search: The Narrative of Lieutenant Frederick Schwatka, U.S.A., 1878–1880, Seeking the Records of the Lost Franklin Expedition.* Ed. Edouard A. Stackpole. Chester, CT: Pequot P for Mystic Seaport, 1977.

Scoresby, William. *An Account of the Arctic Regions with a History and Description of the Northern Whale Fishery.* 2 vols. Edinburgh: Constable, 1820.

———. *Journal of a Voyage to the Northern Whale Fishery.* Edinburgh, 1823.

Scott, Joanna. *The Closest Possible Union.* New York: Ticknor & Fields, 1988.

Scott, Raphael. *The Voyages, Adventures, and Miraculous Escapes of Raphael Scott.* Philadelphia: Barclay, 1850.

Scott, Sir Walter. *The Pirate.* Edinburgh: Constable, 1821.

Seaborn, Captain Adam [pseud.]. *Symzonia.* New York: Seymour, 1820.

Searls, Hank. *The Hero Ship.* New York: World, 1969.

———. *Jaws 2.* Based on the screenplay by Howard Sackler and Dorothy Tristan. Garden City, NY: Doubleday, 1978.

———. *Overboard.* New York: Norton, 1977.

———. *Sounding.* New York: Ballantine, 1982.

Sea Stories. New York: Putnam, 1858.

Semmes, Lieut. Raphael. *Memoirs of Service Afloat During the War Between the States.* Baltimore: Kelly, 1868. Seacaucus, NJ: Blue & Grey P, 1987.

———. *Service Afloat and Ashore During the Mexican War.* Cincinnati: Moore, 1851.

Seymour, Frederick H. *A Lark in the Water; or, A Canoe Trip.* Detroit: Free P, 1880.

Shakespeare, William. *The Tempest. The Riverside Shakespeare,* Boston: Houghton, 1974.

Shange, Ntozake. *Sassafrass, Cypress & Indigo.* New York: St. Martin's, 1982.

Shaw, Elijah. *A Short Sketch of the Life of Elijah Shaw, Who Served for Twenty-two Years in the Navy of the United States. . . .* Rochester, NY: Strong and Dawson, 1843.

Sheppard, Francis Henry. *Love Afloat: A Story of the American Navy.* New York: Sheldon, [1875].

Sherburne, Andrew. *Memoirs of Andrew Sherburne: A Pensioner of the Navy of the Revolution.* Utica, NY: Williams, 1828.

Sherburne, Jacob. *A Narrative of the Principal Incidents in the Life and Adventures of Captain Jacob Sherburne. . . .* Castine, ME: Bond, 1829.

Shippen, Edward. *Thirty Years at Sea.* Philadelphia: Lippincott, 1879. New York: Arno, 1979.

"The Shipwrecked Coaster." *Boston Pearl and Literary Gazette* 17 Jan. 1835: 149–51.

Sigourney, Lydia. *Poems.* New York: Leavitt & Allen, 1850.

———. *Poems.* New York: Leavitt & Allen, 1860.

———. *Poems for the Sea.* Hartford: Parsons, 1850.

Simms, William Gilmore. *Lyrical and Other Poems.* Charleston: Ellis & Neufville, 1827.

———. *The Maroon; A Legend of the Caribes and Other Tales.* Philadelphia: Lippincott, 1855.

———. "A Picture of the Sea." *Southern Literary Gazette* Dec. 1828: 208–15.

———. *Poems Descriptive, Dramatic, Legendary and Contemplative.* 2 vols. New York: Redfield, 1853.

———. "A Sea Piece." *The Atlantic Club Book.* Vol. 1. New York: Harper, 1834. 264–79.

Sinclair, Captain Robert. *Winds Over Lake Huron.* Hicksville, NY: Exposition P, 1977.

Sirna, Anthony, and Allison Sirna, eds. *The Wanderings of Edward Ely: A Mid-Nineteenth Century Seafarer's Diary.* New York: Hastings House, 1954.

Skelton, Joan. *The Survivor of the Edmund Fitzgerald.* Moonbeam, Ontario: Penumbra, 1985.

Sleeper, John Sherburne. *See* Martingale, Hawser [pseud.].

Slocum, Joshua. *Sailing Alone Around the World.* New York: Century, 1900. New York: Dover, 1956.

———. *Voyage of the Liberdade.* Boston: Robinson & Stephenson, 1890.

Slonczewski, Joan. *A Door into Ocean.* New York: Arbor House, 1986.

Smallfull, Jerry [pseud.]. "Adventures of a Reefer." *Military and Naval Magazine* May 1835: 222–26.

Smith, Elizabeth Oakes. "Jack Spanker and the Mermaid." *Graham's Magazine* Aug. 1843: 68–71.

———. *The Poetical Writings of Elizabeth Oakes Smith.* New York: Redfield, 1845.

Smith, James L. *Autobiography of James L. Smith, Including, Also, Reminiscences of Slave Life. . . .* Norwich, CT: Bulletin Co., 1881. New York: Negro UP, 1969.

Smith, Captain John. *A Sea Grammar.* London, 1627.

———. "The Sea Marke." *Advertisements for the Unexperienced Planters of New England, or Anywhere.* London, 1631.

Smith, Moses. *Naval Scenes in the Last War; or, Three Years on Board the Frigate Constitution. . . .* Boston: Gleason's, 1846.

Smith, Thomas W. *A Narrative of the Life, Travels and Sufferings of Thomas W. Smith.* Boston: Hill, 1844.

Smith, Venture. *The Life and Adventures of Venture, a Native of Africa, But Resident Above Sixty Years in the United States of America.* 1798. Rpt. in *Five Black Lives: The Autobiographies of Venture Smith,* . . . . Middletown: Wesleyan UP, 1971.

Smollett, Tobias. *The Adventures of Roderick Random.* London, 1748. Ed. Paul-Gabriel Bouce. Oxford: Oxford UP, 1979.

Snider, Charles Henry Jeremiah. *The Story of the "Nancy" and Other Eighteen-Twelvers.* Toronto: McClelland, 1926.

Snow, Elliot, ed. *The Sea, the Ship and the Sailor.* Salem, MA: Marine Research Society, 1925. Rpt. as *Adventures at Sea in the Great Age of Sail: Five Firsthand Narratives.* New York: Dover, 1986. [Includes: "A Narrative of the Adventures of Capt. Charles H. Barnard of New York . . ."; "A Journal of a Voyage in the Brig *Spy,* of Salem (1832–1834), John B. Knights, Master"; and "A Narrative of Events in the Life of John Bartlett of Boston, Massachusetts, in the Years 1790–1793. . . ."]

Spears, John Randolph. *The Fugitive: A Tale of Adventure in the Days of Clipper Ships and Slavers.* New York: Scribner's, 1899.

———. *The Port of Missing Ships, and Other Stories of the Sea.* New York: Macmillan, 1897.

Spofford, Harriet Prescott, Louise Imogen Guiney, and Alice Brown. *Three Heroines of New England Romance.* Illus. Edward H. Garrett. Boston: Little, 1894.

Stanton, Dick [pseud.]. "Off the Cape." *Burton's Gentleman's Magazine* Apr. 1839: 233–36.

Stedman, Edmund Clarence, ed. *An American Anthology, 1787–1900: Selections Illustrating the Editor's Critical Review of American Poetry in the Nineteenth Century.* 2 vols. Cambridge: Riverside, 1900.

Steere, Richard. *The Daniel Catcher: The Life of the Prophet Daniel, in a Poem.* Boston, 1713.

———. *A Monumental Memorial of Marine Mercy.* Boston, 1684.

———. "On a Sea-Storm Nigh the Coast." *Seventeenth-Century American Poetry.* Ed. Harrison T. Meserole. New York: Anchor, 1968, p. 265.

Steinbeck, John. *Sea of Cortez.* New York: Viking, 1941.

Stevens, Wallace. *The Collected Poems of Wallace Stevens.* New York: Knopf, 1954.

Stewart, C. S. *Brazil and La Plata.* New York: Putnam, 1856.

———. *The Private Journal of a Voyage to the Pacific Ocean and Residence at the Sandwich Islands, in the Years 1822, 1823, 1824, and 1825.* New York: Haven, 1828.

———. *A Visit to the South Seas, in the United States Ship Vincennes, During the Years 1829 and 30.* 2 vols. New York: Haven, 1831.

Stockton, Frank Richard. *The Adventures of Captain Horn.* New York: Scribner's, 1895.

Stoddard, Elizabeth. *The Morgesons.* New York: Carleton, 1862. Rpt., ed. Lawrence Buell and Sandra A. Zagarell. Philadelphia: U of Pennsylvania P, 1984.

———. *Temple House.* New York: Carleton, 1867.

Stone, William Leete. "The Dead of the Wreck." *The Atlantic Souvenir for 1831.* Philadelphia: Carey, Lea, and Carey, 1830. 164–93.

———. "The Spectre Fire-Ship." *Knickerbocker* May 1834: 361–70.

Stories of the Sea. New York: Scribner's, 1893.

[Stout, Benjamin]. *A Narrative of the Loss of the Ship Hercules*. Hudson, NY, 1798.

Stowe, Harriet Beecher. "Deacon Pitkin's Farm." *Christian Union* 3 Nov. 1875: 357–58; 10 Nov.: 377–78; 17 Nov.: 397–99; 24 Nov.: 417–19. *Stories, Sketches and Studies*. Vol. 14 of *The Writings of Harriet Beecher Stowe*. Boston: Houghton, 1896.

———. *The Pearl of Orr's Island*. Boston: Ticknor, 1862. Hartford: Stowe-Day, 1979.

———. *Uncle Tom's Cabin; or, Life Among the Lowly*. Boston: Jewett, 1852.

Stowe, Phineas. *Ocean Melodies, and Seaman's Companion*. Boston: Stowe, 1849.

Strachey, William. "A True Reportory of the Wracke, and Redemption of Sir Thomas Gates Knight. . . ." *Purchas His Pilgrimes*. Ed. Samuel Purchas. Vol. 4. London: Stansby, 1625.

Styron, William. *The Confessions of Nat Turner*. New York: Random, 1967.

Sullivan, A. M. *The Bottom of the Sea: A Poem. . . .* Chicago: Dun & Bradstreet, 1966.

Sumner, Charles Allen. *'Round the Horn: A Christmas Yarn*. San Francisco: Bacon, [1855?].

Swayze, Fred. *The Rowboat War on the Great Lakes, 1812–1815*. Toronto: Macmillan of Canada, 1965.

Swell Life at Sea; or, Fun, Frigates and Yachting: A Collection of Nautical Yarns. From the log-book of a youngster of the mess. New York: Stringer & Townsend, 1854.

Synge, John Millington. *Riders to the Sea*, 1904.

Tate, Allen. *Collected Poems, 1919–1976*. New York: Farrar, Straus & Giroux, 1977.

Taylor, Fitch W. *A Voyage Round the World . . . in the United States Frigate Columbia*. New York: Appleton, 1840.

Taylor, J. Bayard. *At Home and Abroad: A Sketch-Book of Life, Scenery, and Men*. New York: Putnam, 1860.

———. *Eldorado; or, Adventures in the Path of Empire*. 2 vols. New York: Putnam, 1850.

———. *Views Afoot; or, Europe Seen with Knapsack and Staff*. New York: Wiley, 1846.

Taylor, Nathaniel W., M.D. *Life on a Whaler; or, Antarctic Adventures in the Isle of Desolation*. Ed. Howard Palmer. New London, CT: New London County Historical Society, 1929.

Thacher, Anthony. ". . . A Letter . . . Consarning His Grat Deliverance out of the Deapes of the Sea." *In the Trough of the Sea*. Ed. Donald P. Wharton. Westport, CT: Greenwood, 1978. 56–64.

Thalatta: A Book for the Sea-Side. Ed. Samuel Longfellow and Thomas Wentworth Higginson. Boston: Ticknor, Reed, and Fields, 1853.

Thaxter, Celia. *Among the Isles of Shoals*. Boston: Osgood, 1873.

———. *The Poems of Celia Thaxter*. Preface, Sarah Orne Jewett. Boston: Houghton, 1896.

Thomas, R., ed. *Interesting and Authentic Narratives of the Most Remarkable Shipwrecks. . . .* New York: Strong, 1836.

Thomes, William Henry. *Running the Blockade; or, U.S. Secret Service Adventures*. Boston: Lee, Shepard, and Dillingham, 1875.

———. *A Slaver's Adventures on Land and Sea*. Boston: Lee, Shepard, and Dillingham, [1872].

Thompson, John. *The Life of John Thompson, A Fugitive Slave; Containing His History of 25 Years in Bondage, and His Providential Escape.* Written by Himself. Worcester: Thompson, 1856. New York: Negro UP, 1986.

Thoreau, Henry David. *Cape Cod.* Boston: Ticknor and Fields, 1865.

———. *Collected Poems of Henry Thoreau.* Ed. Carl Bode. Baltimore: Johns Hopkins UP, 1964.

———. *Walden; or, Life in the Woods.* Boston: Ticknor and Fields, 1854.

———. *A Week on the Concord and Merrimack Rivers.* Boston: Munroe, 1849.

———. *The Writings of Henry David Thoreau.* 20 vols. Boston: Houghton, 1906.

Timrod, Henry. *Collected Poems, a Variorum Edition.* Ed. Edd Winfield Parks and Aileen Wills Parks. Athens: U of Georgia P, 1965.

Titus, Harold. *Spindrift, a Novel of the Great Lakes.* New York: Doubleday, 1925.

Tolson, Melvin B. *A Gallery of Harlem Portraits.* Columbia: U of Missouri P, 1979.

———. *Harlem Gallery; Book I, The Curator.* Boston: Twayne, 1965.

———. *Libretto for the Republic of Liberia.* New York: Twayne, 1953.

Toomer, Jean. *The Collected Poems of Jean Toomer.* Chapel Hill: U of North Carolina P, 1987.

Traven, B. *The Death Ship: The Story of an American Sailor.* New York: Knopf, 1934.

A Treasury of Afro-American Folklore. Ed. Harold Courlander. New York: Crown, 1976.

Trowbridge, J. T. "At Sea." *Atlantic Monthly.* Jan. 1859: 69.

A True Account of the Loss of the Ship Columbia. Portsmouth, NH, 1792.

A True and Particular Narrative of the Late Tremendous Tornado; or, Hurricane at Philadelphia and New York. Boston, 1792.

A True Declaration of the Estate of the Colonie in Virginia. London, 1610.

Tuckerman, Frederick Goddard. *Poems.* Boston: Wilson, 1860.

Twain, Mark. *Adventures of Huckleberry Finn.* New York: Webster, 1885. Ed. Walter Blair and Victor Fischer. Berkeley: U of California P, 1985.

———. *Following the Equator: A Journey Around the World.* Hartford: American, 1897.

———. *The Innocents Abroad; or, The New Pilgrim's Progress.* Hartford: American, 1869.

———. *The Mysterious Stranger.* Ed. William M. Gibson. Berkeley: U of California P, 1969.

———. *Roughing It.* Hartford: American, 1872. Ed. Franklin R. Rogers. Berkeley: U of California P, 1972.

———. *Which Was the Dream? And Other Symbolic Writings of the Late Years.* Ed. John S. Tuckey. Berkeley: U of California P, 1966.

Tyler, Royall. *The Algerine Captive.* Walpole, NH, 1797. Ed. Don L. Cook. New Haven: College and University P, 1970.

Vail, James E. "The Sea Voyage." *Ladies Companion* Oct. 1837: 285–87.

Very, Jones. *Poems and Essays.* Boston: Houghton, 1886.

Vidal, Gore. *Williwaw.* New York: Dutton, 1946.

Vonnegut, Kurt. *Cat's Cradle.* New York: Delacorte, 1963.

———. *Galápagos.* New York: Delacorte, 1985.

Voyage and Venture; or, The Pleasures and Perils of a Sailor's Life. Philadelphia: Peck & Bliss, 1854.

"Voyaging." *New-England Magazine* Dec. 1834: 447–48.

Vukelich, George. "The Bosun's Chair." *Arts in Society* (1960): 45–64.

———. *Fisherman's Beach.* New York: St. Martin's, 1962.

Wade, Robert L. "The Doomed Ship." *Knickerbocker* Nov. 1843: 403–11.

Walling, William. *The Wonderful Providence of God.* 2d ed. Boston, 1730.

Warneford, Lieutenant [Archibald Clavering Gunter]. *The Adventures of a Naval Officer: A Narrative.* New York: Home Publishing, 1898.

Warren, T. Robinson. *Dust and Foam; or, Three Oceans and Two Continents.* New York: Scribner, 1859.

Warriner, Francis. *The Cruise of the United States Frigate Potomac Round the World During the Years 1831–34. . . .* New York: Leavitt, 1835.

Washburn, James. *A True and Concise Account of the Voyage and Sufferings of James Washburn Jun.* Boston: Spear, 1822.

Waterhouse, Benjamin. *A Journal of a Young Man of Massachusetts, Late a Surgeon on Board an American Privateer.* Boston: Rowe, 1816.

Watkins, Paul. *Calm at Sunset, Calm at Dawn.* Boston: Houghton, 1989.

Watson, Elkanah. *Men and Times of the Revolution.* Ed. Winslow C. Watson. New York: Dana, 1856.

Watts, Isaac. *The Psalms of David, Imitated in the Language of the New-Testament.* London, 1719.

[Weld, Horatio Hastings]. "A Chapter on Whaling." *New-England Magazine* May 1835: 445–49.

Wells, Louisa Susannah. *The Journal of a Voyage from Charlestown, to London Undertaken During the American Revolution by a Daughter of an Eminent American Loyalist.* New York, 1906. New York: Arno, 1968.

Wendt, Frederick W. *Ocean Sketches.* New York: Colonial, 1897.

West, Morris. *The Navigator.* New York: Morrow, 1976.

Wetmore, Claude Hazeltine. *Sweepers of the Sea: The Story of a Strange Navy.* Indianapolis: Bowen-Merrill, 1900.

Wharton, Edith. *The House of Mirth.* New York: Scribner's, 1905.

White, Andrew. *A Relation of the Colony of the Lord Baron of Baltimore: A Narrative of the First Voyage to Maryland.* Trans. N. C. Brooks. Baltimore: Maryland Historical Society, 1847.

Whitecar, William B., Jr. *Four Years Aboard the Whaleship.* Philadelphia: Lippincott, 1860.

Whitman, Walt. *Leaves of Grass.* New York, 1855. Ed. Sculley Bradley and Harold W. Blodgett. New York: Norton, 1973.

Whittemore, Reed. *Poems, New and Selected.* Minneapolis: U of Minnesota P, 1967.

———. *The Self-Made Man.* New York: Macmillan, 1959.

Whittier, John Greenleaf. *The Complete Poetical Works of John Greenleaf Whittier.* Ed. Horace Scudder. Cambridge ed. Boston: Houghton, 1894.

Whitwell, William. *A Discourse, Occasioned by the Loss of a Number of Vessels, with Their Mariners Belonging to the Town of Marblehead.* Salem, 1770.

The Whole Book of Psalmes Faithfully Translated into English Metre. Cambridge, MA, 1640.

Wideman, John Edgar. *Hurry Home.* New York: Harcourt, 1970.

Wilbur, Richard. *New and Collected Poems.* San Diego: Harcourt, Brace, Jovanovich, 1988.

Wilkes, Lieut. Charles. *Narrative of the United States Exploring Expedition During the Years 1838, 1839, 1840, 1841, 1842.* 5 vols. Philadelphia: Lea and Blanchard, 1845.

Williams, Eliza Azelia. "Journal of a Whaleing Voyage to the Indian and Pacific Oceans. . . ." *One Whaling Family.* Ed. Harold Williams. Boston: Houghton, 1964. 3–204.

Williams, Henry Llewellyn. *See* Mizzen, Mat [pseud.].

Williams, James. *Blow the Man Down!* Ed. Warren F. Kuehl. New York: Dutton, 1959.

Williams, William. *Mr. Penrose: The Journal of Penrose, Seaman.* Ed. John Eagles. 4 vols. London: Murray, 1815. Ed. David H. Dickason. Bloomington: Indiana UP, 1969.

Williams, William Carlos. *The Collected Earlier Poems.* New York: New Directions, 1951.

———. *The Collected Later Poems.* New York: New Directions, 1950. Revised 1963.

Willis, Nathaniel P. *Health Trip to the Tropics.* New York: Scribner, 1854.

———. *Summer Cruise in the Mediterranean on Board an American Frigate.* New York: Scribner, 1853.

Wilson, Sloane. *Ice Brothers: A Novel.* New York: Arbor, 1979.

———. *Pacific Interlude.* New York: Arbor, 1982.

Winslow, William Henry. *Cruising and Blockading.* Pittsburgh: Weldin, 1885.

Winters, Yvor. *Collected Poems.* Denver: Alan Swallow, 1952. Revised 1960.

———. *The Journey.* Ithaca, NY: Dragon, 1931.

Woodworth, Samuel. *The Champions of Freedom; or, The Mysterious Chief, a Romance of the Nineteenth Century.* New York: Charles Baldwin, 1816.

Woolson, Constance Fenimore. *Castle Nowhere: Lake Country Sketches.* New York: Harper, 1875.

———. "Margaret Morris." *Appleton's Journal* 13 Apr. 1872: 394–99.

Wouk, Herman. *The Caine Mutiny: A Novel of World War II.* Garden City, NY: Doubleday, 1951.

———. *War and Remembrance.* Boston: Little, 1978.

———. *The Winds of War.* Boston: Little, 1971.

Wright, John C. *Lays of the Lakes.* Boston: Gorham P, 1911.

Wunsch, Josephine [pseud.]. *See* Gillette, Virginia M.

"A Yankee Tar's Adventures with the Flying Dutchman." *Burton's Gentleman's Magazine* 1 (1837): 331–33.

Yeats, W. B. *The Collected Poems of W. B. Yeats.* New York: Macmillan, 1933. 2d ed., 1950 [later poems added].

A Younker [Josiah Cobb]. *A Green Hand's First Cruise, Roughed Out From the Log-Book of Memory, of Twenty-Five Years Standing: Together with a Residence of Five Months in Dartmoor.* 2 vols. Boston: Otis, 1841.

Zollers, George D. *Thrilling Incidents on Sea and Land: The Prodigal's Return.* Mount Morris, IL: Brethren's, 1892.

Secondary Bibliography

Adams, Charles H. "Cooper's Sea Fiction and *The Red Rover.*" *Studies in American Fiction* 16.2 (1988): 155–68.

——. *The Guardian of the Law: Authority and Identity in James Fenimore Cooper.* University Park: Pennsylvania State UP, 1990.

Adams, Percy G. *Travel Literature and the Evolution of the Novel.* Lexington: U of Kentucky P, 1983.

Adler, Joyce Sparer. *War in Melville's Imagination.* New York: New York UP, 1981.

Adler, Thomas P. "Beyond Synge: O'Neill's *Anna Christie.*" *Eugene O'Neill Newsletter* 12.1 (1988): 34–39.

Albion, Robert G. *Naval and Maritime History: An Annotated Bibliography.* 4th ed. Mystic, CT: Marine Historical Association, 1972.

——. *Square Riggers on Schedule.* Princeton: Princeton UP, 1938.

Albion, Robert G., William A. Baker, and Benjamin W. Labaree. *New England and the Sea.* Middletown, CT: Marine Historical Association, 1972.

Allen, Priscilla. "*White-Jacket:* Melville and the Man-of-War Microcosm." *American Quarterly* 25 (1973): 32–47.

Almy, Robert F. "J. N. Reynolds: A Brief Biography with Particular Reference to Poe and Symmes." *Colophon* 2 n.s. (1937): 227–45.

Alvarez, Alfred. *The Shaping Spirit.* London: Chatto, 1968.

Anderson, Charles R., ed. *Journal of a Cruise to the Pacific Ocean, 1842–1844, in the Frigate United States, with Notes on Herman Melville.* Durham, NC: Duke UP, 1937.

——. *Melville in the South Seas.* New York: Columbia UP, 1939.

——. "A Reply to Herman Melville's *White-Jacket* by Rear-Admiral Thomas O. Selfridge, Sr." *American Literature* 7 (1935): 123–44.

Ashley, Clifford W. *The Yankee Whaler.* Boston: Houghton, 1926.

Askins, Justin. "Melville's Natural World: A Voyage to the Metaphysical." *DAI* 47 (1987): 4388A. City U of New York.

Astro, Richard. "Steinbeck's *Sea of Cortez* (1941)." *A Study Guide to Steinbeck: A Handbook to His Major Works.* Ed. Tetsumaro Hayashi. Metuchen, NJ: Scarecrow, 1974. 168–86.

Astro, Richard, ed. *Literature and the Sea: Proceedings of a Conference Held at the Marine Science Center Newport, Oregon, May 8, 1976.* Corvallis: Oregon State U, 1976.

Ault, Nelson A. "The Sea-Imagery in Herman Melville's *Clarel.*" *Research Studies of the State College of Washington* 27 (1959): 115–27.

Axelrod, Steven Gould. *Robert Lowell: Life and Art.* Princeton: Princeton UP, 1978.

Babcock, C. Merton. "Herman Melville's Whaling Vocabulary." *American Speech* 29 (1954): 161–74.

——. "Melville's Backwoods Seamen." *Western Folklore* 10 (1951): 126–33.

——. "Melville's Proverbs of the Sea." *Western Folklore* 11 (1952): 254–65.

Bachelard, Gaston. *Water and Dreams: An Essay on the Imagination of Matter.* Trans. Edith R. Farrell. Dallas: Dallas Institute of Humanities and Culture, 1983.

Bailey, J. O. "An Early American Utopian Fiction." *American Literature* 14 (1942): 285–93.

———. Introduction. *Symzonia: A Voyage of Discovery (1820)*. Gainesville, FL: Scholars' Facsimiles & Reprints, 1965.

———. "Sources of Poe's *The Narrative of Arthur Gordon Pym*, 'Hans Pfaal,' and Other Pieces." *PMLA* 57 (1942): 513–35.

Baker, Carlos. "The Ancient Mariner." *Hemingway: The Writer as Artist*. Princeton: Princeton UP, 1972.

Bald, R. C. *John Donne: A Life*. New York: Oxford, 1970.

Bandy, William T. "A New Light on the Source of Poe's 'A Descent into the Maelstrom.'" *American Literature* 24 (1953): 534–37.

Baskett, Sam S. "Fronting the Atlantic: *Cape Cod* and 'The Dry Salvages.'" *New England Quarterly* 56 (1983): 200–219.

———. "Sea Change in *The Sea Wolf*." *American Literary Realism* 24 (1992): 5–22.

Bauer, K. Jack. *A Maritime History of the United States: The Role of America's Seas and Waterways*. Columbia: U of South Carolina P, 1988.

Beach, Joseph Warren. "Hart Crane and *Moby Dick*." *Western Review* 20 (1956): 183–96.

Beck, Horace. *Folklore and the Sea*. Middletown, CT: Wesleyan UP, 1973.

Behrman, Cynthia F. *Victorian Myths of the Sea*. Athens, Ohio: Ohio UP, 1977.

Bender, Bert. "*Far Tortuga* and American Sea Fiction Since *Moby-Dick*." *American Literature* 56 (1984): 227–48.

———. "The Influence of Darwin in American Literature of the Sea." *Proteus* 6.2 (1989): 30–37.

———. "Jack London in the Tradition of American Sea Fiction." *American Neptune* 46 (1986): 188–99.

———. "Joshua Slocum and the Reality of Solitude." *American Transcendental Quarterly* 6 (1992): 59–71

———. *Sea Brothers: The Tradition of American Sea Fiction from "Moby-Dick" to the Present*. Philadelphia: U of Pennsylvania P, 1988.

Bercovitch, Sacvan. *The Puritan Origins of the American Self*. New Haven: Yale UP, 1975.

Bernhard, Frank. "Dickinson's 'I Started Early . . .': An Anatomy; or, Whatever Happened to Emily's Dog." *Dickinson Studies* 76 (1990): 14–20.

Berthold, Dennis. "Cape Horn Passages: Literary Conventions and Nautical Realities." *Literature and Lore of the Sea*. Ed. Patricia Ann Carlson. Amsterdam: Rodopi, 1986. 40–50.

———. "Deeper Soundings: The Presence of *Walden* in Joshua Slocum's *Sailing Alone Around the World*." *Literature and Lore of the Sea*. Ed. Patricia Ann Carlson. Amsterdam: Rodopi, 1986. 161–75.

Black, Ronald J. "The Paradoxical Structure of the Sea Quest in Dana, Poe, Cooper, Melville, London, and Hemingway." *DAI* 40 (1980): 5862A-63A. Wayne State U.

Bloom, Harold. *Wallace Stevens: The Poems of Our Climate*. Ithaca: Cornell UP, 1976.

Bloom, Harold, ed. *Hart Crane*. New York: Chelsea, 1987.

———. *Marianne Moore*. New York: Chelsea, 1987.

Bonetti, Kay. "An Interview with Peter Matthiessen." *Missouri Review* 12 (1989): 109–24.

Bonner, Willard H. "Captain Thoreau: Gubernator to a Piece of Wood." *New England Quarterly* 39 (1966): 26–46.

———. *Harp on the Shore: Thoreau and the Sea*. Ed. George R. Levine. Albany: State U of New York P, 1985.

———. "Mariners and Terreners: Some Aspects of Nautical Imagery in Thoreau." *American Literature* 34 (1963): 507–19.

Bontemps, Arna, and Langston Hughes, eds. *Book of Negro Folklore*. New York: Dodd, 1958.

Bowman, Larry G. *Captive Americans: Prisoners During the American Revolution*. Athens: Ohio UP, 1976.

Brennen, Joseph X. "Stephen Crane and the Limits of Irony." *Criticism* 11 (1969): 183–200.

Briand, Paul, Jr. *In Search of Paradise: The Nordhoff-Hall Story*. New York: Duell, 1966.

Brodtkorb, Paul, Jr. *Ishmael's White World: A Phenomenological Reading of "Moby-Dick."* New Haven: Yale UP, 1965.

Brogunier, Joseph. "Walking My Dog in 'Sand Dunes.'" *Journal of Modern Literature* 16 (1990): 648–50.

Bronk, William. "Walt Whitman's Marine Democracy." *Vectors and Smoothable Curves*. San Francisco: North Point, 1983. 131–62.

Brophy, Robert J. *Robinson Jeffers: Myth, Ritual and Symbol in His Narrative Poems*. Cleveland: Case Western Reserve UP, 1973.

Brown, Bill. "Interlude: The Agony of Play in 'The Open Boat.'" *Arizona Quarterly* 45.5 (1989): 23–46.

Brown, Mrs. Helen E. *A Good Catch; or, Mrs. Emerson's Whaling Cruise*. Philadelphia: Presbyterian Board of Pub., [1884].

Brunvand, Jan Harold. "Sailors' and Cowboys' Folklore in Two Popular Classics." *Southern Folklore Quarterly* 29 (1965): 266–83.

Bryant, John. "Melville and Charles F. Briggs: *Working a Passage* to *Billy Budd*." *English Language Notes* 22.4 (1985): 48–54.

Bryant, John, ed. *A Companion to Melville Studies*. Westport, CT: Greenwood, 1986.

Bryant, Samuel W. *The Sea and the States: A Maritime History of the American People*. New York: Crowell, 1947.

Burhans, Clinton S., Jr. "Spindrift and the Sea: Structural Patterns and Unifying Elements in *Catch-22*." *Twentieth Century Literature* 19 (1973): 239–49.

Burnett, Gary. "The Identity of 'H': Imagism and H.D.'s *Sea Garden*." *Sagetrieb* 8.3 (1989): 55–75.

Busch, Frederick. "Icebergs, Islands, Ships Beneath the Sea." *A John Hawkes Symposium: Design and Debris*. New York: New Directions, 1977. 50–63.

Butterfield, Stephen. *Black Autobiography*. Amherst: U of Massachusetts P, 1974.

Butterick, George F. *A Guide to the Maximus Poems of Charles Olson.* Berkeley: U of California P, 1978.

Byrd, Don. *Charles Olson's Maximus.* Urbana: U of Illinois P, 1980.

Calhoun, Richard J., and Robert W. Hill. *James Dickey.* Boston: Twayne, 1983.

Callens, Johan. "Memories of the Sea in Shepard's Illinois." *Modern Drama* 29 (1986): 403–15.

Cameron, Kenneth Walter. "Thoreau on the Limitations of Great Circle Sailing." *American Transcendental Quarterly* 14 (1972): 70–71.

Campbell, Joseph. *Hero with a Thousand Faces.* 2d ed. Princeton: Princeton UP, 1968.

Candido, Anne Marie. "A New Sea in a New Frontier: American Leadership in *The Pathfinder.*" *Etudes Anglaises* 44 (1991): 285–95.

Cannon, Agnes D. "Melville's Use of Sea Ballads and Songs." *Western Folklore* 23 (1964): 1–16.

Canright, Stephen. "The Black Man and the Sea." *South Street Reporter* 7.4 (1973–74): 5–10.

Carlson, Patricia Ann, ed. *Literature and Lore of the Sea.* Amsterdam: Rodopi, 1986.

Carothers, Robert L., and John L. Marsh. "The Whale and the Panorama." *Nineteenth Century Fiction* 26 (1971): 319–28.

Carpenter, Frederic I. *Robinson Jeffers.* New York: Twayne, 1962.

Chapelle, Howard I. *The Search for Speed Under Sail.* New York: Norton, 1967.

Chase, George David. *Sea Terms Come Ashore.* Orono: U of Maine P, 1942.

Chase, Richard. *The American Novel and Its Tradition.* New York: Doubleday, 1967.

Church, Albert. *Whale Ships and Whaling.* New York: Norton, 1938.

Clagett, John H. "Cooper and the Sea: Naval History in the Writings of James Fenimore Cooper." Diss. Yale U, 1954.

———. "The Maritime Works of James Fenimore Cooper as Sources for Sea Lore, Sea Legend, and Sea Idiom." *Southern Folklore Quarterly* 30 (1966): 323–31.

Clare, Miriam. "The Sea and Death in *Leaves of Grass.*" *Walt Whitman Review* 10 (1964): 14–16.

Cline, Walter. "Dana at the Point, Discrepancies in the Narrative." *Historical Society of Southern California Quarterly* 32 (1950): 127–32.

Colcord, Joanna C. *Sea Language Comes Ashore.* New York: Cornell Maritime P, 1945.

———. *Songs of American Sailormen.* New York: Norton, 1938.

Colcord, Lincoln. "Notes on 'Moby Dick.'" 1922. Rpt. in *The Recognition of Herman Melville.* Ed. Hershel Parker. Ann Arbor: U of Michigan P, 1967.

———. Rev. of *Lightship,* by Archie Binns. *New York Herald Tribune Books* 2 Sept. 1934: 1–2.

———. Rev. of *Ships and Women,* by Bill Adams. *New York Herald Tribune Books* 11 Apr. 1937: 6.

Connolly, James Brendan. *The Port of Gloucester.* New York: Doubleday, 1940.

Cooksey, Philip Neil. "A Thematic Study of James Fenimore Cooper's Nautical Fiction." Diss. Louisiana State U, 1977.

Cooper, Jane Roberta. *Reading Adrienne Rich: Reviews and Revisions, 1951–1981.* Ann Arbor: U of Michigan P, 1984.

Cowan, Bainard. *Exiled Waters: "Moby-Dick" and the Crisis of Allegory.* Baton Rouge: Louisiana State UP, 1982.

Cowart, David. "Fantasy and Reality in *Mrs. Caliban*." *Critique: Studies in Contemporary Fiction* 30 (1989): 77–83.

Cowell, Pattie. "'Sailing with sealed orders': Herman Melville's *White Jacket* and B. Traven's *The Death Ship*." *B. Traven: Life and Work.* Ed. Ernst Schurer and Philip Jenkens. University Park: Pennsylvania State UP, 1987. 316–25.

Cox, Edward G. *A Reference Guide to the Literature of Travel.* 2 vols. Seattle: U of Washington P, 1935–38.

Cox, James M. "Richard Henry Dana's *Two Years Before the Mast:* Autobiography Completing Life." *The Dialectic of Discovery: Essays on the Teaching and Interpretation of Literature Presented to Lawrence E. Harvey.* Lexington, KY: French Forum, 1984. 160–78.

Cramer, Timothy R. "Testing the Waters: Contemplating the Sea in ED's Poem 520 and Kate Chopin's *The Awakening*." *Dickinson Studies* 83 (1992): 51–56.

Creighton, Margaret Scott. "Fraternity in the American Forecastle, 1830–1870." *New England Quarterly* 63 (1990): 531–57.

———. "The Private Life of Jack Tar: Sailors at Sea in the Nineteenth Century." *DAI* 46 (1986): 2729A. Boston U.

Cressy, David. "The Vast and Furious Ocean: The Passage to Puritan New England." *New England Quarterly* 57 (1984): 511–32.

Cross, Richard K. "The Seer and the Revealed: Reflections on Eberhart's Maine." *Richard Eberhart: A Celebration.* Ed. Sidney Lee, Jay Parini, M. Robin Barone. Hanover, NH: Kenyon Hill, 1980.

Cuddy, Lois A. "Eliot and Huck Finn: River and Sea in 'The Dry Salvages.'" *T. S. Eliot Review* 3 (1976): 3–12.

Curry, David Park. *Childe Hassam: An Island Garden Revisited.* New York: Norton/ Denver Art Museum, 1990.

Cutler, Carl C. *Greyhounds of the Sea: The Story of the American Clipper Ship.* New York: Halcyon House, 1930.

Dameron, J. Lasley. "Another Source for Poe's 'City in the Sea.'" *Poe Studies* 22 (1989): 43–44.

Davie, Donald. "John Ledyard: The American Traveler and His Sentimental Journies." *Eighteenth-Century Studies* 4 (1970): 57–70.

Davis, Merrell R. *Melville's Mardi: A Chartless Voyage.* New Haven: Yale UP, 1952.

Day, Cyrus. "Stephen Crane and the Ten-foot Dinghy." *Boston University Studies in English* 3 (1957): 193–213.

Day, Martin S. "Travel Literature and the Journey Theme." *Forum* 12.2 (1974): 37–47.

Day, Robert A. "Image and Idea in Voyage II." *Criticism* 7 (1965): 224–34.

DeFalco, Joseph M. "'Bimini' and the Subject of Hemingway's *Islands in the Stream*." *Topic: A Journal of the Humanities* 31 (1977): 41–51.

———. "Hemingway's Islands and Streams: Minor Tactics for Heavy Pressure." *Hemingway in Our Time.* Ed. Richard Astro and Jackson J. Benson. Corvallis: Oregon State UP, 1974. 39–51.

————. "Metaphor and Meaning in Poe's *Pym.*" *Topic: A Journal of the Humanities* 30 (1976): 54–67.

————. "Psyche and Sea: The Waste Land Era in America." *Literature and the Sea.* Ed. Richard Astro. Corvallis: Oregon State UP, 1976. 31–38.

De Pauw, Linda Grant. *Seafaring Women.* Boston: Houghton, 1982.

Dillingham, William B. *An Artist in the Rigging: The Early Work of Herman Melville.* Athens: U of Georgia P, 1972.

————. *Melville's Later Novels.* Athens: U of Georgia P, 1986.

————. *Melville's Short Fiction, 1853–1856.* Athens: U of Georgia P, 1977.

Dimock, Wai-Chee. "*White Jacket:* Authors and Audiences." *Nineteenth Century Fiction* 36 (1981): 296–317.

Doerflinger, William M. *Shantymen and Shantyboys.* New York: Macmillan, 1951.

Doreski, Carole Kiler. "'Back to Boston': Elizabeth Bishop's Journeys from the Maritimes." *Colby Library Quarterly* 24 (1988): 151–61.

Dorson, Richard M., ed. *American Negro Folktales.* Greenwich, CT: Fawcett, 1967.

Dow, George Francis. *Slave Ships and Slaving.* Salem, MA: Marine Research Society, 1927.

————. *Whale Ships and Whaling: A Pictorial History of Whaling During Three Centuries.* Salem, MA: Marine Research Society, 1925.

Downey, Alice. "Writing Among the Isles of Shoals: Celia Thaxter as Woman and Writer." *A Stern and Lovely Scene: A Visual History of the Isles of Shoals.* Durham: U of New Hampshire Art Galleries, 1978. 85–105.

Druce, Robert. "*Jaws:* A Case Study." *Dutch Quarterly Review of Anglo-American Letters* 13 (1983): 72–86.

Druett, Joan. "Those Female Journals." *Log of Mystic Seaport* 40 (1989): 115–25.

Dyer, Joyce. "Symbolism and Imagery in *The Awakening.*" *Approaches to Teaching Chopin's "The Awakening."* Ed. Bernard Koloski. New York: Modern Language Association, 1988. 126–31.

Eby, Cecil D., Jr. "William Starbuck Mayo and Herman Melville." *New England Quarterly* 35 (1962): 515–20.

Edelman, Lee. "'Voyages.'" *Hart Crane.* Ed. Harold Bloom. New York: Chelsea, 1986. 255–91.

Egan, Hugh. "Introduction." *Proceedings of the Naval Court Martial in the Case of Alexander Slidell Mackenzie.* Delmar, NY: Scholars' Facsimiles and Reprints, 1992.

————. "Gentlemen-Sailors: The First-Person Sea Narratives of Dana, Cooper, and Melville." Diss. U of Iowa, 1983.

————. "'One of Them': The Voyage of Style in Dana's *Two Years Before the Mast.*" *American Transcendental Quarterly* 2 (1988): 177–90.

Ekambaram, E. J. "Sea Imagery in American Poetry." *Literary Half-Yearly* 23 (1982): 115–22.

Farr, James. "Black Odyssey: The Seafaring Tradition of Black Americans." New York: Peter Lang, 1989.

Federbush, Sherri. "The Journal of Eliza Brock at Sea on the Lexington." *Historic Nantucket* 30 (1982): 13–17.

Fields, Annie. *Authors and Friends.* Boston: Houghton, 1896.

Fiess, Edward. "Byron's Dark Blue Ocean and Melville's Rolling Sea." *English Language Notes* 3 (1966): 274–78.

Fink, Ernst O. " 'She Springs Her Luff!' Coopers Einsatz der Fachsprache im Seeroman." *Die Amerikanische Literatur in der Weltliteratur: Themen und Aspecte.* Ed. Claus Uhlig and Volker Bischoff. Berlin: Schmidt, 1982. 110–27.

Fogel, Daniel Mark. *Henry James and the Structure of the Romantic Imagination.* Baton Rouge: Louisiana State UP, 1981.

Fogle, Richard H. "*Billy Budd*—Acceptance or Irony." *Tulane Studies in English* 8 (1958): 107–13.

Foulke, Robert. "Life in the Dying World of Sail, 1870–1910." *Literature and Lore of the Sea.* Ed. Patricia Ann Carlson. Amsterdam: Rodopi, 1986. 72–115.

———. "The Literature of Voyaging." *Literature and Lore of the Sea.* 1–13.

Fowler, William M., Jr. *Rebels Under Sail; The American Navy During the Revolution.* New York: Scribner's, 1976.

Frank, Armin Paul. "The Long Withdrawing Roar: Eighty Years of the Ocean's Message in American Poetry." *Forms and Functions of History in American Literature: Essays in Honor of Ursula Brumm.* Ed. Winfried Frick, Jurgen Peper, and Willi Paul Adams. Berlin: Schmidt, 1981. 71–90.

Frank, Bernhard. "Dickinson's 'I Started Early . . .': An Anatomy; or, Whatever Happened to Emily's Dog." *Dickinson Studies* 76 (1990): 14–20.

Frank, Stuart. " 'Cheer'ly Men': Chantying in *Omoo* and *Moby-Dick.*" *New England Quarterly* 58 (1985): 68–82.

———. *Herman Melville's Picture Gallery: Sources and Types of the "Pictorial" Chapters of "Moby-Dick."* Fairhaven, MA: Lefkowicz, 1986.

———. " 'The King of the Southern Sea' and 'Captain Bunker': Two Songs in *Moby-Dick.*" *Melville Society Extracts* 63 (1985): 4–7.

———. "Melville in the South Seas and *The Friend.*" *Melville Society Extracts* 83 (1990): 1, 4–6.

Franklin, Rosemary F. "The Seashore Sketches in *Twice-Told Tales* and Melville." *English Language Notes* 21.4 (1984): 57–63.

Franklin, Wayne. *Discoverers, Explorers, Settlers: The Diligent Writers of Early America.* Chicago: U of Chicago P, 1979.

FreeHand, Julianna. *A Seafaring Legacy.* New York: Random, 1981.

Frye, Northrop. *Anatomy of Criticism: Four Essays.* Princeton: Princeton UP, 1957.

Fukuchi, Curtis. " 'The Only Firmament': 'Sea-Room' in Emerson's *English Traits.*" *American Transcendental Quarterly* 1 (1987): 197–209.

Gale, Robert L. *Richard Henry Dana.* New York: Twayne, 1969.

Gerstenberger, Donna. " 'The Open Boat': Additional Perspective." *Modern Fiction Studies* 17 (1971–72): 557–61.

Gidmark, Jill. *A Melville Sea Dictionary.* Westport, CT: Greenwood, 1982.

Gilbert, Sandra M. "The Second Coming of Aphrodite: Kate Chopin's Fantasy of Desire." *Kenyon Review* 5.3 (1983): 42–56. Rpt. as intro. to *The Awakening and Selected Stories*, by Kate Chopin. New York: Penguin, 1984.

Gilman, William H. *Melville's Early Life and Redburn.* New York: New York UP, 1951.

Goetzmann, William H. *New Lands, New Men: America and the Second Great Age of Discovery.* New York: Viking, 1986.

Goheen, Cynthia J. "Rebirth of the Seafarer: Sarah Orne Jewett's *The Country of the Pointed Firs.*" *Colby Library Quarterly* 23 (1987): 154–64.

Goldberg, Joseph P. *The Maritime Story: A Study in Labor-Management Relations.* Cambridge: Harvard UP, 1958.

Grazia, Emilio de. "The Great Plain: Rolvaag's New World Sea." *South Dakota Review* 20.3 (1982): 35–49. Rpt. *Literature and Lore of the Sea.* Ed. Patricia Ann Carlson. Amsterdam: Rodopi, 1986. 244–55.

———. "Poe's Other Beautiful Woman." *Literature and Lore of the Sea.* 176–84.

Greenberg, Robert M. "Cetology: Center of Multiplicity and Discord in *Moby-Dick.*" *Emerson Society Quarterly* 27 (1981): 1–13.

———. "The Three-Day Chase: Multiplicity and Coherence in *Moby-Dick.*" *Emerson Society Quarterly* 29 (1983): 91–98.

Greer, Ann L. "Early Development in America, 1825–1850, of Travel Books as Literature." Diss. U of Southern California, 1955.

Gregory, Eileen. "Rose Cut in Rock: Sappho and H.D.'s *Sea Garden.*" *Contemporary Literature* 27 (1986): 525–52.

Griffiths, John W. *Treatise on Marine and Naval Architecture.* London: Philip, 1856.

Grossman, James. *James Fenimore Cooper.* New York: Sloane, 1949.

Grove, James P. "Pastoralism and Anti-Pastoralism in Peter Matthiessen's *Far Tortuga.*" *Critique* 21 (1979): 15–29.

Guerra, Jonnie G. "Dickinson's 'I Started Early, Took My Dog.'" *Explicator* 50 (1992): 78–80.

Guilds, John C., Jr. "Poe's 'Manuscript Found in a Bottle': A Possible Source." *Notes and Queries* n.s. 3.7 (1956): 452.

Guzlowski, John Z. "No More Sea Changes: Hawkes, Pynchon, Gaddis, and Barth." *Critique: Studies in Modern Fiction* 23.2 (1981): 48–59. Rpt. as "No More Sea Changes: Four American Novelists' Responses to the Sea." *Literature and Lore of the Sea.* Ed. Patricia Ann Carlson. Amsterdam: Rodopi, 1986. 232–43.

Hamilton, William. "Melville and the Sea." *Soundings: A Journal of Interdisciplinary Studies* 62 (1979): 417–29.

Harlow, Frederick Pease. *Chanteying Aboard American Ships.* Barre, MA: Barre P, 1962.

Hart, James D. "The Education of Richard Henry Dana, Jr." *New England Quarterly* 9 (1936): 3–25.

———. "An Eyewitness of Eight Years Before the Mast." *New Colophon* 3 (1950): 128–31.

———. "Melville and Dana." *American Literature* 9 (1937): 49–55.

———. *The Popular Book: A History of America's Literary Taste.* New York: Oxford UP, 1950.

Haverstick, Iola S. "A Note on Poe and *Pym* in Melville's *Omoo.*" *Poe Newsletter* 2 (1969): 37.

Haverstick, Iola, and Betty Shepard, eds. *The Wreck of the Whaleship Essex: A Narrative Account by Owen Chase*. New York: Harcourt, 1965.

Hayford, Harrison. "Hawthorne, Melville, and the Sea." *New England Quarterly* 19 (1946): 435–52.

———. "The Sailor Poet of *White-Jacket*." *Boston Public Library Quarterly* 3 (1951): 221–28.

Hayford, Harrison, ed. *The Somers Mutiny Affair*. Englewood Cliffs, NJ: Prentice Hall, 1959.

Heaman, Robert J., and Patricia B. Heaman. "Hemingway's Fabulous Fisherman." *Pennsylvania English* 12 (1985): 29–33.

Heck, Francis S. "*The Old Man and the Sea* and Rimbaud's *Le Bateau ivre*: 'Solidaire'/ 'Solitaire.'" *North Dakota Quarterly* 49 (1981): 61–67.

Hedgpeth, Joel W. "Genesis of the *Sea of Cortez*." *Steinbeck Quarterly* 6.3 (1973): 74–80.

———. "Since the Days of Aristotle." *Literature and the Sea*. Ed. Richard Astro. Oregon State U, 1976. 19–24.

Heffernan, Thomas F. "Melville the Traveler." *A Companion to Melville Studies*. Ed. John Bryant. Westport, CT: Greenwood, 1986. 35–61.

———. *Stove by a Whale: Owen Chase and the Essex*. Middletown, CT: Wesleyan UP, 1981.

Heflin, Wilson. "Melville's Celestial Navigation, and Dead Reckoning." *Melville Society Extracts* 29 (1977): 3.

———. "New Light on Herman Melville's Cruise in the *Charles and Henry*." *Historic Nantucket* 22 (1974): 6–27.

Helms, Randel. "Another Source for Poe's *The Narrative of Arthur Gordon Pym*." *American Literature* 41 (1970): 572–75.

Higgins, Joseph. *The Whale Ship Book*. New York: Rudder, 1927.

Hill, Douglass B., Jr. "Richard Henry Dana, Jr., and 'Two Years Before the Mast.'" *Criticism* 9 (1967): 312–25.

Hinz, Evelyn J., and John J. Teunissen. "*Islands in the Stream* as Hemingway's Laocoön." *Contemporary Literature* 29 (1988): 26–48.

Hitchcock, Champion Ingraham. *The Dead Men's Song: Being the Story of a Poem and a Reminiscent Sketch of Its Author Young Ewing Allison*. Louisville: privately printed, 1914.

Hoffman, Steven K. "Sailing into the Self: Jung, Poe and 'MS. Found in a Bottle.'" *Tennessee Studies in Literature* 26 (1981): 66–74.

House, Kay Seymour. "The Unstable Element." *Cooper's Americans*. Columbus: Ohio State UP, 1965. 181–203.

Howard, Leon. *Herman Melville: A Biography*. Berkeley and Los Angeles: U of California P, 1951.

Hughes, James M. "Inner and Outer Seas in Dickinson, Dana, Cooper, and Roberts." *Literature and Lore of the Sea*. Ed. Patricia Ann Carlson. Amsterdam: Rodopi, 1986. 202–11.

————. "Popular Imagery of the Sea: Lore Is a Four-Letter Word." *Literature and Lore of the Sea.* 215–24.

————. "Whitman and Longfellow: The Breaking Up of Vessels." *West Hills Review: A Walt Whitman Journal* 8 (1988): 63–69.

Huguenin, Charles A. "The Truth About the Schooner Hesperus." *New York Folklore Quarterly* 16 (1960): 48–53.

Hunter, George M. "The Novels of the Sea." *Bookman* 31 (1910): 316–18.

Huntress, Keith. "Another Source for Poe's *The Narrative of Arthur Gordon Pym.*" *American Literature* 16 (1944): 19–25.

————. *A Checklist of Narratives of Shipwrecks and Disasters at Sea to 1860.* Ames: Iowa State UP, 1979.

————. "Melville's Use of a Source for *White-Jacket.*" *American Literature* 17 (1945): 66–74.

International Seamen's Union of America. *The Red Record: A Brief Resume of Some of the Cruelties Perpetrated upon American Seamen at the Present Time.* Supplement to the *Coast Seamen's Journal.* San Francisco, [1897?].

Irwin, John. *American Hieroglyphics.* New Haven: Yale UP, 1980.

Jaffé, David. *The Stormy Petrel and the Whale: Some Origins of "Moby-Dick."* Baltimore: Jaffé, 1976.

James, C. L. R. *Mariners, Renegades, and Castaways.* New York: James, 1953.

Johnson, Barbara. "The Lure of the Whaling Journal." *Manuscripts* 23 (1971): 159–77.

Jones, Daryl. "Mark Twain's Symbols of Despair: A Relevant Letter." *American Literary Realism* 15 (1982): 266–68.

Jones, Howard Mumford. *O Strange New World.* New York: Viking, 1964.

Jung, C. G. *Symbols of Transformation: An Analysis of the Prelude to a Case of Schizophrenia.* Trans. R. F. C. Hull. London: Routledge, 1956.

Kligerman, Jack. "Style and Form in James Fenimore Cooper's *Homeward Bound* and *Home as Found.*" *Journal of Narrative Technique* 4 (1974): 45–61.

Krob, Carol. "Columbus of Concord: *A Week* as a Voyage of Discovery." *Emerson Society Quarterly* 21 (1975): 215–21.

Kuehl, Warren F. *Blow the Man Down!* New York: Dutton, 1959.

Lang, Hans-Joachim, and Benjamin Lease. "The Authorship of *Symzonia:* The Case for Nathaniel Ames." *New England Quarterly* 48 (1975): 241–52.

Lang, Hans-Joachim, and Fritz Fleischmann. " 'All this Beauty, All this Grace': Longfellow's 'The Building of the Ship' and Alexander Slidell Mackenzie's 'Ship'." *New England Quarterly* 54 (1981): 104–18.

La Rue, Robert. "Whitman's Sea: Large Enough for *Moby Dick.*" *Walt Whitman Review* 12 (1966): 51–59.

Lawrence, D. H. *Studies in Classic American Literature.* New York: Selzer, 1923.

Lederer, Richard. "Lost Metaphors of Land and Sea." *Verbatim* 11 (1984): 1–2.

Leitz, Robert C., III. "The *Moran* Controversy: Norris's Defense of His 'Nautical Absurdities.' " *American Literary Realism* 15 (1982): 119–24.

Lemoine, Bernadette. "Lafcadio Hearn et la mer." *Cahiers Victoriens et Edwardiens* 23 (1986): 157–76.

Lenz, William E. "Narratives of Exploration, Sea Fiction, Mariners' Chronicles, and the Rise of American Nationalism: 'To Cast Anchor on that Point Where All Meridians Terminate.'" *American Studies* 32.2 (1991): 41–61.

———. "Poe's *Arthur Gordon Pym* and the Narrative Techniques of Antarctic Gothic." *CEA Critic* 53.3 (1991): 30–38.

Levine, Robert S. "Benito Cereno." *Conspiracy and Romance*. Cambridge, Eng.: Cambridge UP, 1989. 165–230.

Lewis, Charles Lee. *Books of the Sea: An Introduction to Nautical Literature*. Annapolis: Naval Institute P, 1943.

———. "Edgar Allan Poe and the Sea." *Southern Literary Messenger* 7 (1941): 5–10.

Leyda, Jay. *The Melville Log: A Documentary Life of Herman Melville, 1819–1891*. 2d ed. 2 vols. New York: Gordian P, 1969.

Leyris, Pierre. "Herman Melville: Poemes Marins." *Poèsie (Poèsie)* 36 (1986): 17–36.

Liebowitz, Herbert A. *Hart Crane: An Introduction to the Poetry*. New York: Columbia UP, 1968.

Ljungquist, Kent. "Poe and the Sublime: His Two Short Sea Stories in the Context of an Aesthetic Tradition." *Criticism* 17 (1975): 131–51.

Loving, Jerome M. "Melville's Pardonable Sin." *New England Quarterly* 47 (1974): 262–78.

Lucid, Robert F. "The Composition, Reception, Reputation, and Influence of *Two Years Before the Mast*." Diss. U of Chicago, 1958.

———. "The Influence of *Two Years Before the Mast* on Herman Melville." *American Literature* 31 (1959): 243–56.

———. "Two Years Before the Mast as Propaganda." *American Quarterly* 12 (1960): 392–403.

Lytle, Andrew. "'The Open Boat': A Pagan Tale." *The Hero with the Private Parts*. Baton Rouge: Louisiana State UP, 1966. 60–75.

McElrath, Joseph R., Jr. "A Critical Edition of Frank Norris's *Moran of the Lady Letty*." *DAI* 34 (1973): 5981A. U of South Carolina.

McFarland, Philip. *Sea Dangers: The Affair of the Somers*. New York: Schocken, 1985.

McFee, William. Introduction. *Great Sea Stories of Modern Times*. Ed. William McFee. New York: McBride, 1953.

———. Introduction. *Two Years Before the Mast*. New York: Limited Editions, 1947.

———. *Swallowing the Anchor*. London: Heinemann, 1925.

McKeithan, D. M. "Two Sources of Poe's *The Narrative of Arthur Gordon Pym*." *University of Texas Studies in English* 13 (1933): 116–37.

MacMechan, Archibald. "The Best Sea Story Ever Written." *Queen's Quarterly* 7 (1899): 120–30.

McWilliams, John P., Jr. *Political Justice in a Republic: James Fenimore Cooper's America*. Berkeley: U of California P, 1972.

Macy, Obed. *History of Nantucket and the Whale Fishery.* New York: Hilliard, 1835.

Madden, Fred. "'A Descent into the Maelstrom': Suggestions of a Tall Tale." *Studies in the Humanities* 14 (1987): 127–38.

Madigan, Francis V., Jr. "Mermaids in the Basement: Emily Dickinson's Sea Imagery." *Greyfriar: Siena Studies in Literature* 23 (1982): 39–56.

Madison, R. D. "Cooper's Columbus." *James Fenimore Cooper: His Country and His Art.* Ed. George A. Test. Oneonta & Cooperstown: State U of New York College, 1985. 75–85.

———. "Cooper's *The Wing-and-Wing* and the Concept of the Byronic Pirate." *Literature and Lore of the Sea.* Ed. Patricia Ann Carlson. Amsterdam: Rodopi, 1986. 119–32.

———. "Getting Under Way with James Fenimore Cooper." *James Fenimore Cooper and His Country; or, Getting Under Way.* Papers from the 1978 Conference at State University of New York College at Oneonta and Cooperstown, NY. Oneonta: State of New York College, 1979. 45–54.

———. "Introduction." *The History of the Navy of the United States of America.* Ed. R. D. Madison. Delmar: Scholars' Facsimiles & Reprints P, 1988.

———. "Redburn's Seamanship and Dana's Guide Book." *Melville Society Extracts* 57 (1984): 13–15.

Malloy, Mary. "The Old Sailor's Lament: Recontextualizing Melville's Reflections on the Sinking of 'The Stone Fleet.'" *New England Quarterly* 64 (1991): 633–42.

Martin, Robert A. "Gatsby and the Dutch Sailors." *American Notes and Queries* 12 (1973): 61–63.

Martin, Robert K. *Hero, Captain, and Stranger: Male Friendship, Social Critique, and Literary Form in the Sea Novels of Herman Melville.* Chapel Hill: U of North Carolina P, 1986.

Mathewson, Stephen. "Against the Stream: Thomas Hudson and Painting." *North Dakota Quarterly* 57 (1989): 140–45.

Maufort, Marc. "Exoticism in Melville's Early Sea Novels." *Archiv Für Liturgiewissenschaft* 11 (1991): 65–75.

———. "Mariners and Mystics: Echoes of *Moby Dick* in O'Neill." *Theatre Annual* 43 (1988): 31–52.

Mead, Joan Tyler. "'Spare me a few minutes i have Something to Say': Poetry in Manuscripts of Sailing Ships." *Literature and Lore of the Sea.* Ed. Patricia Ann Carlson. Amsterdam: Rodopi, 1986. 23–31.

———. "Traditions and Functions of the Songs in 'Midnight, Forecastle.'" *Melville Society Extracts* 83 (1990): 1–5.

Merrin, Jeredith. "Re-Seeing the Sea: Marianne Moore's 'A Grave' as a Woman Writer's Re-Vision." *Marianne Moore: Woman and Poet.* Ed. Patricia C. Willis. Orono: National Poetry Foundation, U of Maine, 1990. 155–67.

Metwalli, Ahmed M. "Americans Abroad: The Popular Art of Travel Writing in the Nineteenth Century." *America: Exploration and Travel.* Ed. Steven E. Kagle. Bowling Green, OH: Bowling Green UP, 1979. 68–82.

Miller, Lewis H., Jr. "On Keeping Watch: Robert Frost and Marianne Moore." *College Language Association Journal* 27 (1984): 411–18.

Miller, Nathan. *Sea of Glory: The Continental Navy Fights for Independence, 1775–1785.* New York: McKay, 1974.

Miller, Pamela A. *And the Whale Is Ours: Creative Writing of American Whalemen.* Boston and Sharon, MA: Godine and the Kendall Whaling Museum, 1979.

———. "What the Whalers Read." *Pages* 1 (1976): 242–47.

Milne, Gordon. *Ports of Call: A Study of the American Nautical Novel.* Lanham, MD: UP of America, 1986.

Miner, Roy Waldo. "Marauders of the Sea." *National Geographic Magazine* Aug. 1935: 185–207.

Mistri, Zenobia. "Absurdist Contemplations of a Sperm in John Barth's "Night-sea Journey." *Studies in Short Fiction* 25 (1988): 151–52.

Moffat, Mary Jane, and Charlotte Painter, eds. *Revelations: Diaries of Women.* New York: Random, 1974.

Monteiro, George, and Barton Levi St. Armand. "The Experienced Emblem: A Study of the Poetry of Emily Dickinson." *Prospects* 6 (1981): 187–280.

———. "The Pilot-God Trope in Nineteenth-Century American Texts." *Modern Language Studies* 7 (1977): 42–51.

Morgan Robertson the Man. New York: Metropolitan Magazine, 1915.

Morison, Samuel Eliot. Rev. of *An Instrument of the Gods*, by Lincoln Colcord. *Yale Review* 14 (1924): 195–96.

———. *The Maritime History of Massachusetts.* Boston: Houghton, 1921. Boston: Northeastern UP, 1979.

Morrison, Gail M. "A Manuscript Fragment." *Mississippi Quarterly* 34 (1981): 340–41.

Morrison, Kristin. "Conrad and O'Neill as Playwrights of the Sea." *Eugene O'Neill Newsletter* 2 (1978): 3–5.

Mortland, Donald F. "Lincoln Colcord: At Sea and at Home." *Colby Library Quarterly* 19 (1983): 125–43.

Muhlestein, Daniel K. "Crane's 'The Open Boat.'" *Explicator* 45.2 (1987): 42–43.

Mulvey, Christopher. *Anglo-American Landscapes: A Study of Nineteenth-Century Anglo-American Travel Literature.* Cambridge: Cambridge UP, 1983.

Nathanson, Tenney. "Whitman's Tropes of Light and Flood: Language and Representation in the Early Editions of *Leaves of Grass.*" *Emerson Society Quarterly* 31 (1985): 116–34.

Neeser, Robert W. *American Naval Songs & Ballads.* New Haven: Yale UP, 1938.

Newberry, I. "'The Encantadas': Melville's *Inferno.*" *American Literature* 38 (1966): 49–68.

O'Brien, Ellen J. "That 'Insular Tahiti': Melville's Truth-Seeker and the Sea." *Literature and Lore of the Sea.* Ed. Patricia Ann Carlson. Amsterdam: Rodopi, 1986. 193–201.

O'Hanlon, Redmond. *Joseph Conrad and Charles Darwin: The Influence of Scientific Thought on Conrad's Fiction.* Edinburgh: Salamander, 1984.

Oliver, Paul. *The Meaning of the Blues.* New York: Collier, [1963].

Olson, Charles. *Call Me Ishmael.* New York: Reynal, 1947.

Parker, Hershel. *The Recognition of Herman Melville.* Ann Arbor: Michigan UP, 1967.

Patteson, Richard F. "Holistic Vision and Fictional Form in Peter Matthiessen's *Far Tortuga.*" *Bulletin of the Rocky Mountain Modern Language Association* 37 (1983): 70–81.

Paul, Sherman. *The Shores of America.* Urbana: U of Illinois P, 1958.

Peck, H. Daniel. "A Repossession of America: The Revolution in Cooper's Trilogy of Nautical Romances." *Studies in Romanticism* 15 (1976): 589–609.

Peña, Horacio. "Aproximaciones a Rubén Darío y Walt Whitman: El Mar en Darío y Whitman." *Káñina* 8 (1984): 165–76.

Perez, Betty L. "The Form of the Narrative Section of *Sea of Cortez:* A Specimen Collected from Reality." *Steinbeck Quarterly* 9.2 (1976): 36–44. Rpt. in *Steinbeck's Travel Literature.* Steinbeck Monograph Series 10. Ed. Tetsumaro Hayashi, 1980. 47–55.

Philbrick, Thomas. "Cooper and the Literary Discovery of the Sea." *Canadian Review of American Studies* 20.3 (1989): 35–45.

———. "Cooper and the Literary Discovery of the Sea: Papers from the Bicentennial Conference, 1989." *James Fenimore Cooper: His Country and His Art.* Ed. George A. Test. Oneonta: State U of New York, English Dept., 1991.

———. Introduction. *Two Years Before the Mast.* By Richard Henry Dana, Jr. Ed. Philbrick. New York: Penguin, 1981.

———. *James Fenimore Cooper and the Development of American Sea Fiction.* Cambridge: Harvard UP, 1961.

———. "Language and Meaning in Cooper's *The Water-Witch.*" *Emerson Society Quarterly* 60 (1970): 10–16.

———. Rev. of *La Mer et le Roman Américain dans le Première Moitié du Dix-Neuvième Siècle,* by Jeanne-Marie Santraud. *American Literature* 45 (1973): 456.

Plimpton, George. "The Craft of Fiction in *Far Tortuga.*" Interview with Peter Matthiessen. *Paris Review* 60 (1974): 79–82.

Pollin, Burton. *Edgar Allan Poe: The Imaginary Voyages.* Boston: Twayne, 1981.

———. "The Narrative of Benjamin Morrell: Out of the 'Bucket' and Into Poe's *Pym.*" *Studies in American Fiction* 4 (1976): 157–72.

———. "Poe's Use of Material from Barnardin de Saint-Pierre's *Etudes.*" *Romance Notes* 12 (1971): 331–38.

Pommer, Henry F. "Herman Melville and the Wake of the *Essex.*" *American Literature* 20 (1948): 290–304.

Porte, Joel. "Henry Thoreau and the Reverend Poluphloisboios Thalassa." *The Chief Glory of Every People.* Ed. Matthew Bruccoli. Carbondale: Southern Illinois UP, 1973. 191–210.

Pratt, John Clark. "Neither Tea nor Comfortable Advice: The Elements in Literature." *Literature and the Sea.* Ed. Richard Astro. Corvallis: Oregon State U, 1976. 13–18.

Probert, K. G. "Nick Carroway and the Romance of Art." *English Studies in Canada* 10 (1984): 188–208.

Pryse, Marjorie. "Women 'At Sea': Feminist Realism in Sarah Orne Jewett's 'The Foreigner.'" *American Literary Realism* 15 (1982): 244–52.

Pudaloff, Ross J. "The Gaze of Power: Cooper's Revision of the Domestic Novel, 1835–1850." *Genre* 17 (1984): 275–95.

Putney, Martha S. *Black Sailors: Afro-American Merchant Seamen and Whalemen Prior to the Civil War.* Westport, CT: Greenwood, 1987.

Quinlan, Kieran. "Sea and Sea-Shore in 'Song of Myself': Whitman's Liquid Theme." *Literature and Lore of the Sea.* Ed. Patricia Ann Carlson. Amsterdam: Rodopi, 1986. 185–92.

Rampersad, Arnold. *Melville's Israel Potter: A Pilgrimage and a Progress.* Bowling Green, OH: Bowling Green U Popular P, 1969.

Ranta, Jerrald. "Marianne Moore's Sea and the Sentence." *Essays in Literature* 15 (1988): 245–57.

Reck, Tom S. "Melville's Last Sea Poetry: *John Marr and Other Sailors.*" *Forum* 12 (1974): 17–22.

Reed, Jeremy. "Hart Crane and the Riddle of the Sailor in 'Cutty Sark.'" *Modern American Poetry.* Ed. R. W. Butterfield. Totowa, NJ: Barnes and Noble, 1984. 127–41.

Rhea, R. L. "Some Observations on Poe's Origins." *U of Texas Studies in English* 10 (1930): 135–46.

Rice, Julian C. "The Ship as Cosmic Symbol in *Moby Dick* and *Benito Cereno.*" *Centennial Review* 16 (1972): 138–54.

Ridgely, J. V. "The Continuing Puzzle of *The Narrative of Arthur Gordon Pym.*" *Poe Newsletter* 3.1 (1970): 5–6.

Riesenberg, Felix. "'Communists' at Sea." *Nation* 23 Oct. 1937: 432–33.

Ringe, Donald A. *James Fenimore Cooper.* New York: Twayne, 1962.

———. "The Moral Geography of Cooper's 'Miles Wallingford' Novels." *Hudson Valley Regional Review* 21 (1985): 52–68.

Rink, Mary Terese. "The Sea in Lowell's 'Quaker Graveyard in Nantucket.'" *Renascence* 20 (1967): 39–43.

Robillard, Douglas. "'I Laud the Inhuman Sea': Melville as Poet in the 1880's." *Essays in Arts and Sciences* 5 (1976): 193–206.

Roppen, Georg. "Melville's Sea Shoreless, Indefinite as God." *Americana-Norvegica, Vol. 4: Norwegian Contributions to American Studies Dedicated to Sigmund Skard.* Ed. Brita Seyersted. Oslo: Universitetsforlaget, 1973. 137–81.

Rosenberry, Edward H. "The Problem of *Billy Budd.*" *PMLA* 80 (1965): 489–98.

Russell, W. Clark. "Sea Stories." *Contemporary Review* 46 (1884): 343–63.

Ryan, Paul R. "The *Titanic:* Lost and Found (1912–1985)." *Oceanus* 28.4 (1985–86): 4–14.

Sabo, William J. "The Ship and Its Related Imagery in 'Inscriptions' and 'Song of Myself.'" *Walt Whitman Review* 24 (1978): 118–23.

Sadler, Lynn Veach. "The Sea in Selected American Novels of Slave Unrest." *Journal of American Culture* 10.2 (1987): 43–48.

St. Armand, Barton Levi. "Dickinson's 'Red Sea.'" *Explicator* 43.3 (1985): 17–20.

Samson, John. *White Lies: Melville's Narratives of Facts.* Ithaca: Cornell, 1989.

Santraud, Jeanne-Marie. *Les Chantres Passionnés du Grand Océan.* Paris: Presses de l'Université de Paris—Sorbonne, 1989.

———. *La Mer et le Roman Américain dans le Première Moitié du Dix-Neuvième Siècle.* Paris: Didier, 1972.

Sawyer, Edith A. "A Year of Cooper's Youth." *New England Magazine* 43 (1907–8): 498–504.

Scharnhorst, Gary. "Another Night-Sea Journey: Poe's 'MS. Found in a Bottle.'" *Studies in Short Fiction* 22 (1985): 203–8.

Schenck, Celeste M. "'Every Poem an Epitaph': Sea-Changes in Whitman's 'Out of the Cradle . . .' and Crane's 'Voyages.'" *Ariel: A Review of International English Literature* 16.1 (1985): 3–25.

Schirmer, Gregory A. "Becoming Interpreters: The Importance of Tone in Crane's 'The Open Boat.'" *American Literary Realism* 15 (1982): 221–31.

Schwendinger, Robert J. "The Language of the Sea: Relationships Between the Language of Herman Melville and Sea Shanties of the Nineteenth Century." *Southern Folklore Quarterly* 37 (1973): 53–73.

Scudder, Harold H. "Melville's Benito Cereno and Captain Delano's Voyages." *PMLA* 43 (1928): 502–32.

Sealts, Merton M., Jr. *Melville's Reading.* Madison: U of Wisconsin P, 1988.

Seed, David. "Sargassos of the Imagination: *The Day of the Locust.*" *NMAL: Notes on Modern American Literature* 9 (1985): Item 12.

Seelye, John. "'Andsome Is as 'Andsome Does." *Literature and the Sea.* Ed. Richard Astro. Corvallis: Oregon State U, 1976. 7–12.

Sherman, Sarah Way. *Sarah Orne Jewett: An American Persephone.* Hanover: UP of New England, 1989.

Sherwood, John C. "Vere as Collingwood: A Key to *Billy Budd.*" *American Literature* 35 (1964): 476–84.

Shuffelton, Frank. "Going Through the Long Vaticans: Melville's 'Extracts' in *Moby-Dick.*" *Texas Studies in Literature and Language* 25 (1983): 528–40.

Shurr, William H. "Mysticism and Suicide: Anne Sexton's Last Poetry." *Soundings* 68 (Fall 1985): 335–56.

Singh, B. M. "Look at the Sea." *Rajasthan Studies in English* 10 (1977): 22–26.

Skallerup, Harry R. *Books Afloat and Ashore: A History of Books, Libraries, and Reading Among Seamen During the Age of Sail.* Hamden, CT: Archon, 1974.

Slethaug, Gordon E. "Floating Signifiers in John Barth's *Sabbatical.*" *Modern Fiction Studies* 33 (1987): 647–55.

Sloan, Edward W., III. "The Sailor and the Steamer: Literary Reflections on a Maritime Revolution." *Trinity* [College—Hartford, CT] *Reporter* May 1975: 2–7.

Smith, Harold F. *American Travellers Abroad: A Bibliography of Accounts Published Before 1900*. Carbondale-Edwardsville: Southern Illinois UP, 1969.

Smith, Laura A. *The Music of the Waters*. London: Paul, Trench, 1888.

Smith, Myron J., Jr. *American Naval Bibliography Series*. 5 vols. Metuchen, NJ: Scarecrow, 1973–74.

Smith, Myron J., Jr., and Robert L. Weller. *Sea Fiction Guide*. Metuchen, NJ: Scarecrow, 1976.

Smyth, William D. "Water: A Recurring Image in Frederick Douglass' *Narrative*." *CLA Journal* 34 (1990): 174–87.

Snow, Edward R. *Women of the Sea*. New York: Dodd, 1962.

Sojka, Gregory S. "Art and Order in *Islands in the Stream*." *Hemingway: A Revaluation*. Ed. Donald R. Noble. Troy, NY: Whitson, 1983. 263–80.

Spengemann, William C. *The Adventurous Muse: The Poetics of American Fiction, 1789–1900*. New Haven: Yale UP, 1977.

Spiller, Robert E. *James Fenimore Cooper*. Minneapolis: U of Minnesota P, 1974.

Springer, Haskell. "Call Them All Ishmael?: Fact and Form in Some Nineteenth-Century Sea Narratives." *Literature and Lore of the Sea*. Ed. Patricia Ann Carlson. Amsterdam: Rodopi, 1986. 14–22.

———. "The Nautical *Walden*." *New England Quarterly* 57 (1984): 84–97.

Stackpole, Edouard A. *The Sea Hunters*. Philadelphia: Lippincott, 1953.

Stanonik, Janez. *Moby Dick: The Myth and the Symbol*. Ljubljana, Yugoslavia: Ljubljana UP, 1962.

Stanton, William. *The Great United States Exploring Expedition of 1838–1842*. Berkeley: U of California P, 1975.

Stanwood, Paul Grant. "The Triumph of the Sea: Three Poems of John Peale Bishop (1892–1944)." *Jahrbuch Für Amerikastudien* 8 (1963): 212–18.

Starbuck, Alexander. *History of the American Whale Fishery*. Waltham, MA, 1876. New York: Argosy-Antiquarian, 1964.

Stein, Paul. "Cooper's Later Fiction: The Theme of 'Becoming.'" *South Atlantic Quarterly* 70.1 (1971): 77–78.

Stein, Roger B. "Copley's *Watson and the Shark* and Aesthetics in the 1770s." *Discoveries and Considerations*. Ed. Calvin Isreal. Albany: SUNY, 1976.

———. "Picture and Text: The Literary World of Winslow Homer." *Winslow Homer: A Symposium*. Ed. Nicolai Cikovsky, Jr. Studies in the History of Art 26. Washington, DC: National Gallery of Art, 1990. 32–59.

———. "Pulled Out of the Bay: American Fiction in the Eighteenth Century." *Studies in American Fiction* 2.1 (1974): 13–36.

———. "Seascape and the American Imagination: The Puritan Seventeenth Century." *Early American Literature* 7.1 (1972): 17–37.

———. *Seascape and the American Imagination*. New York: Whitney Museum of American Art/Clarkson N. Potter, 1975.

———. "Thomas Smith's Self-Portrait: Image/Text as Artifact." *Art Journal* 44 (1984): 316–27.

Steiner, George. *After Babel: Aspects of Language and Translation.* New York: Oxford UP, 1975.

A Stern and Lovely Scene: A Visual History of the Isles of Shoals. Durham: U Art Galleries, U of New Hampshire, 1978.

Stiffler, Randall. "The Sea Poems of W. S. Merwin." *Modern Poetry Studies* 11 (1983): 247–66.

Stone, Albert E. "The Sea and the Self: Travel as Experience and Metaphor in Early American Autobiography." *Genre* 7 (1974): 279–306.

Stout, Janis P. *The Journey Narrative in American Literature: Patterns and Departures.* Westport, CT: Greenwood, 1983.

Strauss, W. Patrick. *Americans in Polynesia, 1783–1842.* East Lansing: Michigan State UP, 1963.

Swann, Thomas Burnett. *The Classical World of H.D..* Lincoln: U of Nebraska P, 1962.

Tanner, Tony. "Henry James: The Candid Outsider." *The Reign of Wonder: Naivety and Reality in American Literature.* New York: Harper, 1967.

Taylor, Paul S. *The Sailors' Union of the Pacific.* 1923. New York: Arno, 1971.

Terramorsi, Bernard. "Des Tenants et des Aboutissants du Fantastique." *Du Fantastique en Litterature: Figures et Figurations.* Ed. Max Duperray. Aix-en-Provence: U de Provence, 1990. 169–88.

Thaxter, Rosamond. *Sandpiper: The Life of Celia Thaxter.* 1962. Rpt. as *Sandpiper: The Life and Letters of Celia Thaxter. . . .* Francistown, NH: M. Jones, 1963.

Thomas, Russell. "Melville's Use of Some Sources in 'The Encantadas.'" *American Literature* 3 (1932): 432–56.

Tintner, Adeline R. "The Sea of Asof in 'The Turn of the Screw' and Maurice Barrès's *Les Déracinés.*" *Essays in Literature* 14.1 (1987): 139–43.

Trimpi, Helen P. "Melville's Use of Demonology and Witchcraft in *Moby-Dick.*" *Journal of the History of Ideas* 30 (1969): 543–62.

"A Trio of Sailor Authors." *Dublin University Magazine* 47 (1856): 47–54.

Turner, Arlin. "Sources of Poe's 'A Descent into the Maelstrom.'" *JEGP* 46 (1947): 298–301.

Tymieniecka, Anna-Teresa, ed. *Poetics of the Elements in the Human Condition: The Sea.* Analecta Husserliana 19 (1985).

Tynan, Daniel. "J. N. Reynolds' *Voyage of the Potomac:* Another Source for *The Narrative of Arthur Gordon Pym.*" *Poe Studies* 4.1 (1971): 35–37.

Vallier, Jane E. *Poet on Demand: The Life, Letters and Works of Celia Thaxter.* Camden, ME: Peter Randall/Down East, 1982.

Vincent, Howard P. *The Tailoring of Melville's White-Jacket.* Evanston: Northwestern UP, 1970.

———. *The Trying-Out of Moby-Dick.* Boston: Houghton, 1949.

Vitzthum, Richard C. *Land and Sea: The Lyric Poetry of Philip Freneau.* Minneapolis: U of Minnesota P, 1978.

Walker, Warren S. "The Gull's Way." *James Fenimore Cooper: An Introduction and Interpretation.* New York: Barnes and Noble, 1962. 64–85.

Wallace, James D. *Early Cooper and His Audience.* New York: Columbia UP, 1986.

Wallace, Robert K. "'The Sultry Creator of Captain Ahab': Herman Melville and J. M. W. Turner." *Turner Studies* 5.2 (1985): 2–20.

Ward, J. A. "The Function of the Cetological Chapters in *Moby-Dick.*" *American Literature* 28 (1956): 164–83.

Watson, Charles. *The Novels of Jack London: A Reappraisal.* Madison: U of Wisconsin P, 1983.

Watters, R. E. "Melville's Isolatoes." *PMLA* 60 (1945): 1138–48.

Weaver, Raymond M. *Herman Melville: Mariner and Mystic.* New York: Doran, 1921.

Welland, Dennis. "The Dark Voice of the Sea: A Theme in American Poetry." *American Poetry.* Ed. Irvin Ehrenpreis. Stratford-upon-Avon Studies 7. London: Arnold, 1965.

Werth, Lee F. "The Paradox of Single-Handed Sailing (Case Studies in Existentialism)." *Journal of American Culture* 10.1 (1987): 65–77.

Westbrook, Perry D. *Acres of Flint: Writers of Rural New England, 1870–1900.* Washington: Scarecrow, 1951. Rev. ed. *Acres of Flint: Sarah Orne Jewett and Her Contemporaries.* Metuchen, NJ: Scarecrow, 1981.

Whall, W. B. *Sea Songs and Shanties.* Glasgow: Brown, 1927.

Wharton, Donald P. "Anne Bradstreet and the *Arbella.*" *Critical Essays on Anne Bradstreet.* Ed. Pattie Cowell and Ann Stanford. Boston: G. K. Hall, 1983.

———. "Hudson's Mermaid: Symbol and Myth in Early American Sea Literature." *Early American Literature and Culture: Essays Honoring Harrison T. Meserole.* Ed. Kathryn Zabelle Derounian-Stodola. Newark: U of Delaware P, 1992.

———. *Richard Steere: Colonial Merchant Poet.* University Park: Pennsylvania State UP, 1976.

Wharton, Donald P., ed. *In the Trough of the Sea: Selected American Sea-Deliverance Narratives, 1610–1766.* Westport, CT: Greenwood, 1979.

Whitehill, Walter Muir. "Cooper as Naval Historian." *James Fenimore Cooper: A Reappraisal.* Cooperstown: New York State Historical Association, 1954. 468–79.

Whiting, Emma, and Henry Beetle Hough. *Whaling Wives.* Boston: Houghton, 1952.

Willett, Ralph W. "Nelson and Vere: Hero and Victim in *Billy Budd, Sailor.*" *PMLA* 82 (1967): 370–76.

Williams, Harold, ed. *One Whaling Family.* Boston: Houghton, 1964.

Williams, William Appleman. "Angles of Vision: or, As We Say at Sea, Shots for a Fix." *Literature and the Sea.* Ed. Richard Astro. Corvallis: Oregon State U, 1976. 45–49.

Wilmerding, John. *American Marine Painting.* 2d ed. New York: Abrams, 1987.

Wilson, G. R., Jr. "Saints and Sinners in the Caribbean: The Case for *Islands in the Stream.*" *Studies in American Fiction* 18.1 (1990): 27–40.

Winterich, John T. "Two Years Before the Mast by Richard Henry Dana, Jr." *Georgia Review* 9 (1955): 459–61.

Wright, Louis B., ed. *The Elizabethans' America.* Cambridge: Harvard, 1965.

Young, Vernon. "Same Sea, Same Dangers: W. S. Merwin." *American Poetry Review* 7 (1978): 4–5.

Zacharias, Greg W. "The Marine Metaphor, Henry James, and the Moral Center of *The Awkward Age*." *Philological Quarterly* 69 (1990): 91–105.

Ziff, Larzer. *Literary Democracy: The Declaration of Cultural Independence in America*. New York: Viking, 1981.

Zigerell, James. "Crane's 'Voyages II.' " *Explicator* 13.2 (1954).

Zoellner, Robert. "The Conceptual Metaphor." *Literature and the Sea*. Ed. Richard Astro. Corvallis: Oregon State U, 1976. 39–43.

———. "James Fenimore Cooper's Sea Novels: His Social Theories as Expressed Symbolically Through the Gentleman-leader of the Microcosmic Ship on the Sea-frontier." Diss. U of Wisconsin, 1962.

———. *The Salt-Sea Mastodon*. Berkeley: U of California P, 1973.

Contributors

BERT BENDER, professor of English at Arizona State University, has written widely on American authors, and is himself the author, most notably, of *Sea Brothers: The Tradition of American Sea Fiction from "Moby-Dick" to the Present* (1988). Chapters from his forthcoming study of the response to Darwin in American fiction from 1871 to 1926 have appeared in *American Literature* 63, *Journal of American Studies* 26, and *Prospectus* 18.

DENNIS BERTHOLD, professor of English at Texas A&M University, has published articles on art and landscape in Brown, Hawthorne, and Melville, and on such nautical topics as Cape Horn and Joshua Slocum. He is co-author of *Hawthorne's American Travel Sketches* (1989) and co-editor of *Dear Brother Walt: The Letters of Thomas Jefferson Whitman* (1984). He regularly teaches a course in sea literature.

VICTORIA BREHM is assistant professor of English at Grand Valley State University in Allendale, Michigan. She edited the first collection of Great Lakes maritime literature, *Sweetwater, Storms, and Spirits: Stories of the Great Lakes* (1990), and has contributed essays on the subject to *Inland Seas, American Neptune,* and *American Literary Realism.* She holds a master's license for Great Lakes and inland waters from the U.S. Coast Guard.

JOSEPH DEFALCO, professor of English at Marquette University, has published *The Hero in Hemingway's Short Stories* (1963, 1968) and *Collected Poems of Christopher Pearse Cranch* (1971, 1986), as well as a number of articles on various topics in American literature. His maritime concerns have been addressed in articles such as "Psyche and Sea: The Waste Land Era in America," "Metaphor and Meaning in Poe's *Pym*," " 'Bimini' and the Subject of Hemingway's *Islands in the Stream*," and in scholarly conference papers on Hemingway, Poe, Cooper, Melville, and Whitman.

HUGH EGAN is associate professor of English at Ithaca College and currently chair of his department. He has published on Dana, Cooper, and Irving. Most recently, he wrote the introduction to the 1844 *"Somers" Affair* trial transcript, *Proceedings of the Naval Court Martial in the Case of Alexander Slidell Mackenzie* (1992).

JOSEPH FLIBBERT is professor of English at Salem State College, Massachusetts, where he teaches graduate and undergraduate courses on literature of the sea and participates in the marine studies interdisciplinary minor. He is the author of

Melville and the Art of Burlesque and has written articles on various authors, including Hawthorne, Melville, and Dickens. Among his professional activities, Flibbert has lectured on nautical aspects of American works, and edited the papers of a conference on Massachusetts and the sea.

ROBERT D. MADISON, associate professor of English at the U.S. Naval Academy, teaches American literature and literature of the sea. An expert in nautical song, he has been Chantey Interpreter at Mystic Seaport Museum. His editions include *Newfoundland Summers: The Ballad Collecting of Elisabeth Bristol Greenleaf* (1982), James Fenimore Cooper's *History of the Navy* (1988), David Porter's *Journal of a Cruise* (1986), and Robert Southey's *Life of Nelson* (1990). He is a contributing scholar to the Northwestern-Newberry edition of *The Writings of Herman Melville.*

DOUGLAS ROBILLARD recently retired from the University of New Haven. His publications include *Poems of Melville* (1976) and Jack London and Anna Strunsky's *Kempton-Wace Letters* (1990), as well as articles on Melville, Henry James, Edith Wharton, and Conrad Aiken.

JOHN SAMSON is associate professor of English at Texas Tech University. A former editor of *The Eighteenth Century: Theory and Interpretation,* he is the author of *White Lies: Melville's Narratives of Facts* (1989), and articles on American and British fiction. He is currently president of the Society for Eighteenth-Century American Studies.

ELIZABETH SCHULTZ is Chancellor's Club Teaching Professor of English at the University of Kansas. A former Fulbright senior lecturer in Japan, she has also taught at Tuskeegee Institute. Among her many articles are pieces on Ralph Ellison, other African-American authors, Herman Melville, and Japanese whaling. Her book *"Unpainted to the Last": "Moby-Dick" and Twentieth-Century American Art* will be published by the University Press of Kansas in 1995.

HASKELL SPRINGER, professor of English at the University of Kansas, has held posts in Rio de Janeiro and at the Sorbonne, and has taught sea literature at sea and ashore. Among his publications are *Studies in Billy Budd* (1970) and *The Sketch Book of Geoffrey Crayon, Gent.* (1978), as well as nautically focused articles on personal writings by both men and women, and on Thoreau's *Walden.* He currently serves as Area Chair for Sea Literature for both the American Culture and Popular Culture associations.

ROGER B. STEIN is professor of the history of art at the University of Virginia, though he has taught both American literature and history of American art over the years, including at maritime institutes at Mystic Seaport and the University of Rhode Island. He has been curator and consultant to museum exhibitions and has published on seventeenth- and eighteenth-century literary seascape, Thomas Smith, John Singleton Copley, Winslow Homer, and other visual artists of the sea.

DENNIS WELLAND is professor emeritus of American literature at the University of Manchester. A founder-officer of the British Association for American Studies, he was also founder-editor of the *Journal of American Studies.* His publications include *Wilfred Owen: A Critical Study* (1960, 1978), *Arthur Miller* (1961, now in third edition as *Miller the Playwright*), and *Mark Twain in England* (1975), as well as other books and articles on English and American literature.

DONALD P. WHARTON is president of Plymouth State College, New Hampshire. His publications on sea literature include *Richard Steere: Colonial Merchant Poet* (1978), *In the Trough of the Sea: Selected American Sea-Deliverance Narratives, 1610–1766* (1979), "Anne Bradstreet and the *Arbella*" in *Critical Essays on Anne Bradstreet* (1983), and "Hudson's Mermaid: Symbol and Myth in Early American Sea Literature" in *Early American Literature and Culture* (1992).

Index

Anna Christie (O'Neill), 294, 296
Antarctica, 30, 74–75, 79, 81, 89–90, 312–13
Antimodernism, 177
Anti-romanticism, 199, 262
"Apology for Bad Dreams" (Jeffers), 267
Appledore House, 201, 202, 204
Arabian Nights (trans. Burton), 319, 320
Archaeologist of Morning (Olson), 284
Arctic, 81, 90, 165–67, 203–4, 213
Arctic Explorations . . . in the Years 1853, '54 (Kane), 55, 90, 165
"Argonaut, The" (Freneau), 61
"Armada of Thirty Whales" (Hoffman), 272–73
Armory Show, 289
Arnold, Matthew, 118, 260–61, 283
"Artist of the Beautiful, The" (Hawthorne), 121
Ashcan realism, 179
Ashton, Philip, 43
Ashton's Memorial (Barnard), 43
Asia: in nineteenth-century narratives, 88, 91
"As I Ebb'd with the Ocean of Life" (Whitman), 119
"As if the Sea Should part" (Dickinson), 157
"Aspern Papers, The" (James), 196
Astor, John Jacob, 4, 71
Astoria (Irving), 4, 24, 71, 81
"At Ithaca" (H.D.), 261
Atlantic High (Buckley), 325
Atlantic Monthly, 201
Atlantis (Simms), 111
"At Melville's Tomb" (Crane), 186
"At Sea" (Trowbridge), 193
"At the Fishhouses" (Bishop), 274–75, 281
Auden, W. H., 79, 113, 123, 291–92, 293, 297, 302
Augusta Victoria (painting, Jacobsen), 173–75
"Aunt Jemima of the Ocean Waves" (Hayden), 249–50
Autobiography, 242, 244–46, 324. *See also* Personal narratives
Awakening, The (Chopin), 21, 154, 206–8

Bailey, Capt. Joseph, 44
Balano, Dorothea, 19
Balboa, Vasco Núñez de, 71
Baldwin, James, 247
"Ballad of Carmilhan, The" (Longfellow), 113, 115

"Ballad of Nantucket, A" (Aldrich), 114
"Ballad of Old Zeb" (song, Kaplan), 108
"Ballad of Sir John Franklin" (Willis), 117
"Ballad of Yarmouth Castle" (song, Lightfoot), 108
Ballou, Maturin Murray, 79
Balmer, Edwin, 225
Baltimore (Maryland), 11
Banjo (McKay), 244, 254
"Barbados" (Marshall), 252
Barber, John W., 168–69
Bard, James, 173
Bard, John, 173
Barge of State, The (sculpture, MacMonnies), 182–83, 198
Barker, Benjamin, 79
Barlow, Joel, 149
Barnard, John, 43, 53
"Barrett's Privateers" (song, Rogers), 108
Barth, John, 14, 318–20, 326
Bartholomew, Benjamin, 42
"Bartleby, the Scrivener" (Melville), 139
Batchelor, John Calvin: *The Birth of the People's Republic of Antarctica*, 30, 312–13
"Bathers, the" (Crane), 286
Battle of Lake Erie, 170
Battle of Midway, 315
"Battle of Stonington, The" (Freneau), 102
"Battle of the Kegs, The" (song, anonymous), 101
Battle-Pieces and Aspects of the War (Melville), 140, 141, 191
Bay Mild (Kintziger), 230
Bay Psalm Book, 100, 101
Beach, Cmdr. Edward, 315
Beach Boys, 108
"Beacon, The" (Wilbur), 280–81
Beast (Benchley), 309
"Becalmed" (Longfellow), 120–21
Beckoning Waters, The (Carse), 227
Bedford Incident, The (Rascovich), 315
Belland, F. W.: *The True Sea*, 30, 313
Below Zero (painting, Homer), 166
Benchley, Peter, 22, 308–9, 326
Bender, Bert, ix
"Benito Cereno" (Melville), 138, 143, 205, 209, 241, 257, 322
Bennett, F. D., 93, 96
Bennett, William James, 170
Beowulf, 310, 312